Tools, Methodologies and Techniques Applied to Sustainable Supply Chains

Tools, Methodologies and Techniques Applied to Sustainable Supply Chains

Special Issue Editor

Jorge Luis García-Alcaraz

MDPI • Basel • Beijing • Wuhan • Barcelona • Belgrade • Manchester • Tokyo • Cluj • Tianjin

Special Issue Editor
Jorge Luis García-Alcaraz
Autonomous University of
Ciudad Juárez
Mexico

Editorial Office
MDPI
St. Alban-Anlage 66
4052 Basel, Switzerland

This is a reprint of articles from the Special Issue published online in the open access journal *Sustainability* (ISSN 2071-1050) (available at: https://www.mdpi.com/journal/sustainability/special_issues/Sustainable_Supply_Chains_Tools).

For citation purposes, cite each article independently as indicated on the article page online and as indicated below:

LastName, A.A.; LastName, B.B.; LastName, C.C. Article Title. *Journal Name* **Year**, *Article Number*, Page Range.

ISBN 978-3-03928-318-7 (Pbk)
ISBN 978-3-03928-319-4 (PDF)

© 2020 by the authors. Articles in this book are Open Access and distributed under the Creative Commons Attribution (CC BY) license, which allows users to download, copy and build upon published articles, as long as the author and publisher are properly credited, which ensures maximum dissemination and a wider impact of our publications.

The book as a whole is distributed by MDPI under the terms and conditions of the Creative Commons license CC BY-NC-ND.

Contents

About the Special Issue Editor . . . . **vii**

Preface to "Tools, Methodologies and Techniques Applied to Sustainable Supply Chains" . . **ix**

José Manuel Velarde, Susana García, Mauricio López and Alfredo Bueno-Solano
Implementation of a Mathematical Model to Improve Sustainability in the Handling of Transport Costs in a Distribution Network
Reprinted from: *Sustainability* **2020**, *12*, 63, doi:10.3390/su12010063 . . . . **1**

Ernesto A. Lagarda-Leyva and Angel Ruiz
A Systems Thinking Model to Support Long-Term Bearability of the Healthcare System: The Case of the Province of Quebec
Reprinted from: *Sustainability* **2019**, *11*, 7028, doi:10.3390/su11247028 . . . . **11**

Marco A. Miranda-Ackerman, Catherine Azzaro-Pantel, Alberto A. Aguilar-Lasserre, Alfredo Bueno-Solano and Karina C. Arredondo-Soto
Green Supplier Selection in the Agro-Food Industry with Contract Farming: A Multi-Objective Optimization Approach
Reprinted from: *Sustainability* **2019**, *11*, 7017, doi:10.3390/su11247017 . . . . **25**

Elías Escobar-Gómez, J.L. Camas-Anzueto, Sabino Velázquez-Trujillo, Héctor Hernández-de-León, Rubén Grajales-Coutiño, Eduardo Chandomí-Castellanos and Héctor Guerra-Crespo
A Linear Programming Model with Fuzzy Arc for Route Optimization in the Urban Road Network
Reprinted from: *Sustainability* **2019**, *11*, 6665, doi:10.3390/su11236665 . . . . **45**

Leonardo Rivera-Cadavid, Pablo Cesar Manyoma-Velásquez and Diego F. Manotas-Duque
Supply Chain Optimization for Energy Cogeneration Using Sugarcane Crop Residues (SCR)
Reprinted from: *Sustainability* **2019**, *11*, 6565, doi:10.3390/su11236565 . . . . **63**

Jania Astrid Saucedo Martinez, Abraham Mendoza and Maria del Rosario Alvarado Vazquez
Collection of Solid Waste in Municipal Areas: Urban Logistics
Reprinted from: *Sustainability* **2019**, *11*, 5442, doi:10.3390/su11195442 . . . . **79**

Roman Rodriguez-Aguilar, Jose Antonio Marmolejo-Saucedo and Brenda Retana-Blanco
Prices of Mexican Wholesale Electricity Market: An Application of Alpha-Stable Regression
Reprinted from: *Sustainability* **2019**, *11*, 3185, doi:10.3390/su11113185 . . . . **95**

Riccardo Accorsi, Giulia Baruffaldi, Riccardo Manzini and Chiara Pini
Environmental Impacts of Reusable Transport Items: A Case Study of Pallet Pooling in a Retailer Supply Chain
Reprinted from: *Sustainability* **2019**, *11*, 3147, doi:10.3390/su11113147 . . . . **109**

Sasha Shahbazi, Martin Kurdve, Mats Zackrisson, Christina Jönsson and Anna Runa Kristinsdottir
Comparison of Four Environmental Assessment Tools in Swedish Manufacturing: A Case Study
Reprinted from: *Sustainability* **2019**, *11*, 2173, doi:10.3390/su11072173 . . . . **123**

Martin Krajčovič, Viktor Hančinský, Ľuboslav Dulina, Patrik Grznár, Martin Gašo and Juraj Vaculík
Parameter Setting for a Genetic Algorithm Layout Planner as a Toll of Sustainable Manufacturing
Reprinted from: *Sustainability* **2019**, *11*, 2083, doi:10.3390/su11072083 . . . . **143**

José Roberto Mendoza-Fong, Jorge Luis García-Alcaraz, José Roberto Díaz-Reza, Emilio Jiménez-Macías and Julio Blanco-Fernández
The Role of Green Attributes in Production Processes as Well as Their Impact on Operational, Commercial, and Economic Benefits
Reprinted from: *Sustainability* **2019**, *11*, 1294, doi:10.3390/su11051294 . . . . **169**

Chih-Hung Yuan, Yenchun Jim Wu and Kune-muh Tsai
Supply Chain Innovation in Scientific Research Collaboration
Reprinted from: *Sustainability* **2019**, *11*, 753, doi:10.3390/su11030753 . . . . **193**

Xiaodong Zhu, Lingfei Yu, Ji Zhang, Chenliang Li and Yizhao Zhao
Warranty Decision Model and Remanufacturing Coordination Mechanism in Closed-Loop Supply Chain: View from a Consumer Behavior Perspective
Reprinted from: *Sustainability* **2018**, *10*, 4738, doi:10.3390/su10124738 . . . . **205**

Maria-Lizbeth Uriarte-Miranda, Santiago-Omar Caballero-Morales, Jose-Luis Martinez-Flores, Patricia Cano-Olivos, Anastasia-Alexandrovna Akulova
Reverse Logistic Strategy for the Management of Tire Waste in Mexico and Russia: Review and Conceptual Model
Reprinted from: *Sustainability* **2018**, *10*, 3398, doi:10.3390/su10103398 . . . . **227**

Verónica Duque-Uribe, William Sarache and Elena Valentina Gutiérrez
Sustainable Supply Chain Management Practices and Sustainable Performance in Hospitals: A Systematic Review and Integrative Framework
Reprinted from: *Sustainability* **2019**, *11*, 5949, doi:10.3390/su11215949 . . . . **253**

About the Special Issue Editor

Jorge Luis García-Alcaraz is a full-time professor at Autonomous University of Ciudad Juárez (Mexico). He received a MSc in Industrial Engineering from Technological Institute of Colima (Mexico), a PhD in Industrial Engineering from Technological Institute of Ciudad Juárez (Mexico), a PhD in Innovation in Product Engineering and Industrial Process from University of La Rioja (Spain) and a PhD in Sciences and Industrial Technologies for University of Navarre (Spain). His main research areas are Multicriteria decision making applied to lean manufacturing, production process modeling, supply chain management and statistical inference. He is founding member of the Mexican Society of Operation Research and active member in the Mexican Academy of Industrial Engineering. Currently, he is a recognized as National Researcher level III by the Mexican National System of Researcher and the National Council for Science and Technology. He wrote more that 150 papers indexed in Scopus, ten books with Springer and IGI GLOBAL editorial, among others.

ORCID: 0000-0002-7092-6963. Scopus Author ID: 55616966800. ResearcherID: N-9124-2013.

Preface to "Tools, Methodologies and Techniques Applied to Sustainable Supply Chains"

Nowadays, supply chains and production systems are globalized and distributed geographically around the world, since the raw materials for a product can be extracted in one country, processed in a second, assembled or converted into final product in a third, and distributed and marketed in other countries. Due to material and products flow, some authors indicate that 60% of the total cost of some products is associated with the supply chain and logistics, which have the largest environmental impact. Several researchers have therefore focused on minimizing the costs and environmental impacts of supply chain and logistics.

This book reports a set of tools, methodologies, and techniques that managers are using to improve the supply chain and that allow them to generate a competitive advantage to retain the position of their company in the globalized market with low-cost products while being socially responsible. The 15 chapters explain the application of these tools, methodologies, and techniques in the supply chain, illustrating the focus of managers on cost reduction, partner integration, use of information and communication technologies, algorithms that optimize resources, human resources involvement, and information flow among partners. Every chapter reports an application from various sectors such as the automotive, aerospace, agricultural, and healthcare industries.

Chapter 1 is entitled "Implementation of a Mathematical Model to Improve Sustainability in the Handling of Transport Costs in a Distribution Network" by Velarde et al. They report a mathematical model using mixed-integer linear programming for minimizing cost in the vehicle routing problem from a distribution center to some customers. The model considers different capacities in the distribution network with a starting point for fulfilling each customer's demand, the vehicle carrying capacity, work schedule, and the sustainable use of resources. The model proposes the amount of equipment suitable to satisfy the demand and improves customer service, and optimizing human and economic resources.

In Chapter 2, Lagarda and Leyva report "A Systems Thinking Model to Support Long-Term Bearability of the Healthcare System: The Case of the Province of Quebec", an application to the Canadian healthcare system. They integrate universities, hospitals, doctors, the Ministry of Health and Social Services of Québec, and society in a dynamic system model to find the relationships among these entities, integrating experts from all them.

In Chapter 3, Miranda-Ackerman et al. report "Green Supplier Selection in the Agro-Food Industry with Contract Farming: A Multi-Objective Optimization Approach". They describe a genetic algorithm joined with a multicriteria suppler selection model applied to the agro-food industry, integrating a life cycle assessment, environmental collaborations, and contract farming to produce social and environmental benefits. The main contribution in this chapter is that several scenarios are generated for sharing environmental risk.

Chapter 4 is entitled "A Linear Programming Model with Fuzzy Arc for Route Optimization in the Urban Road Network" by Escobar-Gomez et al. They outline a fuzzy model based on shortest path problem (SPP) to optimize route distances, the amount of fuel used, and travel times. The major contribution in this model is that it integrates uncertainty in demand and delivery time and was applied to a Mexican city.

In Chapter 5, Rivera Cadavid et al. report "Supply Chain Optimization for Energy Cogeneration Using Sugarcane Crop Residues (SCRs)" with a Colombian case study regarding the incentives for the implementation of energy projects with non-conventional sources, specifically related to the use of residual biomass generated by sugar cane supply chains. They report a mixed-integer programming model to decide which plots to harvest on a given day. Although no additional energy is generated, the model shows that it is feasible to replace all coal used in the boilers with sugarcane crop residues for power cogeneration.

Chapter 6 is entitled "Collection of Solid Waste in Municipal Areas: Urban Logistics", by Saucedo-Martinez et al. The chapter proposes improving the planning territories for urban cleaning, weeding, and collection of solid waste in municipal areas using two mixed-integer linear programming models. The model was applied to a Mexican city and integrates variables such as the amount of waste, frequency, and service coverage.

In Chapter 7, Rodriguez-Aguilar et al. report the "Prices of Mexican Wholesale Electricity Market: An Application of Alpha-Stable Regression". The authors propose a model to estimate prices in the Mexican wholesale electric market, integrating gradual increases in the number of competitors and the geographic and technical characteristics of electric power generation. They conclude that prices in electricity distribution fluctuate due to seasonality, the availability of fuel, congestion problems in the electrical network, as well as other risks such as natural hazards.

Chapter 8 is entitled "Environmental Impacts of Reusable Transport Items: A Case Study of Pallet Pooling in a Retailer Supply Chain" by Accorsi et al. The chapter focuses on pallet management for transport operations. The authors analyze a combination of the pooler's management strategies with different retailer network configurations results in different scenarios, which are assessed and compared through a what-if analysis. The logistical and environmental impacts generated by the pallet distribution activities are quantified for each scenario through tailored software incorporating geographic information system (GIS) and routing functionalities.

In chapter 9, Shahbazi et al. report "Comparison of Four Environmental Assessment Tools in Swedish Manufacturing: A Case Study" with four environmental assessment tools commonly used among Swedish manufacturing companies: green performance mapping (GPM), environmental value stream mapping (EVSM), waste flow mapping (WFM), and life cycle assessment (LCA) to help practitioners and scholars to understand the different features of each tool. The conclude that some overlap and differences exist between the tools and a given tool may be more appropriate for a situation depending on the specific context.

Chapter 10 is entitled "Parameter Setting for a Genetic Algorithm Layout Planner as a Toll of Sustainable Manufacturing", by Krajčovič et al. The chapter describes a genetic algorithm layout planner (GALP) used to optimize the spatial arrangement in manufacturing and logistics system design applied to industrial layout. The structure was integrated into the design process of manufacturing systems. The main contribution in this chapter is related to parameters identification for an adequate industrial layout.

In chapter 11, Mendoza-Fong et al. report "The Role of Green Attributes in Production Processes as Well as Their Impact on Operational, Commercial, and Economic Benefits" using a second-order structural equation model to identify the relationship among the green attributes before and after an industrial production process, the operating benefits, the commercial benefits, and the economic benefits. The authors conclude that companies focused on increasing their greenness level must monitor and evaluate the green attributes in their production process to guarantee benefits and make fast decisions, if required, due to deviations.

Chapter 12 is entitled "Supply Chain Innovation in Scientific Research Collaboration" by

Hung-Yuan et al., who conducted a content analysis followed by a social network analysis to systematically review the supply chain innovation (SCI) in the last two decades. They conclude that SCI research was originally concentrated in the United States and did not receive much attention in Europe and Asia until more recently. An analysis of collaboration networks indicates that an SCI research community has just started to form, with the United Kingdom at the center of the international collaborative network.

In Chapter 13, Zhu et al. report their work "Warranty Decision Model and Remanufacturing Coordination Mechanism in Closed-Loop Supply Chain: View from a Consumer Behavior Perspective". They discuss and compare decision variables such as remanufacturing product pricing, extended warranty service pricing, and warranty period to measure the supply chain system profit. The findings indicate that consumers' decision-making significantly affirms the dual marginalization effect of the supply chain system while significantly affecting the supply chain warranty decision.

Chapter 14 is entitled "Reverse Logistic Strategy for the Management of Tire Waste in Mexico and Russia: Review and Conceptual Model, by Uriarte-Miranda et al. They studied tire waste management from economic, environmental, and social contexts with a reverse logistic (RL) model to improve the process in Russia and Mexico. The model considers regulations and policies in each country to assign responsibilities regarding RL processes for the management of tire waste.

In Chapter 15, Duque-Uribe et al. report their paper "Sustainable Supply Chain Management Practices and Sustainable Performance in Hospitals: A Systematic Review and Integrative Framework". The chapter presents a systematic literature review and develops a framework for identifying the supply chain management practices that may contribute to sustainable performance in hospitals. The proposed framework is composed of 12 categories of management practices, which include strategic management and leadership, supplier management, purchasing, warehousing and inventory, transportation and distribution, information and technology, energy, water, food, hospital design, waste, and customer relationship management.

Jorge Luis García-Alcaraz
Special Issue Editor

Article

Implementation of a Mathematical Model to Improve Sustainability in the Handling of Transport Costs in a Distribution Network

José Manuel Velarde [1], Susana García [1], Mauricio López [1] and Alfredo Bueno-Solano [2],*

[1] Department of Industrial Engineering, Instituto Tecnológico de Sonora, Navojoa, Sonora 85800, Mexico; jose.velarde@itson.edu.mx (J.M.V.); susana.garciavil@gmail.com (S.G.); mlopeza@itson.edu.mx (M.L.)
[2] Department of Industrial Engineering, Instituto Tecnológico de Sonora, Cd. Obregon, Sonora 85000, Mexico
* Correspondence: abuenosolano@itson.edu.mx

Received: 30 October 2019; Accepted: 12 December 2019; Published: 19 December 2019

Abstract: This work considers the application of a mathematical model using mixed-integer linear programming for the vehicle routing problem. The model aims at establishing the distribution routes departing from a distribution center to each customer in order to reduce the transport cost associated with these routes. The study considers the use of a fleet of different capacities in the distribution network, which presents the special characteristic of a star network and which must meet different efficiency criteria, such as the fulfillment of each customer's demand, the vehicle carrying capacity, work schedule, and sustainable use of resources. The intention is to find the amount of equipment suitable to satisfy the demand, thus improving the level of customer service, optimizing the use of both human and economic resources in the distribution area, and leveraging maximum vehicle capacity usage. The MILP mixed-integer linear programming mathematical model of the case study is presented, as well as the corresponding numerical study.

Keywords: MIPL; VRP; star network; optimal plant location

1. Introduction

As a result of globalization, the distances that products travel from the producer to the final consumer have increased, so ensuring the constant flow of products in safe and quality environments have become a critical challenge from the point of view of logistics sustainability. In this sense, a tool that has been a valuable ally for proper planning of distribution networks is the vehicle routing problem (VRP), which belongs to the class of combinatorial optimization problems that have been studied extensively over the past several decades due to their multiple applications in everyday life [1]. The VRP is focused on the proper and sustainable management of the elements that comprise the distribution system, a problem faced by thousands of companies and enterprises dedicated to the collection and/or delivery of goods in a distribution network.

Depending on each variant of the VRP, there can be different approaches to finding a solution, so the objectives and restrictions to be found in practice are very broad, offering the opportunity to apply different models and algorithms in the search for a solution that guarantees to secure the lowest cost possible based on the efficient use of the resources available in the distribution area. There are currently numerous research studies being carried out by different authors, in which they propose different solution algorithms of this problem [2]. In the literature, the classic VRP model is characterized by the optimal design of routes, including the construction and selection of routes from a central depot to each customer and to which this vehicle must return upon completion of its distribution route. In this environment, it is necessary to fulfill different conditions or restrictions, such as visiting each customer only once to meet the demand (known), as well as respecting the different capacity restrictions on each type of vehicle if not heterogeneous, vehicle capacity, maximum distance covered, and working hours, with the main objective of finding the lowest cost when selecting distribution routes [3]. The VRP has been the subject of research for many years due to the great scientific interest in its multiple applications to everyday problems, considered NP-complete problems, in which for it is not possible to find an optimal result in due time [4].

Currently, many organizations focus their efforts on improving their different processes in order to be socially responsible, reduce costs, and increase profits. The area with the greatest interest in cost reduction in recent years has been logistics, especially in transport, which represents a direct impact upon the cost of the final product [5]. It is essential to have a method that provides good management of the planning and programming of the different distribution and transport activities that aids in the suitable selection of the distribution route to each customer to minimize the costs associated with this process. To achieve this, it is necessary to use different tools, such as exact optimization, which employs mathematical models in the search of a solution that guarantees to attain the exact solution for small to midsize problems.

The aim of this research is to find a sufficiently robust and actionable solution to practical problems that seeks to streamline the different logistical processes of its distribution system via minimization of the costs associated with the system of transport to its main customers, considering that products are required in specific schedules (within time windows), vehicle capacities, and business hours. These must timely and quality solutions for the midsize instances operated by the company. To do this, a real case study is proposed to address the vehicle routing problem with hard time windows, i.e., each customer has a specific time window within which to be attended to meet demand. To achieve this, we propose a mathematical model based on mixed-integer linear programming, which contemplates the special case of the star distribution network, in which there are only direct routes between the distribution center and one of the customers for each vehicle trip.

2. Literature Review

Numerous works in the literature address the VRP, as well as multiple classifications, such as VRP with limited capacity, time windows, simultaneous deliveries and collections, and others.

A study into the distinct classifications of the VRP is presented in reference [6], specifying the differences in the methods for modeling the objective functions of each variant of the problem, as well as the diverse restrictions. One of the most widely used variants in the VRP is the capacitated version of the problem. This variant considers that the destinations will be visited only once to deliver or collect products and its objective function is based on the minimization of the total distance covered by each of the vehicles. Another variant of this problem would be to consider simultaneous collections and deliveries for each customer, as well as restrictions on the product delivery or collection times for each customer.

The authors of reference [7] propose that, due to the particular evolution of the VRP, it is necessary to include different aspects that were not previously considered, for example, multi-objective programming, since it is imperative to contemplate the need not only for reduction strategies but also for compliance with time windows and customer satisfaction. The aforementioned work also affirms that it is essential to find a balance between the environmental cost and the economic cost.

The research in reference [8] describes, in general terms, the main characteristics of the VRP in use since its formulation in 1959, whose objective is to establish a set of routes to each geographically dispersed customer in one zone, complying with a series of limitations to minimize cost. In the past 10 years, great advances have been made regarding resolution techniques for large problems. It is also noted that the use of information technologies, such as global positioning systems, radio-frequency identification, and the use of high-capacity computerized information processing, favor the development of new techniques and models for obtaining efficient solutions in less time for large problems.

The use of heterogeneous vehicle fleets has given rise to the Heterogeneous Vehicle Routing Problem (HVRP) [9], which aims to minimize the total distance covered on each route by each vehicle, satisfying each customer's demand and including capacities in non-homogeneous vehicles and costs. It resolves the problem by applying a metaheuristic algorithm based on a taboo search, which works by accepting infeasible solutions with a penalty to give the search diversity, reaching quick and efficient solutions in comparison to the traditional method used by the company.

The work developed by the authors of reference [10] presents the formulation of a mathematical model applied to the transport problem of two enterprises seeking to reduce operating costs in the logistics area, intending to improve their level of customer service and competitiveness.

In reference [11], an exact programming model is proposed for programming deliveries from a central depot to each customer (meeting customer demand), looking to minimize the costs related to moving products throughout the distribution network. The results given by the algorithm show that for medium problems, quick, reliable, and valid solutions are obtained. Likewise, reference [12] proposes the Fleet Size and Mix Vehicle Routing Problem with Time Windows (FSMVRPTW) for variations of the VRP and establishes that significant savings can be obtained by conjointly employing a knowledge base and computer-based and operations research tools.

Reference [13] presents the VRP, considering multiple depots, classifying it as NP-hard. It presents a grouping technique (clustering) to generate initial solutions with a local search algorithm, an iterated location search (ILS) to obtain a quick solution that simultaneously seeks to establish the possible routes of a set of vehicles. The main objective is to determine the total distance covered by the vehicles in each route and to minimize that distance, considering the particular characteristics of the system in addition to the capacity of each depot and each vehicle. The performance of the proposed methodology is feasible and effective for resolving the problem in terms of the quality of the computational responses and the times obtained; a comparison was carried out with some testing instances in the literature.

In reference [14], an alternative methodology was implemented to resolve the problem of flower collection and transportation, which used a model reflecting the stochastic behavior of the demand, where the solution method includes clusters for the collection points. This methodology required a model including the correct route design, the proper assignation of routes to trucks, and a regression model to obtain the equation of the total system cost. In reference [15], the use of two heuristics is considered for resolving the vehicle routing problem, which considers a flexible demand in the mix of collection and delivery services, with restrictions on the maximum route duration, the main difference in the VRP with simultaneous collections and deliveries.

In reference [16], two well-known strategies were implemented in delivery routes in urban areas—the first is the application of the capacitated VRP, and the second is the problem of loading plan optimization. This is based on the use of an approximation with a hybrid method of the two strategies and with the concept of robustness introduced into the route to guarantee a level of predefined service, according to vehicle performance.

In the work developed by the authors of reference [12] for the case of large problems where computing time is of the essence when obtaining a solution, the use of Lagrangian relaxation is proposed to generate lower bounds [17], which necessitates adding a penalty term to the objective function to avoid violating relaxed restrictions. When resolving the Lagrangian problem, a lower bound is returned for original optimal objective value minimization problems. Another technique can be applied later—for example, that of ant colony algorithms, simulated annealing, genetic algorithms, taboo search, and/or artificial neural networks—to seek a better solution to the problem [18,19].

The work presented in reference [20] uses a discrete simulation to represent and analyze transport and distribution process performance in construction material mining enterprises. To this end, subjects such as construction materials, discrete simulation, transport, and distribution are covered. As a result of the article, the discrete simulation allows quantitatively analyzing transport and distribution performance, making it possible to measure the amounts mobilized, the efficiency of the process, and the use of resources.

A distribution network involves everything related to material delivery or collections; crucial elements to consider are the plant's capacity, its distribution centers, and its transport fleet, to satisfy the demands of each customer in due time and manner. The efficient use of the logistical resources and the human capital are significant factors for meeting the objectives and challenges proposed in the logistics activities of an organization, with the result that the costs associated are minimal. The growing competition in today's world stage, the introduction of new products with very short lifecycles, and growing customer demand have driven industrial organizations and enterprises to invest in improving their current logistics systems. The above added to the different changes in the systems of communication and technologies applied to transport systems (which aid and streamline movement) have contributed to the continuous development of logistics systems administration and management [5].

Research into logistics is very varied and extensive, as can be observed in works such as reference [21], in which the concept of logistics is managed from different perspectives, one of these being a business that establishes that suppliers must-have products to offer the customers, and that these must be provided in due time, manner, number, quality, undamaged, and at a minimal cost. Logistics addresses the flow of materials, of finished products, and the information associated with same (the flow of merchandise and the flow of information being developed in parallel), from the supplier to the customer, with the required quality, at the right place and right time, and a minimal cost. Logistics is that part of the process of the supply chain that plans, implements, and controls the flow and storage of products and services and related information, from the point of origin to the point of consumption, efficiently and at the lowest possible cost, to satisfy customer requirements [22]. Logistics can also be called business logistics, emphasizing rapid customer response systems, distribution or delivery channel administration, industrial logistics, physical international distribution, supply chain administration, and currently, on a value network [23]. Distribution channels allow marketing efforts to become a reality and are one of the main pillars in satisfying the end consumer [24].

1. Objective

The objective consists of minimizing the total integrated transport cost by the sum of the costs of the trip, which includes the costs generated, whether due to arriving late or early to the customer, the cost of the fuel used to move from the central depot to each customer and vice versa, and meeting the demands of each customer while considering time windows, vehicle capacity, and distribution center restrictions, and duration of the workday.

2. Justification

In reference [25], the importance of the transport and its impact on the company's logistics distribution systems are weighted, so the proper use of the available logistical resources is at the center of most distribution route design problems. The cost associated with the transportation process represents between 10% and 20% of the total cost of the products [26].

The work presented in reference [27] establishes that land transport represents 75% of available transport utilized, making it the most widely used, followed by rail with 17%, maritime with 7%, and air with 1%. Among its main advantages are its door-to-door service, its flexibility due to a wide variety of adapted vehicles of all load types and volumes, and the speed and facility with which it can be coordinated with other means of transport. The main disadvantage is that it must be limited to the weight and volume of the load to be moved.

3. Problem Statemen

In many organizations, the planning of transport and distribution activities represents a serious decision-making problem. This situation has become increasingly important due to the contribution of distribution costs in the total product costs.

Many enterprises require a fleet of vehicles for the collection and/or delivery of products within a distribution network. The efficient programming and use of said fleet is the main problem in the majority of transport problems. Distribution area managers specifically ask themselves, how many, from which plant, and what capacity should the vehicles be to satisfy the demand of each customer at a minimal cost? This question is difficult to answer due to the substantial number of possible combinations among the mix of fleet and routes; this is at the heart of how the proposed model has been implemented in this work [28].

There are currently two main variables that the business of cargo transport (VRP) focuses on. The first is the cost of fuel and the reduction in contaminant particle emissions into the environment. Therefore, cargo truck manufacturers have focused their attention over the last decade on technological development and innovation to achieve greater fuel consumption efficiency, seeking to gain a competitive advantage over rivals in the market that benefits the enterprise's profitability. To increase truck carrying capacity, automotive equipment for more than 30 tons is being designed, with different interconnected equipment for greater load capacity and consolidation, thus maximizing the movement of goods in long hauls and reducing the transport cost [29].

3. Methodology

The mathematical model proposed in reference [28] was the foundation for the implementation of the practical problem addressed in this study. These works present an optimization model similar to the problem under study. Different modifications were made to adapt it to the problem addressed, which are based on the following:

- The transport network has a star configuration (Figure 1), in which only one round trip per customer is allowed to meet the demand;
- The time windows are closed, because unloading can only take place within an established timeframe, and the vehicle may not arrive either before or after the allocated time window, as this generates additional costs;
- This problem is typical in those cases where the first and final destinations are the same point and not the customer;
- The vehicles are assigned to a specific plant which forces you to return to this plant, that is to say a vehicle must start and finish its journey in the plant to which it is assigned.

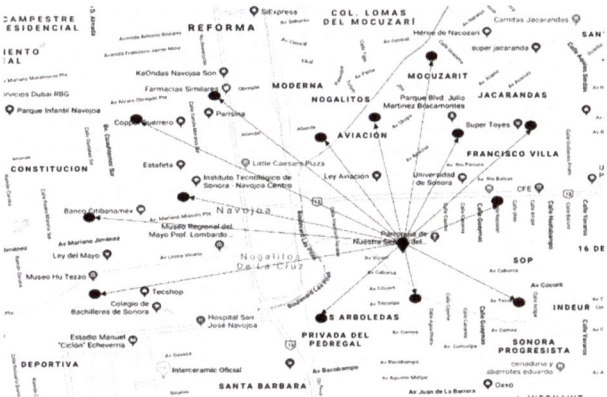

Figure 1. This figure represents a star network topology.

One of the main differences between the model proposed in this real case study and the classic vehicle routing problems [2] consists of the following: the assumption that, on one hand, the transport network has a star topology in those cases in which only direct routes are allowed, i.e., there can only be trips between the central depot and the customers, and, on the other hand, each customer can be visited several times (which can be by the same vehicle) in order to satisfy the demand.

Formulation of the model: sets, indices, and parameters used to characterize the model

I = Set of clients
P = Set of plants
J = Set of travels
K = Set of vehicles
i = Index corresponding to customers $i \in I = \{0, 1, 2, 3, \ldots, I\}$
p = Index corresponding to plants $p \in P = \{1, 2, 3, \ldots, P\}$
j = Index corresponding to travels $j \in J = \{1, 2, 3, \ldots, J\}$
k = Index corresponding to vehicles $k \in K = \{1, 2, 3, \ldots, K\}$
E_i = Window start time for customer $i \in I$
L_i = Window end time for customer $i \in I$
C_{ip} = Cost of a trip to customer i from plant p; $i \in I$, $p \in P$
Cd_{ip} = Cost for arriving late to customer i from plant p; $i \in I$, $p \in P$
Ce_{ip} = Cost for arriving early to customer i from plant p; $i \in I$, $p \in P$
D_i = Customer i demand; $i \in I$
θ_{ip} = Travel time to customer i from plant p; $i \in I$, $p \in P$
Q_{kp} = Capacity of vehicle k from plant p; $k \in K$, $p \in P$

Decision variables

$$X_{ipjk} = \begin{cases} 1 & \text{If customer } i \text{ is visited by vehicle } k \text{ on trip } j \text{ from plant } p \\ 0 & \text{otherwise.} \end{cases}$$

S_{pjk} : Schedule in which each vehicle k must leave to complete each trip j from each plant p
W^+_{ipjk} : Vehicle k waiting time at customer i on trip j from plant p
W^-_{ipjk} : Vehicle k delay time at customer i on trip j from plant p

Objective Function and Restrictions

The objective of the problem consists of determining a set of routes to be followed considering a fleet of heterogeneous vehicles that depart from one or more central depots or warehouses intending to satisfy the demand of various geographically dispersed customers, minimizing the total cost of product transport.

The resulting mathematical model in mixed-integer linear programming is presented below.

Equation (1) calculates the total transport cost, which includes the cost for visiting the customer, as well as the costs for deviations or non-compliance within the time windows for the different customers in the distribution network.

Objective function:

$$\text{Min} \sum_{ipjk} C_{ip} X_{ipjk} + \sum_{ipjk} Ce_{ip} W^+_{ipjk} + \sum_{ipjk} Cd_{ip} W^-_{ipjk} \tag{1}$$

To ensure that for any trip, a vehicle is exactly in the plant (customer (0)) or visiting a customer, the following restriction is established (2):

$$X_{0pjk} + \sum_{i \neq 0} X_{ipjk} = 1, \ p \in P, j \in J, k \in K \tag{2}$$

An obligatory restriction in the majority of routing problems is that of meeting customer demand, which is established with Restriction (3), expressed by the following equation:

$$\sum_{pjk} Q_{kp} X_{ipjk} \geq d_i \ i \in I, \ I \cup \{0\} \tag{3}$$

If a vehicle is used, the time-of-use conditions must be established as well as the departure time for each trip to be made. It is also established that a trip can only begin once vehicle k has returned to the depot after its previous trip. Restriction (4) is established for this purpose via the following expression:

$$S_{p, j+1, k} \geq S_{pjk} + \sum_{i} 2\theta_{ip} X_{ipjk}, \; p \in P, j \in J, k \in K \tag{4}$$

This ensures that if on any trip j, the customer is not visited ($X_{ipjk} = 0$) then there is no cost for arriving early, given that ($W^{+}_{ijpk} = 0$), otherwise, the time violation is calculated as standard, depending on Restriction (5) given by the following expression:

$$W^{+}_{ipjk} \geq E_{ip} X_{ipjk} - \left(S_{pjk} + \theta_{ip} X_{ipjk}\right) \; \forall \; i \neq 0, \; p \in P, j \in J, k \in K \tag{5}$$

To estimate the cost of arriving after each customer's closing time window, Restriction (6) is established, given by the following expression:

$$W^{-}_{ipjk} \geq \left(S_{pjk} + \theta_{ip} X_{ipjk}\right) - L_i - M\left(1 - X_{ipjk}\right) \; \forall \; i \neq 0, \; p \in P, j \in J, k \in K \tag{6}$$

Here, M refers to a large positive integer to full fill $(M \geq S_{pjk} - L_i)$. It must be clarified that in (5) and (6) $W^{+}_{ijpk} > 0$ only for $S_{pjk} + \theta_{ip} < E_i$, while $W^{-}_{ijpk} > 0$ only for $S_{pjk} + \theta_{ip} > L_i$, so for $E_i \leq L_i$ always $W^{+}_{ipjk}, W^{-}_{ipjk} = 0$.

Depending on the characteristics of the variables defined in the model, binary (0,1) and continuous variables are established. This situation is established through Restriction (7), given by the following expression:

$$X_{ipjk} \in \{0, 1\}, \; W^{+}_{ipjk}, W^{-}_{ipjk}, S_{pjk} \geq 0 \tag{7}$$

4. Results

The present real case study addressed the vehicle routing problem with hard time windows, i.e., each customer has a specific time window within which to be attended to meet demand. To achieve this, a mathematical model based on mixed-integer linear programming was used, which contemplates the special case of the star distribution network, in which there are only direct routes between the distribution center and one of the customers for each vehicle trip.

An MIPL was obtained, represented by employing Expressions (1)–(7). The problem consists of the selection of transportation means (vehicles) within a distribution network composed of a single plant or distribution center, with vehicles of two distinct capacities and restrictions upon the windows of attention to each customer and the fulfillment of each customer's demand. The solution must provide the number of vehicles necessary to meet the demand at a minimal cost or identify if the current fleet can face increasing customer demand.

In relation to the fulfillment of the time windows, although the model contemplates the option that the vehicle can reach the customer sooner or later by means of a penalty in the direct cost to the objective function, this situation is not presented for the practical case here addressed, since due to the nature of the VRP where it is sought to minimize total transportation costs, considering this penalty (which is very high) as an option that would result in an increase in the total cost of planning the route to follow.

The results obtained from the experimentation with the mathematical model applied are presented. An AMPL language was used for model implementation as well as CPLEX optimization software version 12.8 executed on a MacBook Pro with macOS High Sierra Version 10.13.4, with a 2.2 Ghz quad-core Intel Core i7 processor with 16 GB of RAM. The real case is performance in Navojoa Sonora, México. The model was used to resolve the practical problem of 22 customers, with a fleet of eight 48-ton capacity vehicles and two 20-ton capacity vehicles, with the demand and time windows known.

The intention is to obtain enough information to make the best decision that guarantees to reduce the costs associated with the distribution system; one important decision is the size and configuration of the fleet. It is important to consider that the results given by the model indicate that the current configuration is not suitable since there is the equipment of a greater capacity than is needed, as the trips normally made do not use 100% of the vehicle carrying capacity, leading to a deficit in the use of the equipment. The percentage of equipment use considering the demand is approximately 65%. Currently, 25 trips are made in total to meet customer demand, distributed as shown in Table 1.

Table 1. The number of trips carried out per vehicle.

Vehicle	1	2	3	4	5	6	7	8	9	10
Current Trips for vehicle	3	2	2	2	3	2	2	2	4	3

It can be seen that some of the vehicles make a greater number of trips than the rest, maintaining a standard deviation of 0.71 in the number of trips made, leading to the current discontent among equipment operators.

The total costs of distribution programming reported by the model ascend to 91,868 per operating day with the current vehicle fleet configuration. It is important to consider a more equitable distribution of each team's trips to establish uniform workloads for each operator. Therefore, Table 2 shows the trips carried made by each vehicle with the proposal for improvement in the number of pieces of equipment available, which was obtained by reducing the number of 40-ton capacity vehicles to be used from eight to two and increasing the number of 20-ton capacity vehicles from two to eight. Obtaining the same total transport cost with the difference in carrying capacity, which was increased to 78%. The standard deviation of the number of trips carried out by each vehicle was also improved to 0.52 of the total trips made.

Table 2. The number of trips carried out per vehicle (proposed).

Vehicle	1	2	3	4	5	6	7
Proposed Trips for vehicle	2	5	4	4	3	4	3

Table 3 shows the trips made by each vehicle in a second proposal for improvements to the system, which consists of reducing to seven units in operation, considering a total of five 20-ton capacity units and two 40-ton capacity units. Although the total operating cost per day increases by 9980 per day, the equipment utilization rate of the total vehicle load capacity in improved to 87%.

Table 3. The number of trips carried out per vehicle (proposal 2).

Vehicle	1	2	3	4	5	6	7
Proposed Trips for vehicle	2	5	4	4	3	4	3

5. Conclusions

It is of utmost importance to consider that the problem addressed considers atypical characteristics of a classic VRP, such as considering the star network and the fact that the same customer can be visited more than once to meet demand.

It was possible to satisfactorily resolve the real problem here addressed by applying a mathematical optimization model in addition to establishing two proposals for improvement to the fleet configuration, considering its size (amounts) and composition (capacities) suitable to the company's available logistical resources.

First, the configuration maintains the same total number of vehicles but considers two 40-ton capacity vehicles and eight 20-ton capacity vehicles, resulting in the same total cost as the original but with a better workload distribution for the equipment operators and above all improving the equipment utilization rate without increasing the total distribution route programming cost.

Second, the reduction to seven units of the total number of vehicles available for use is recommended, considering two 40-ton capacity vehicles and five 20-ton capacity vehicles, where the total vehicle load utilization rate of 87% is considered the main contribution.

These two proposals for improvement are a significant contribution to the problem posed during the study and are considered of great value due to this being a real problem resolved through an exact method of mixed-integer linear programming for a case of 22 customers with capacity, demand, and time window restrictions.

This paper aims to contribute to the ONU sustainable development goal 12 to promote a responsibly consume of resource and energy efficiency, and sustainable use of the infrastructure.

Author Contributions: Conceptualization, J.M.V.-C and S.G.; investigation, J.M.V.-C and S.G.; data curation, M.L.-A; writing—original draft preparation, J.M.V.-C and A.B.-S; writing—review and editing, A.B.-S; visualization, A.B.-S. All authors have read and agreed to the published version of the manuscript.

Funding: This research received no external funding

Acknowledgments: The work team thanks all the participants for their interest and collaboration during the data collection. Thanks also to Technological Institute of Sonora (ITSON) as well as the national laboratory for transportation systems of the National Council of Science and Technology (CONACYT) and the company for their support and facilities for the development of the project; this publication has been funded with resources from PFCE 2019 and the Research Promotion and Support Program (PROFAPI 2019).

Conflicts of Interest: The authors declare no conflicts of interest.

References

1. Baldacci, R.; Battarra, M.; Vigo, D. Routing a heterogeneous fleet of vehicles. In *The Vehicle Routing Problem: Latest Advances and New Challenges*; Springer: Boston, MA, USA, 2008; pp. 3–27.
2. Toth, P.; Vigo, D. *The Vehicle Routing Problem*, 1st ed.; Society for Industrial and Applied Mathematics: Philadelphia, PA, USA, 2002.
3. Bustos, A.; Jiménez, E. Enfasís: Models for better vehicular routing. Available online: http://www.logisticamx.enfasis.com/articulos/69225-modelos-un-mejor-ruteo-vehicular. (accessed on 24 March 2014).
4. Prins, C. A simple and effective evolutionary algorithm for the vehicle routing problem. *Comput. Oper. Res.* **2004**, *31*, 1985–2002. [CrossRef]
5. Ballesteros, D.; Ballesteros, P. Importance of logistics administration. *Sci. Tech.* **2008**, *14*, 217–222.
6. Jaramillo, P.; Rodrigo, J. Tabu search for vehicle routing. *Ind. Eng.* **2012**, *30*, 29–43.
7. Toro, O.; Eliana, M.; Escobar, Z.; Antonio, H.; Granada, E. Literature Review on The Vehicle Routing Problem In The Green Transportation Context. *Luna Azul Mag.* **2016**, 362–387. [CrossRef]
8. Ghannadpour, S.; Noori, S.; Moghaddam, R.; Ghoseiri, K. A multi-objective dynamic vehicle routing problem with fuzzy time windows: Model, solution and application. *Appl. Soft Comput.* **2014**, *14*, 504–527. [CrossRef]
9. Puenayán, D.; Londoño, J.; Escobar, J.; Linfati, R. An algorithm based on a granular taboo search for the solution of a vehicle routing problem considering heterogeneous fleet. *Mag. Eng. Univ. Medellín* **2014**, *13*, 81–89. [CrossRef]
10. Arboleda, J.; López, A.; Lozano, Y. The problem of vehicle routing (VRP) and its application in Colombian medium-sized companies. *Ingenium* **2016**, *10*, 29–36.
11. Raffo, E.; Ruiz, E. Optimization model of the delivery route. *J. Fac. Ind. Eng.* **2005**, *8*, 75–83.
12. Reyes, N. Route programming optimization model for a Peruana logistics company using FSMVRPTW tools. *Ind. Data* **2016**, *16*, 118–123. [CrossRef]
13. Toro, E.; Domínguez, A.; Escobar, A. Performance of grouping techniques to solve the routing problem with multiple deposits. *Technol. Log.* **2016**, *19*, 49–62.
14. Gonzalez, E.; Adarme, W.; Orjuela, J. Stochastic mathematical model for vehicle routing problem in collecting perishable products. *DYNA* **2015**, *82*, 199–206. [CrossRef]

15. Dechampai, D.; Tanwanichkul, L.; Sethanan, K.; Pitakaso, R. A differential evolution algorithm for the capacitated VRP with flexibility of mixing pickup and delivery services and the maximum duration of a route in poultry industry. *J. Intell. Manuf.* **2017**, *28*, 1357–1376. [CrossRef]
16. Guedria, M.; Malhene, N.; Deschamps, J. Urban Freight Transport: From Optimized Routes to Robust Routes. *Transp. Res. Procedia* **2016**, *12*, 413–424. [CrossRef]
17. Guignard, M. En hommage a Joseph-Louis Lagrange. *Oper. Res.* **2007**, *149*, 103–116.
18. Litvinchev, I. Refinement of lagrangian bounds in optimization problems. *Comput. Math. Math. Phys.* **2007**, *47*, 1101–1109. [CrossRef]
19. Litvinchev, I.; Rangel, S. Comparing Lagrangian bounds for a class of generalized assignment problems. *Comput. Math. Math. Phys.* **2008**, *48*, 739–746. [CrossRef]
20. Gómez, M.; Rodrigo, A.; Correa, E.; Alexander, A. Analysis of Transportation and Distribution of Building Materials Using Discrete 3D Simulation. *Earth Sci. Bull.* **2011**, *30*, 39–51.
21. González, N. New freight chains generated by the modal interchange logistics infrastructures. *Transp. Territ. Mag.* **2016**, *14*, 81–108.
22. Ballesteros, D.; Ballesteros, P. Competitive Logistics and Supply Chain Management. *Sci. Tech.* **2004**, *1*, 201–206.
23. Ocampo, P. Logistics and global management. *Sch. Bus. Adm. Mag.* **2009**, *66*, 113–136.
24. Sierra, C.; Moreno, J.; Silva, H. Distribution channels: main features of wholesale distributors of mining extraction construction materials in Barranquilla—Colombia. *TELOS* **2015**, *17*, 512–529.
25. Bermeo, E.; Calderón, J. Design of a transport route optimization model. *Man. Mach.* **2009**, *32*, 52–67.
26. Toth, P.; Vigo, D. *An Overview of Vehicle Routing Problems, The Vehicle Routing Problem*; SIAM: Philadelphia, PA, USA, 2001; pp. 1–26.
27. Segura, R. Land transport, the most used means in Mexico. *Transportes Turismo Magazine*, 6 August 2015. Available online: https://tyt.com.mx/reportajes/el-transporte-terrestre-hace-mas-competitivo-a-mexico/ (accessed on 6 August 2015).
28. Velarde, J.; Castro, L.; González, D.; Cedillo, M.; Fox, J.; Lozano, L.; Litvinchev, I. Fleet optimization and vehicle allocation in a star network. *J. Sci. Eng. Environ.* **2014**, *2*, 21–26.
29. Hnaien, F.; Yalaoui, F.; Mhadhbi, A.; Nourelfath, M. A mixed-integer programming model for integrated production and maintenance. *IFAC PapersOnLine* **2016**, *49*, 556–561. [CrossRef]

© 2019 by the authors. Licensee MDPI, Basel, Switzerland. This article is an open access article distributed under the terms and conditions of the Creative Commons Attribution (CC BY) license (http://creativecommons.org/licenses/by/4.0/).

Article

A Systems Thinking Model to Support Long-Term Bearability of the Healthcare System: The Case of the Province of Quebec

Ernesto A. Lagarda-Leyva [1,*] and Angel Ruiz [2]

1 Industrial Engineering Department, Instituto Tecnológico de Sonora, Cd. Obregón 85000, Mexico
2 Faculté des sciences de l'administration, Département opérations et systèmes de decisión, Université Laval, Quebec City, QC G1V 0A6, Canada; angel.ruiz@fsa.ulaval.ca
* Correspondence: ernesto.lagarda@itson.edu.mx or ernesto.lagarda@me.com

Received: 30 October 2019; Accepted: 4 December 2019; Published: 9 December 2019

Abstract: This paper describes the modeling efforts devoted by the Ministry of Health and Social Services of Québec, Canada (MSSS), to ensure the long-term bearability of their care system. To this end, it studies the relationships between four entities that self-regulate and interact to form the complex care-providing system: (1) universities; (2) hospitals and doctors; (3) the ministry; and (4) society. The first phase of this research focuses on modeling such relationships and relies on the system dynamics methodology to adequately capture the long-term dynamics of the system. The methodology encompasses three phases: (a) determination of the critical variables and parameters of each entity; (b) development of the causal diagram of each entity; and (c) integration of the individual causal diagrams to form the global system diagram. The final casual model illustrates and explains the relationships between all the entities and constitutes an excellent tool to support experts during discussions or focus groups where critical variables that positively or negatively affect the system can be evaluated. We intend to enrich this casual model in a further phase of the project, which will hopefully lead to a simulation and scenario analysis tool that can be used to support managers in their long-term decision-making process.

Keywords: system dynamics; systems thinking; causal models; healthcare services; bearability; sustainability

1. Introduction

The healthcare and social services system offered throughout Québec was created in 1971 following the adoption of a healthcare and social services Act by the National Assembly of Québec [1]. It is a public system, with the State acting as the main insurer and administrator; all residents of the province of Québec are admissible. The plan covers all the basic healthcare services, except some specific treatments such as aesthetic surgery and parallel medical practices known as natural or alternative [2].

In 2012, the entire healthcare expenditure in Québec roughly totaled $43.5 billion. This included both public expenses (including direct expenses covered by the government for the people under its care) and private expenses (amounts claimed from private insurance plans, direct payments—such as contributions to accommodation (CHSLD) and for the purchase of drugs—for example, made by individuals and donations). Public healthcare expenditure, which rose to $30.5 billion in 2012, accounted for 70.2% of the total [1].

Social services expenditure is not included in the estimates of the Canadian Institute for Healthcare Information. It accounts for roughly 12% of the Government of Québec's total healthcare and social services expenditure. Between 2000 and 2012, public and private healthcare expenditure grew annually by an average of 4.9% and 5.8%, respectively [1].

On the other hand, the higher education institutions are responsible for training medical doctors in Québec; four higher education institutions are financed by the State to provide such training. Since the Act to Modify the Organization and Governance of the Healthcare and Social Services Network—in particular by abolishing the regional agencies—came into force on 1 April 2015, the Québec healthcare and social services network have included the following aspects:

(a) Thirteen integrated healthcare and social services centers (CISSS), and nine university healthcare and social services integrated centers (CIUSSS). Only integrated centers located in a healthcare region where a university offers a complete undergraduate medical program or operates a university institute in the social field are entitled to be called CIUSSS.
(b) Seven institutions that were not amalgamated with an integrated center, of which four are university hospital centers (CHU) and three are university institutes (IU).
(c) Five institutions not covered by the Act, which are offering services to an aboriginal and Northern population.
(d) Each institution may offer services in several sites that are physical locations where healthcare and social services are provided.
(e) It should be noted that 17 institutions that were not amalgamated under the Act have been grouped into integrated centers, and are managed by the center's board of directors.
(f) In addition to the services provided by public institutions, society benefits from services such as lodging and long-term care that are provided by private institutions.

Moreover, four integrated university healthcare networks (RUIS) promote collaboration and complementarity and fulfill the combined mission of care, education, and research that is incumbent upon the healthcare institutions and higher education institutions with which they are affiliated. These are the Université Laval, McGill University, Université de Montréal, and Université de Sherbrooke integrated higher education institutions healthcare networks [3].

The future capacity of MSSS to satisfy Quebec's population demand for specialized healthcare depends on the demand (the population) and the available system capacity, which basically depends on the number of available doctors and the number of contracts offered by the MSSS. The number of doctors at a given time t shows a dynamic behavior, that depends on the number of doctors at an initial moment $t_0 < t$ plus the students that graduate from higher education institutions between t_0 and t, minus the doctors that retire between t_0 and t. Since specialized training requires between 10 and 12 years, the MSSS must plan today for the number of students starting medical studies, knowing that they will only join the field ten years from now. Similarly, Quebec's population will change in the next ten years, as it will the current set of doctors. In this context, the aim of this paper is to propose a causal model using the systems thinking approach to support discussions on the challenging task of deciding courses of action such as policies, the number of candidates enrolling medical studies, or the number of contracts to offer, among others, to ensure the long-term bearability of their care system. Bearability concerns the achievement of two of the sustainability pillars: social and environmental. Bearability is targeted in this paper rather than the triple sustainability—social, environmental, and financial pillars—because our goal at this moment is to capture the complex and dynamic relationships between the above-mentioned actors without limiting their choices by external funding constraints. Nonetheless, economic aspects such as funding are of the highest importance for system sustainability and may also impact patients' behavior and cares consumption [4]. The introduction of economic aspects in further healthcare models will be briefly discussed in Section 5.

2. Literature Review

This section first introduces system dynamics methodology and then presents some studies related to this research. Jay Forrester, the creator of system dynamics methodology, defines it as that which studies data feedback characteristics—mainly within industrial activities—in order to prove how organization structure, policy broadening, and delays (in both actions and decisions) interact and impact the success within the organization. Its subject matter is based on examining the interactions among companies' information flows, money, orders, materials, staff, and equipment. Moreover, system dynamic provides a single structure to group the functional areas of top management [5].

From another perspective, Sterman [6] establishes system dynamics as a method used to optimize learning within complex systems; it is a method to develop decision-making simulators by means of specialized computer software. The purpose is to know the dynamic complexity within a system, understand the source of policy resistance, and design more effective strategies.

The first phase in system dynamics methodology encompasses the development of a causal model considering exogenous and endogenous variables, as well as parameters that directly impact the behavior of one or more dynamic variables considered within the entities. Information feedback systems are relevant when a meaning leads to a decisive action, whose result is another influential action in the environment, and therefore, impacts future decisions. The feedback process is ongoing, and, therefore, attracts new results which consequently maintain the process in constant movement by using past data to decide or predict future behavior [5,6]. Thus, a causal diagram is a graphically depicted causal diagram which is created based on the elements of a system and its interrelations. It integrates causal loops to explain certain behaviors and incorporates background data to predict future behaviors. It typically has a greater degree of formality than a linguistic description, but it is less precise than a mathematical equation. This type of diagram is comprised of interrelated elements within the system. It also includes the concept of feedback as circular chains of influence, and its presence explains certain behaviors of the model structure [7–9].

These causal loops help build the foundations of a systemic language that allow for certain system descriptions to be determined and enable the visualization of things which, without these models and loops, could not be appreciated at a glance. The causal loops diagrams are classified in positive and negative based on their self-regulatory behavior system [7,10]. Table 1 shows the configuration of each type of causal loop diagram.

Table 1. Causal loop diagrams.

Causal Loop	Description	Example
Positive Loop (Explosive)	Also known as reinforcing or explosive, and colloquially referred to as "snowball" or "vicious circle". It is characterized by the homogeneity of its positive influences, which are able to increase or decrease equal variables, thus, its name "reinforcing".	
Negative Loop (Balance)	It is considered as a balancing loop due to the fact that it nullifies exterior disruptions, and seeks to move forward with its current elements, that is, cancelling out elements to avoid variations among them.	

Source: Prepared by Sterman [6] with their own information.

Empirical Studies Related to This Research

A comprehensive, non-exhaustive review of recent literature allowed us to identify several works related to the use of system dynamics or systems thinking to support decision making in healthcare or to the study of sustainability in that particular context. The next paragraphs briefly describe their contributions and position our research with respect to them.

The work of Wolstenholme et al. [11] is one of the first to propose system dynamics models as a new way of understanding how complex healthcare systems operate. They demonstrate the value of such modeling approach, providing a safe environment within which stakeholders from across agencies and functions can make explicit their own assumptions, understand the impact that their initiatives might have on other parts of the system, and develop ways of collaborating to achieve maximum benefit for service users. Since then, system dynamics or systems thinking has been successfully applied to the study of public health services. Indeed, the systematic literature review conducted by Carey et al. [12] reports how system-oriented approaches have been applied, and conclude that soft systems modeling techniques are likely to be the most useful addition to public health, and align well with the current debate around knowledge transfer and policy. Marshall et al. [13] reported recommendations on the potential of dynamic simulation modeling methods to support health care decision-makers evaluating interventions to improve the effectiveness and efficiency of health care delivery. Gupta et al. [14] described the steps needed and data required for analysis of supply and demand from a system perspective. They proposed a modeling tool that allows educators and policy makers, in addition to physician specialty organizations, to assess how various factors may affect demand (and supply) of current and emerging health services. Djanatliev et al. [15] enlarged the scope of methods encompassed by system-oriented thinking by discussing the use of hybrid simulation approaches, consisting of system dynamics (SD) and agent-based simulation (ABS), to anticipate the impacts of an innovative product before it is developed.

System-modeling tools have been applied to facilitate a better understanding of the system-wide effects of patient flow-related interventions. Esensoy and Carter [16] presented a multi-panel expert knowledge elicitation approach based on group model building principles to assess changes to rehabilitation patient flows in a community hospital. Vanderby and Carter [17] proposed a system dynamics model that simulates patient flows in a hospital. The model is used to analyze the delays experienced by patients in the emergency department. Several works have focused, like our research, on the study of a system's long-term capacity. Vanderby et al. [18] and later Morgan and Graber-Naidich [19] developed models simulating the evolution of the workforce in the long term. While Vanderby, Carter, Latham and Feindel [18] considers a single specialty at a national level, and provides a tool that helps decision managers planning future workforce, Morgan and Graber-Naidich [19] focused on policies to alleviate the rural care gap in the future. Grida and Zeid [20] proposed a system dynamics model simulating a typical hospital where different types of patients are served using the same limited resources. The model is used to study system capacity, and then to identify bottleneck resources.

Sustainability is still an emerging concept in healthcare, although its importance is rising quickly. To the best of our knowledge, sustainability in healthcare systems has been approached through the development of the social pillar, particularly in studies focusing on patient satisfaction and patient-oriented care. Indore [21] proposed a comprehensive conceptual model to understand and measure variables affecting patient satisfaction in an attempt to assess healthcare quality. It suggests a multi-disciplinary approach combining patient input as well as expert judgment to achieve this challenging task. In Faezipour and Ferreira [22], Faezipour and Ferreira discuss healthcare as a complex system of systems and discusses the challenges related to healthcare sustainability. Indeed, by placing patients at the center of the healthcare system, authors focus on patient satisfaction as a means to develop the social pillar of sustainability. Finally, Faggini et al. [23] studied how some digital enablers (e.g., digital technologies and platforms), boost the sustainability of complex service systems such as healthcare. The paper tries, by analyzing the system dynamics, to explain the role of technologies, in particular digital platforms, empowering actors and making them willing to interact and share their own resources in continually new ways as ongoing value co-creation, which is essential for healthcare system sustainability.

We conclude that, despite of the efforts to analyze the dynamics of the healthcare system in a long-term perspective, additional research must be done to better understand the relationships between the four decisional entities suggested in this paper: (1) higher education institutions; (2) the Ministry or the Government; (3) hospitals or healthcare centers; and (4) the society in demand for services, but also as a source of income for the Government as taxpayers. Moreover, sustainability objectives need to be incorporated in an explicit manner to new models and approaches to study healthcare systems.

3. Materials and Methods

This article applies the first phase of system dynamics methodology: it is related to the conceptualization phase supported by causal diagrams to analyze the influences among the four selected entities.

For its analysis, the first stage of the system dynamics methodology was used, which encompasses three phases:

1. Determination of the critical variables and parameters for each entity;
2. Development of a causal diagram for each entity;
3. Integration of the individual causal loop diagram to form the global system diagram from the selected entities.

4. Results

4.1. Determination of the Critical Variables and Parameters for Each Entity

During this first phase, variables and parameters are selected based on theoretical research conducted through empirical studies and stakeholder experience.

Entity 1: higher education institutions. As described in Table 2, nine initial variables and nine parameters have been determined in order to develop the first causal diagram called the role of higher education institutions.

Table 2. List of variables and parameters for the entity of higher education institutions.

Variable	Definition
Students in	Number of people who wish to enter university every year
Applicants rejected	Number of people not accepted in the University
Students in universities (Doctoral Program)	Students that go through medical training in one of the universities financed by the ministry between 0 and 4–5 years.
Universities' capacity	Number of students that are accepted by the four universities that provide medical training (Laval, Sherbrooke, McGill, Montreal).
Residence familiar medicine	Number of graduates from residence familiar medicine that take two years additional training at university
Residence subspecialities	Student population that take 5 years at universities financed by the Ministry
Subspecialities or complementary training	Number of graduates from medical specialist training that continue their studies for an additional 1 or 2 years at university
Discrepancy	It is the difference between the goal and the number of students that are training as medical doctors at universities.
Total students accepted	Sum of accepted students at the four universities (Laval, Sherbrooke, McGill, UdMontréal)

Table 2. *Cont.*

Variable	Definition
Required parameters	1. % of applicants 2. % students accepted 3. Time for students in higher education institutions = 4–5 years 4. Time for subspecialties or complementary training = 1–2 years 5. Time for residence subspecialties = 5 years 6. Université Laval = 218 * (2017) 7. McGill University = 199 (2017) 8. Université de Montréal = 291 (2017) 9. Université de Sherbrooke = 175 + 2 (2017)

* Two hundred and nine (209) from Québec (which includes up to 8 positions available for applicants to the PhD program and up to 8 positions available for university applicants from the labour market); 3 more positions as per the intergovernmental agreement with New-Brunswick; 1 position available for an applicant of foreign nationality who holds a study permit; 1 potential position for an applicant from another Canadian province or territory, 2 positions for applicants who wish to pursue their postgraduate studies in the field of oral and maxillofacial surgery; and up to 3 positions could be offered to IMGs willing to start with clerkship.

Entity 2: the Ministry. Five initial variables and four parameters have been determined in order to develop the second causal diagram, called Ministry, that describe the funding resources. Table 3 shows this information.

Table 3. List of variables and parameters for the entity of the Ministry.

Variable	Definition
Funds provided by the government	Sum of financial resources from the government and the private sector
Organizations	Set of organizations/companies that provide taxes for ministry
Budget for the universities	Assigned budget each year for universities
Taxes	Incoming financial resources
Services from family doctors care	Financial resources allocated by the ministry for medical training at universities, and funds to pay assigned medical services at hospitals
Parameters	1. Rate of demands (max–min) 2. Family doctors demands 3. $ available for universities per year 4. $ available for hospitals to pay for medical doctors

Entity 3: hospitals. Five initial variables and five parameters have been determined to develop the third causal diagram called hospitals. Table 4 shows this information.

Table 4. List of variables and parameters for the entity of hospitals.

Variable	Definition
Population of doctors in the Province of Quebec (available)	Medical doctors' annual contracts performed by hospitals (Dijk where D = doctors, i = general, j = specialist, k = higher specialty) coming from rural or urban areas or universities
Doctors in the province of Quebec (with activity)	Number of medical doctors in 2019
Doctors in the province of Quebec (without activity)	Number of medical doctors that retire per year
Positions replaced	Number of positions replaced for every retired medical doctor
Service from family doctors care	People that receive medical services a year, grouped into general medicine or per specialist
Parameters	1. Concentration rate of medical doctors in rural areas = 8% 2. Concentration rate of medical doctors in urban areas = 92% 3. Contract rate per time according to type of medical doctor i, j, k where i = general, j = specialist, k= subspecialties 4. Hospitals = number of hospitals in Quebec 5. Doctors per hospital = i, j, k … .n

Entity 4: Quebec society. Six initial variables and six parameters have been determined to develop the four causal diagram called the role of Quebec society. Table 5 shows this information.

Table 5. List of variables and parameters for the entity of Quebec society.

Variable	Definition
Quebec society	Current population in Quebec
Births	Number of people that are born every day
Deaths	Number of people that die every day
Economically active population	Number of people who are employed
Migration	Number of people that leave Quebec
Immigration	Number of people that enter Quebec
Parameters	1. Birth rate 2. Average life expectancy 3. % people that work = 18 years old migration rate 4. Immigration rate

4.2. Causal Diagram Development for Each Entity

In this phase, each of the entities is developed considering the variables and parameters; this leads to a better understanding of the influence that each variable has over the other or others. The parameters generate the dynamic for each of the variables involved. Likewise, each of the entities and their explanation are shown separately in each of the following figures.

The entity of higher education institutions is presented in Figure 1, in which four balancing causal loop diagrams, B1 to B4, can be observed.

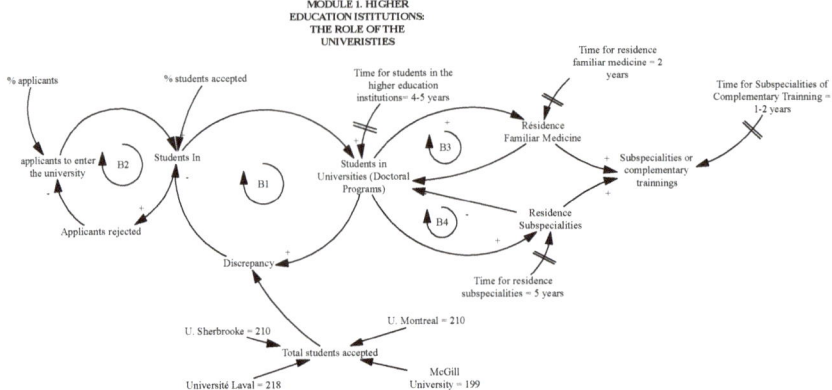

Figure 1. Causal loop diagrams for the Higher Education Institutions entity.

Dynamics of balancing loops B1 and B2 are explained as follows: there is an annual population that applies for admission at university to become part of the population of students that will be accepted; however, some of them will be rejected, so this makes the system become balanced (B2). The dynamic of student entries is trapped in the university: the way in which the system is regulated is the capacity of each of the universities in terms of the total number of students it can accept. The balancing loop B3 and B4 establishes that doctor training goes from one stage to the other; as long as it is possible to go from one level to another, there will be greater interest in becoming part of the university; all the students are in universities (higher education institutions) per 4 or 5 years, then they are in residence familiar medicine or residence subspecialties, and then students in the universities are reduced because they now are in other category.

Figure 2 shows the second entity related to hospitals, viewed as entities that receive medical doctors; in this diagram, causal loop B5 expresses negative or balancing relationships, which comes back to the idea set forth by [23].

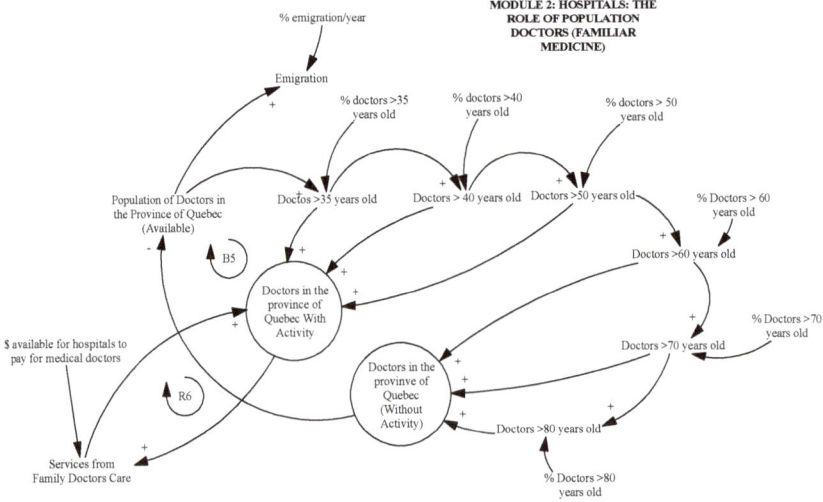

Figure 2. Causal loop diagrams for the hospitals entity.

The loop B5 generates a dynamic that begins with a growth in the population of available doctors. This growth goes through different stages or population groups defined by the age, in which the dynamic leads to the understanding that, as they become older, doctors tend to be less active, which results in a reduction in the number of "net" available doctors. The R6 loop explains a dynamic switch of services, from available doctors to active doctors once the ministry assigns funds to hospitals to enroll an additional workforce.

Figure 3 shows the third entity that explains the dynamic of Quebec society in terms of population growth and decline, the latter either due to natural reasons or as a consequence of family decisions to enter or leave Quebec.

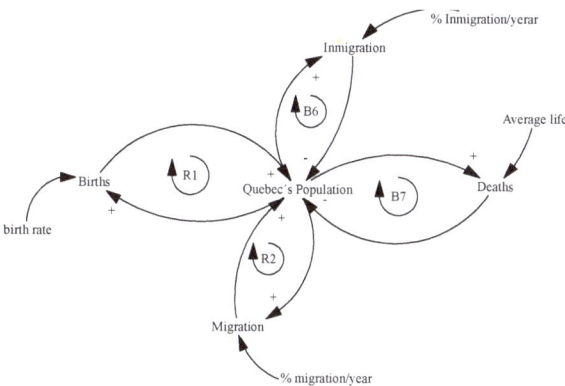

Figure 3. Causal loop diagrams for the entity of society.

There are two reinforcing (R1 and R2) and two balancing (B6 and B7) loops that are explained next. The population of Quebec grows by the effect of two variables: births (R1) and migration, which can be split into doctors arriving from other provinces of Canada, or those who come from other countries (R2). The Quebec population decreases due to natural deaths (B7) or immigration (B6) when people born in Quebec emigrate to other states or countries.

Finally, the entity related to the Ministry and Government is illustrated in Figure 4. It contains four loops: two balancing loops (B8 and B9) and two reinforcing loops (R3 and R4). In order to explain the dynamic of this entity, it has been set forth that as the economically active population increases, there is greater income for the Ministry as a result of tax payments (R3); this allows medical services in family medicine care to be covered by the Ministry, causing a decrease in resources; nevertheless, there are incoming funds on behalf of organizations that contribute to the rise in funds (B8). This sustains the financial resources in the Ministry in terms of family income and organization contributions (R4). The budget for the universities depends on the total students accepted each year; this is the dynamic presented in loop B9.

This analysis, conducted separately for each of the entities, seeks to make clearer the complexity of the whole system. It considers that a system is a set of an entity and at least two interrelated elements and where each element is related to the others, either directly or indirectly [24,25].

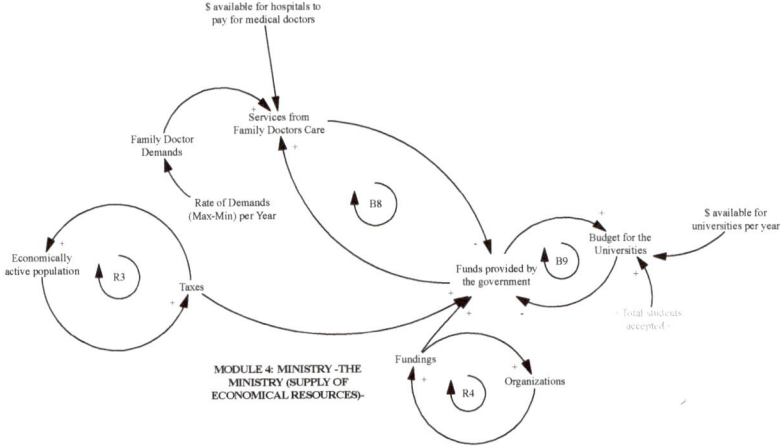

Figure 4. Causal loop diagrams for the Ministry entity.

4.3. Integrating Individual Causal Loop Diagrams to form the Global System Diagram

The last phase is the integration of the four entities, which is presented in Figure 5. The purpose is to provide to all the stakeholders with a global view where they can observe their role, and from that perspective, make the necessary adjustments to the model before moving on to its simulation stage.

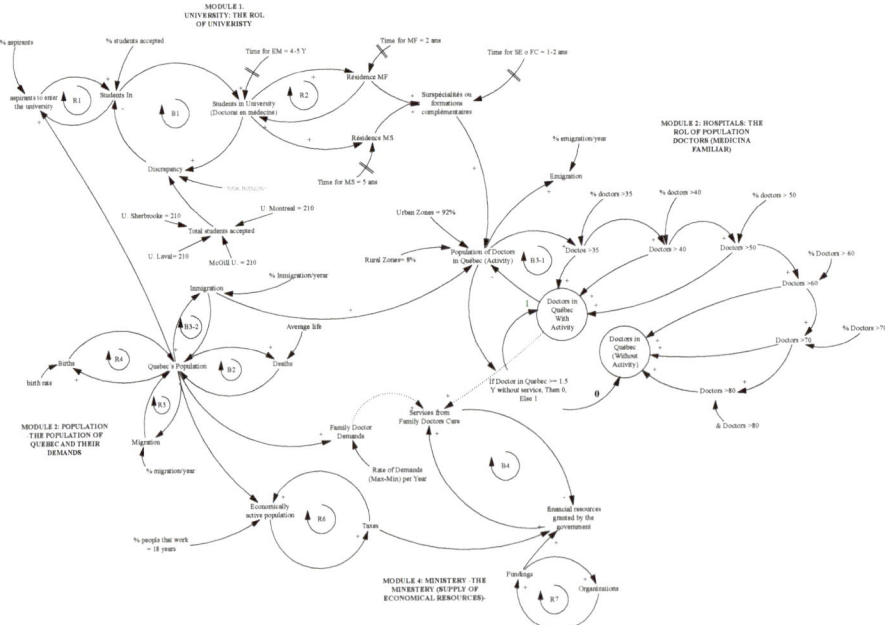

Figure 5. Causal loop diagram for the whole system, including the four entities.

Systems thinking requires the use of causal loop diagrams to improve understanding of the system [26]; in this sense, after having constructed the entities separately, it is fundamental to visualize them as a whole by joining the parts. The model shows the interactions among the four entities, higher education institutions, society, hospitals, and the Ministry. It is important to observe that, in total, there are nine causal balancing loops (Bi, where i = 1 .... 9) and five reinforcing loops (Rj, where j = 1 ... 5).

In order to transform the casual diagram into a simulation model, parameters affecting the relationships as well as initial values for the variables need to be set. A data collection process is therefore required, followed by a validation step before planning numerical experiments that would help to analyze the system's sensitivity to specific decisions or policies. However, casual diagrams support fertile discussions among stakeholders. Also, notice that focal groups must concentrate on each entity, and the discussion must be coordinated by a group of people external to the entity—but having a good knowledge of all of them—in order to achieve favorable results in the final version that will lead to the second phase of the solution development from a quantitative point of view.

5. Discussion

The usage of system dynamics, in their first step related to conceptualization (causal loop diagrams), represents an opportunity for organizations dedicated to providing services to society. In this project, the four selected entities have been studied separately and then merged into a single model. The following considerations are raised for discussion:

1. The model integrates data gathered from official websites, which must be validated by the stakeholders of the areas involved.
2. The model is built from a theoretical and qualitative perspective, based on the fact that the variables are consolidated from empirical studies, and the model is conceptual, incorporating variables and parameters that are explained from the perspective of the system dynamics theory.
3. The model only covers the first phase of the system dynamics, which is related to the conceptualization phase and divided into three stages.
4. Nonetheless, to validate the entire model, it is necessary to question and verify the interrelations among the four entities.

Using systems thinking allows us to conceive causal diagrams of specific behaviors and to connect several of them later to form a more complex system. This process allows a better understanding of the system and its entities. The granularity of the model's components is, however, a decision of the model designer and, since there is no fixed rule to do it, it makes systems thinking a science but also an art.

Finally, it is worth mentioning that both financial aspects and demand modeling for health services and its evolution in forthcoming years by considering the emergent risks or changes in the way that healthcare is provided, as approached by [21], is a key part of healthcare system design and we are considering such a demand model entity in our next version of the model.

6. Conclusions

Based on the analysis of variables and parameters included in the feedback model to observe the complexity of the same in the four entities, that is, the universities as educators that train medical doctor in family medicine, hospitals as spaces where they are able to provide their services to the community integrated by the Quebec society, and the Ministry as the entity that manages financial resources from taxpayers and organizations that are part of society, it can be concluded that:

1. It is possible to understand the complexity of each entities' functions and their role from a systemic perspective supported by cause–effect interrelations provided by causal diagrams.
2. The ministry may allocate funds to universities and hospitals based on the information provided by the scenarios that arise from quantitative analysis with a dynamic simulation that should be developed from the proposed feedback model.
3. The system dynamics methodology in their first step (causal loop diagrams) provides information at every stage in order to better understand the complexity of the entities as a whole.
4. The present work contributes to the Ministry's decision-making process; it is a graphic representation that enables following the cause–effect relationships of each entity.

5. The model may be replicated in other countries that have the four entities analyzed in the Province of Quebec, Canada, with all its required adjustments.

Future Work: The Project has been concluded in its first phase, which is related to the conceptualization of the system under study by using the dynamic systems methodology. The second part is building a simulation model and conducting a sensitivity analysis in order to show future scenarios by implementing "what if" policies and know the pessimist and optimist scenarios after validating the current one. The third phase corresponds to the development of a graphic interphase in which users can run tests by means of certain general-use buttons to observe data, graphs and performance indicators that are worth analyzing in the four entities under study.

Author Contributions: Conceptualization, methodology, software, and formal analysis, E.A.L.-L.; supervision, AR; writing—original draft preparation, E.A.L.-L. and A.R.; writing—review and editing, A.R.

Funding: This research was funded by Federal Program of Mexico (PFCE-2019) and CONACYT, through National Laboratory for Transportation System and Logistics.

Acknowledgments: The authors E.A.L.-L. and A.R., as part of the National Laboratory in Transportation and Logistics- Consolidation Systems, the authors are grateful for the support received through the National Science and Technology Council of Mexico (CONACYT) through the "National Laboratories" program. This Publication was funded with PFCE 2019 as a part of the international stay at Laval University, Canada. Also, with complementary founds provided by PROFAPI-2019-ITSON.

Conflicts of Interest: There is no conflict of interest. The funding sponsors had no role in the design of the study; in the collection, analyses, or interpretation of data; in the writing of the manuscript; and in the decision to publish the results.

References

1. MSSS. © Gouvernement du Québec 2019. Obtenido de Ministère de la Santé et des Services sociaux. Available online: http://www.msss.gouv.qc.ca/en/reseau/systeme-de-sante-et-de-services-sociaux-en-bref/contexte/ (accessed on 17 June 2019).
2. International, Q. Quebec en la cabeza. Available online: https://www.quebecentete.com/es/vivir-en-la-ciudad-de-québec/salud/ (accessed on 19 July 2019).
3. Québec, G.D. *Organization of the Healt and Social Services Network*; Éditeur officiel du Québec: Québec, QC, Canada, 2019.
4. Vuong, Q.; Ho, T.; Nguyen, H. Healthcare consumers' sensitivity to costs: A reflection on behavioural economics from an emerging market. *Palgrave Commun.* **2018**, *4*, 70. [CrossRef]
5. Forrester, J. *Dinámica Industrial, Segunda ed.*; Editorial El Ateneo: Patagones, Argentina, 1981.
6. Sterman, J. *Business Dynamics: Systems Thinking and Modeling for a Complex World*; McGraw Hill: New York, NY, USA, 2000.
7. Aracil, J.; Gordillo, F. *Dinámica de Sistemas*; Alianza: Madrid, Spain, 1997.
8. Senge, P.; Roberts, C.; Ross, R. *La Quinta Disciplina en la Práctica: Estrategias y Herramientas Para Construir la Organización Abierta al Aprendizaje*; Ediciones Granica: Buenos Aires, Argentina, 2006.
9. Guo, H.; Qiao, W.; Liu, J. Dynamic Feedback Analysis of Influencing Factors of Existing Building Energy-Saving Renovation Market Based on System Dynamics in China. *Sustainability* **2019**, *11*, 273. [CrossRef]
10. Lagarda, E. Collection and Distribution of Wheat, Dynamic of the Process of Shipping to International Markets: Case Study. *Int. J. Supply Chain Manag.* **2019**, *8*, 43–57.
11. Wolstenholme, E.; McKelvie, D.; Smith, G.; Monk, D. Using System Dynamics in Modelling Health and Social Care Commissioning in the UK. In Proceedings of the 2004 International System Dynamics Conference, Oxford, UK, 25–29 July 2004.
12. Carey, G.; Malbon, E.; Carey, N.; Joyce, A.; Crammond, B.; Carey, A. Systems science and systems thinking for public health: A systematic review of the field. *BMJ Open* **2015**, *5*. [CrossRef] [PubMed]
13. Marshall, D.; Burgos-Liz, L.; IJzerman, M.J.; Osgood, N.D.; Padula, W.V.; Higashi, M.K.; Wong, P.K.; Pasupathy, K.S.; Crown, W. Applying dynamic simulation modeling methods in health care delivery research-the SIMULATE checklist: Report of the ISPOR simulation modeling emerging good practices task force. *Value Health* **2015**, *18*, 5–16. [CrossRef] [PubMed]

14. Gupta, S.; Black-Schaffer, W.S.; James, M.; Gross, D.; Donald, S.; Kaufman, J.; Knapman, D.; Prystowsky, M.B.; Wheeler, T.M.; Bean, S.; et al. An Innovative Interactive Modeling Tool to Analyze Scenario-Based Physician Workforce Supply and Demand. *Acad. Pathol.* **2015**, *2*. [CrossRef] [PubMed]
15. Djanatliev, A.; Kolominsky-Rabas, P.; Hofmann, B.M.; Aisenbrey, A.; German, R. System Dynamics and Agent-Based Simulation for Prospective Health Technology Assessments. In *Simulation and Modeling Methodologies, Technologies and Applications. Advances in Intelligent Systems and Computing*; Obaidat, M., Filipe, J., Kacprzyk, J., Pina, N., Eds.; Springer: Berlin/Heidelberg, Germany, 2014; Volume 256.
16. Esensoy, A.; Carter, M. Health system modelling for policy development and evaluation: Using qualitative methods to capture the whole-system perspective. *Oper. Res. Health Care* **2015**, *4*, 15–26. [CrossRef]
17. Vanderby, S.; Carter, M. An evaluation of the applicability of system dynamics to patient flow modelling. *J. Oper. Res. Soc.* **2010**, *61*, 1572–1581. [CrossRef]
18. Vanderby, S.; Carter, M.; Latham, T.; Feindel, C. Modelling the future of the Canadian cardiac surgery workforce using system dynamics. *J. Oper. Res. Soc.* **2014**, *65*, 1325–1335. [CrossRef]
19. Morgan, J.; Graber-Naidich, A. Small system dynamics model for alleviating the general practitioners rural care gap in Ontario, Canada. *Socio-Econ. Plan. Sci.* **2019**, *66*, 10–23. [CrossRef]
20. Grida, M.; Mahmoud, Z. A System Dynamics-Based Model to Implement the Theory of Constraints in a Healthcare System. *Simulation* **2019**, *95*, 593–605. [CrossRef]
21. Indore, A.N. Factors affecting patient satisfaction and healthcare quality. *Int. J. Health Care Qual. Assur.* **2009**, *22*, 366–381. [CrossRef]
22. Faezipour, M.; Ferreira, S. A System Dynamics Perspective of Patient Satisfaction in Healthcare. *Procedia Comput. Sci.* **2013**, *16*, 148–156. [CrossRef]
23. Faggini, M.; Cosimato, S.; Nota, F.D.; Nota, G. Pursuing Sustainability for Healthcare through Digital Platforms. *Sustainability* **2019**, *11*, 165. [CrossRef]
24. Van Gigch, J. *Teoría General de Sistemas*, 3rd ed.; Trillas: Veracruz, Mexico, 2012.
25. Ackoff, R. *El Paradigma de Ackoff: Una Administración Sistémica*; Limusa: Mexico City, Mexico, 2002.
26. Richardson, G.; Pugh, A., III. *Introduction to System Dynamics Modeling with Dynamo*; Pegasus Communications: Whaltan, MA, USA, 1981.

© 2019 by the authors. Licensee MDPI, Basel, Switzerland. This article is an open access article distributed under the terms and conditions of the Creative Commons Attribution (CC BY) license (http://creativecommons.org/licenses/by/4.0/).

Article

Green Supplier Selection in the Agro-Food Industry with Contract Farming: A Multi-Objective Optimization Approach

Marco A. Miranda-Ackerman [1],*, Catherine Azzaro-Pantel [2],*, Alberto A. Aguilar-Lasserre [3], Alfredo Bueno-Solano [4] and Karina C. Arredondo-Soto [1]

[1] Facultad de Ciencias Químicas e Ingeniería, Universidad Autónoma de Baja California, Tijuana, Baja California 22390, Mexico; karina.arredondo@uabc.edu.mx
[2] Laboratoire de Génie Chimique, Université de Toulouse, CNRS, INPT, UPS, 31432 Toulouse, France
[3] Division of Research and Postgraduate Studies, Tecnológico Nacional de México/Instituto Tecnológico de Orizaba, Orizaba 94320, Veracruz, Mexico; albertoaal@hotmail.com
[4] Department of Industrial Engineering, Instituto Tecnológico de Sonora, Cd. Obregon 85000, Sonora, Mexico; alfredo.buenos@itson.edu.mx
* Correspondence: miranda.marco@uabc.edu.mx (M.A.M.-A.); catherine.azzaroPantel@ensiacet.fr (C.A.-P.)

Received: 31 October 2019; Accepted: 4 December 2019; Published: 9 December 2019

Abstract: An important contribution to the environmental impact of agro-food supply chains is related to the agricultural technology and practices used in the fields during raw material production. This problem can be framed from the point of view of the Focal Company (FC) as a raw material Green Supplier Selection Problem (GSSP). This paper describes an extension of the GSSP methodology that integrates life cycle assessment, environmental collaborations, and contract farming in order to gain social and environmental benefits. In this approach, risk and gains are shared by both parties, as well as information related to agricultural practices through which the FC can optimize global performance by deciding which suppliers to contract, capacity and which practices to use at each supplying field in order to optimize economic performance and environmental impact. The FC provides the knowledge and technology needed by the supplier to reach these objectives via a contract farming scheme. A case study is developed in order to illustrate and a step-by-step methodology is described. A multi-objective optimization strategy based on Genetic Algorithms linked to a MCDM approach to the solution selection step is proposed. Scenarios of optimization of the selection process are studied to demonstrate the potential improvement gains in performance.

Keywords: green supplier selection; contract farming; life cycle assessment; environmental collaboration; multi-objective optimization; agro-food supply chain

1. Introduction

Environmental awareness has shifted consumer behavior towards more efficient and environmentally-friendly products, including processed foods. This has led food manufacturers to find opportunities by developing strategies targeting eco-friendly consumers and markets, through the use of eco-labelling [1,2]. In order to satisfy these niche markets and continue developing this branding strategy, a shift from conventional food production to a more sustainable one has been progressively pursued. The transformation is that consumer awareness about the "greenness" of products is incentivizing a change towards alternative practices and technologies that may affect the entire agro-food supply chain (ASC).

One of the most important links in the ASC lies in the interface between farms and manufacturers, given that the raw materials needed to produce the current selection of food products at retail stores and markets are sourced. An important part of the environmental impact located at this point is due to agrochemical, water, land and energy use related to farming [3,4]. This is why there is an interest in both sustainable agricultural practices as well as green process design steps further downstream, in order to look at the sustainability of ASC, which leads to the potential improvements that could lie at the interface between farmers and manufacturers.

The study of this interface is sometimes referred to as the supplier selection problem (SSP). This subfield of supply chain management (SCM) has been tackled largely in the dedicated literature, where it is mostly described by taking into consideration a set of criteria, traditionally based on cost, delivery time and quality among many other components [5], to then classify and rank suppliers [6,7]. This paper looks at the SSP of an orange juice producing company that uses suppliers in a collaborative scheme, that additionally includes in their supplier roster, small and medium farmers, under a contract farming model. Furthermore, it proposes the use of the green supplier selection paradigm that incorporates environmental performance of suppliers in the selection process. The objective of this paper is then to show how with synergy made through collaboration, contract farming and a green supplier selection perspective, improvements in the performance of the food supply chain can be achieved in economic, social and environmental terms. It answers the specific research question: What is the result of including partnerships in a supplier selection decision modeling approach in terms of selection criteria within the green supply chain paradigm? This approach will be referred to in what follows as Partnership for Sustainability (PfS) (see Figure 1).

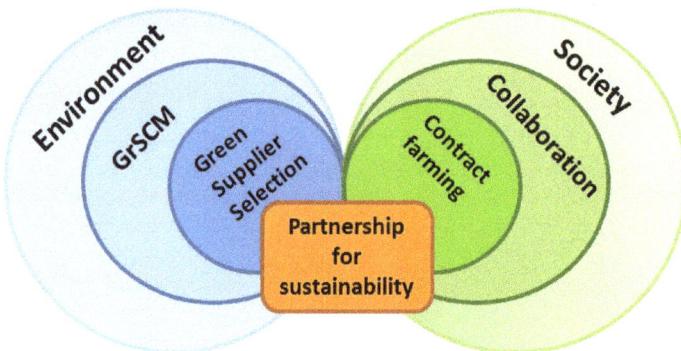

Figure 1. Partnership for sustainability method integration diagram.

In order to illustrate the approach and proposed solution methodology, a case study is developed focusing on the production and supplying of oranges produced in Mexico used for juice production. This case study is based on the locally available technological alternatives which are used at each supplying orchard, taking into account the local technological package and their corresponding production yields and related environmental impacts for the specific citrus fruit producing region. The methodological framework for the solution of this problem is based on the use of a multi-objective optimization strategy through genetic algorithms followed by a multi-criteria decision-making method in order to select a solution.

1.1. Green Supply Chain Management

The backdrop of the proposed approach lies in a promising and somewhat recent paradigm to evaluate and improve the environmental and overall performance of production systems called Green Supply Chain Management (GSCM). This approach integrates two conceptual scopes: Supply Chain Management (SCM) and Life Cycle Assessment (LCA). The former provides a framework to visualize and improve the economic and operational performance of productive systems by modelling the flow of materials, information and money, throughout the links in the production chain, with the end objective of economic profit [8]. The latter, is a technique to aid in the decision-making process by providing a system-oriented approach to evaluate the environmental impacts at some or all the stages in the life cycle of a product [9]. By integrating these two holistic approaches, the economic and operational objectives can be set side by side with sustainability objectives when trying to make decisions on design or improvements of the overall performance of a production system.

Although much progress has been made in the field, there are still opportunities in exploring the application of GSCM in different contexts [10,11]. This work proposes the extension of the current GSCM model to include special issues inherent in agro-food supply chain systems and by looking at the decision-making process of supplier selection in this context, i.e., when suppliers are farms.

1.2. GSCM Modelling and Optimization Approach

The modelling approach assumes that there is a set of products that can be differentiated by a "green" quality attribute based on the technological methods used to produce them. When a product is made with specific quality characteristics, for example "big" oranges, that have a higher demand and/or market price against, say, "small" oranges, this leads the orange farmer to change the production processes configuration and capacity (within physical and cost-benefit limitations), in order to increase the output of the most profitable product i.e., "big" oranges.

Let us consider that a new product quality attribute is now being associated with environmentally produced/processed products. These products are marketed through eco-labels such as: organic, bio, green, eco-friendly, etc. Let us also consider that these new attributes have changed the market value and pricing of the products, given that they are willing to pay more for the story behind the product and for precise information [12]. This type of labelling has been widely used since the 1990s in the U.S. and the 1980s in France [13]. It is interesting to explore the possible effect this could have when modelling and optimizing an ASC. Thus, the approach proposed here takes this green preference into account and integrates it within the current GSCM modelling and optimization methods.

In the literature, different methods have been proposed in order to solve these types of problems, some use MILP (Mixed Integer Linear Programming) [14,15] as a preferable technique for network configuration; others take into account non-linear behaviour that requires MINLP (Mixed Integer Non Linear Programming) capability [16], or a stochastic programming approach to handle uncertainties [17]. A relevant approach that has been widely used to handle the multiple objective nature of GSC models (sometimes in tandem with other modelling approaches such as MILP) is the so-called MOO (Multi-Objective Optimization) technique [14,18]; as well as other Multi-Criteria Decision Methods such as TOPSIS and AHP [19,20]. These final two approaches (i.e., MOO and TOPSIS) were selected to manage the complex decision-making that takes place when working with a SC scope. It may be important to point out that MOO through Genetic Algorithms was used instead of other approaches, such as Wighted-Formula and Lexicographic methods, given that it allows the modeler to take a black box approach, where objectives and constraints are not restricted in their structure [21]. The advantage of ease of use and adoption [22] allows for scalability and integration with other complex models with little or no modifications needed. This last point was critical in its selection due to its integration to a wider reaching green supply chain network design approach [23].

Our work assumes that production practices at all stages of the supply chain can be changed to improve environmental performance, which means the changes in consumer preference for "green" products have to be met by a market demand (as this attribute adds intangible value to the product). In supply chain modelling terms, using the orange juice production as an example, the following set of questions have been formulated:

- Production design problem: What agricultural practices should be used to add "green" value to the product at the raw material production stage e.g., optimal use of pesticides, fertilizers, gasoline powered machine, etc.?

- Product mix problem: What quantity of each quality-type orange should be produced to obtain the desired orange juice quality mix (e.g., "organic", "environmentally-friendly" or "30% less environmental impact")?
- Location-allocation problem: Which and how many orchards should produce environmentally-friendly oranges; which and how many should use intensive agricultural, in order to satisfy demand?
- Supplier selection problem: Which supplier to buy from? How much to buy? Which criteria to use to evaluate supplier performance?

In order to answer these questions, the Partnership for Sustainability approach takes the green supplier selection problem and integrates the benefits from contract farming and an environmental collaborative approach described in the following sections.

1.3. Green Supplier Selection Problem

In the field of supplier selection, there is a wide body of publications looking at many different aspects such as formulation, method and application [24]. These include recent works on green or sustainable supplier selection. Although Green Supplier Selection (GSS) has been named in different ways such as Green Vender, Green Purchasing, and Environmental Purchasing, most definitions coincide in its reach. We consider [25], and define Green Supplier Selection as "the set of purchasing policies held, actions taken and relationships formed in response to concerns associated with the natural environment." It goes on to clarify that "These concerns relate to the acquisition of raw materials, including supplier selection, evaluation and development." This integration is promoted because of the benefits that interaction of the different members of the supply chain produce, not only on sustainability issues, but also on economic and operations performance [26]. A review on Green Supplier Selection has highlighted some interesting research challenges [27]. According to these authors, the integration of Life Cycle Assessment (LCA) in the definition and measurement of environmental criteria offers an interesting methodological framework, since it involves a system-wide approach for environmental impact evaluation. Some other examples of GSS have been published for industries such as electronic and consumer goods [28,29]. A study on the effectiveness of a large-scale Sustainable Supplier program in Mexico was published. Among its conclusions, it suggests that a need to focus on micro and small businesses must be considered as an important objective in further research within the field [30].

This paper locates itself in these groups of research that support the use of LCA as a decision support tool for the supplier selection process and develops the case study in the context of an agro-food chain in order to contribute to the current body of research in the field. It also contributes to the potential improvement from contract farming and its derived environmental collaboration that is further discussed in the following section.

1.4. Contract Farming

In the SCM paradigm as in the GSCM, a central or focal company (FC), as proposed by Seuring and Muller [10], is characterized by being the designer or owner of the product or service offered, governing the supply chain, and having contact with all SC stakeholders including the customers. The FC can also sometimes be the processing or manufacturing company, as is the case of our case study.

We proposed that the FC be the integrator firm within the context of contract farming as described by Rehber [31], under a Management and Income Guaranteeing contract [32], also known as a Production Management Contract (PMC) [33]. They describe the PMC type of contract model as one that both specifies product quality measures that are acceptable to the integrator, and the integrator providing production resources, takes on substantial managerial responsibilities, and supervises the supply chain activities. This form matches well with the green supplier development aspect of the GSCM approach, given that green supplier development draws its importance from the fact that suppliers are often small and do not have the knowhow or resource necessaries to face the environmental issues related to their business process [10,34]. This is why some research has been focused on collaboration, certification and education of suppliers [35–37]. The similarity in the scope and roles that different elements play in GSCM and contract farming models are illustrated in Figure 2.

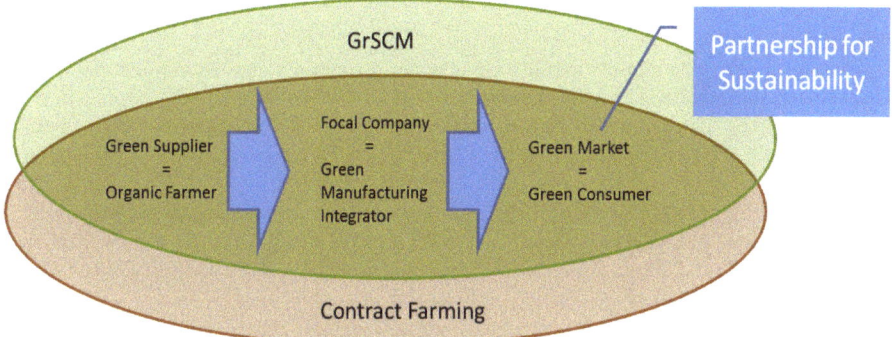

Figure 2. Analogy between Green Supply Chain Management (GSCM) and contract farming elements.

The manufacturer used in the case study is both the Focal Company and Integrator, thus the model we present is adapted to this type of supply chain configuration. In other words, this paper is limited to this situation, i.e., where the focal company is the processing company, and is the principal negotiator with suppliers and distributers or consuming markets.

It is important to note that the details related to the negotiation and contracting stages are out of the scope of this research; however, the contract form that allows for the collaboration and technology transfer, that includes which technological package to use at the suppliers' fields, is part of the general approach.

Contract farming as described in a special report for FAO by Eaton and Chepherd [38], and is the use of contracts to build a foundation of collaborations between the FC and the supplier, where the whole or a fraction of production output from suppliers is bought guaranteeing (not without risk) a reliable source of product, while the supplier has an assured buyer. According to the FAO report, this model is an effective way of taking advantage of the investment power and knowledge base of the FC by transferring technical skills in order to use the synergy reached to improve supplier performance. This, in turn, helps to guarantee both parties a partnership for growth and helps mitigate and spread risk. Most importantly, within the context of this paper, it allows the FC to require and share technological packages or practices to be used at the supplier field in order to obtain conforming and consistent raw materials. It also gives enough flexibility to find the optimal supplier technology package selection.

1.5. Partnership for Sustainability

Partnership for sustainability is derived from the premise documented by Vachon and Klassen [39], in a paper related to the improvement in operational performance based on partnerships in the supply chain. They use the term "green project partnership", to describe "the extent of interaction between a plant and its primary suppliers and major customers in developing and implanting pollution prevention technologies." This paradigm is considered here and implemented in what follows into the food supply chain within the GSCM scope. As pointed out by Vachon and Klassen [39] and corroborated by Albino et al. [40], the integration of suppliers through partnerships has a positive effect of operations performance, and furthermore may stimulate gains in integrated knowledge and knowhow from the supply network [41], which, in turn, can have a positive effect on the development of agricultural practices and supply chain negotiations and distribution issues.

This paper outlines a framework to apply this type of partnership as a step-by-step methodology, to distinguish from green project partnership as a general term that describes partnerships both with suppliers and customers. It will be referred to in the following as Partnership for Sustainability.

2. Materials and Methods

The general approach to solve the SSP consists of implementing a set of steps, where after scouting a list of potential suppliers, a discrimination process is made through basic common sense judgements; this is followed by a measurement and characterization step based on requirements and desirable attributes. Then, a classification or ranking is done in order to target negotiations and contracting (Table 1).

Although this can be a dynamic process, meaning that there are new suppliers being added and old suppliers eliminated from the approved supplier catalogue, it can also be part of the strategic and tactical planning stages for long-term improvements that can be reviewed periodically. This research is situated in the latter approach.

The steps in Table 1 extend from the conventional supplier selection process to form part of a holistic approach to characterize the potential partners in addition to modelling, optimization and selection steps, to end with the execution of a contractual negotiation and agreement. This last step is not yet described in detail since it falls outside of the scope of this paper; steps 3 to 6 are described in detail in this section.

Table 1. Partnership for Sustainability Supplier Selection processes.

No.	Description
1	Pre-selection or scouting of suppliers
2	Short listing based on common sense judgment
3	Supplier characterization
4	Supplier network model
5	Supplier network optimization
6	Supplier network selection
7	Supplier negotiation and contracting

2.1. Partner for Sustainability: Supplier Characterization

The current SSP is framed by looking at a set of characteristics required by the production plant being supplied. The most common characteristics evaluated include quality, cost and delivery performance; although some efforts have been made to include environmental criteria in the supplier selection process, this kind of approach gaining traction in becoming the norm.

It must be also highlighted that the evaluation methods found in the domain literature describe the supplier selection process as a search for the most competitive vender without looking necessarily at the potential benefit of a long-term partnership, although there are some instances. In this approach, the characterization process itself consists of looking at the requirements or criteria most valued by the FC and implementing scoring or measurement systems in order to allocate a value to each supplier based on observed or estimated performance. For this purpose, the Partnership for Sustainability takes the potential capability of the supplier given the field or region characteristics (i.e., land and environment), and the technological package that can be used into account. This means that the question is not only which supplier to choose but also what technology should be matched to obtain the best criteria measurements.

The justification of this approach is also made given that many of the sustainability leaders, at least in the chemical industry, that manage ecology and social sustainability beyond their company boundary and view managing supplier relations as part of their fundamental strategy [35].

The output from each region, from each orchard and even with the same technology, can differ. This is why an initial estimation of the output and cost from each orchard or the regional location of the supplying orchards must be made. This information is then analyzed and processed in order to have useful operational information. This may be difficult for some agro-food products but is already used in orange production in many regions of Mexico. These types of characterizations can be made by collaboration through a confederation of growers, trade group, sponsorship of the FC or by government and independent research bodies.

The characterization of the performance of the fields can be either an internal exercise of the different production fields or it can be an experimental one carried out by expert bodies. In the second alternative, a third body has to collect and integrate data into information that can be used by all stakeholders. This can be carried out by the use of experimental fields in order to characterize the surrounding environment and local soil production.

The information that is necessary to apply this methodology is: field production yield (expressed in tons per hectare per year) and operations cost (energy cost, agrochemicals input cost, etc.). Other important indicators may also be considered such as land and irrigation cost, but are not included in the scope of the case study of this investigation.

The most widely used technique to measure the environmental impact in the current literature on GSCM is using LCA, which matches well with the holistic systems approach of SCM. LCA can provide information on the effect of each step depending on the depth of the analysis, from process and product design to industrial systems design and even at more strategic scales, such as GSCM. For the SSP, a collaborative scenario is investigated in which information and knowhow are shared between the Focal Company and the suppliers, this is also known as Supplier Development [34]; the objective is to use the synergy created through the flow of information that helps the Focal Company make better decisions. This explains why the FC has the insight to perform a system-wide analysis such as LCA. To collect and analyze the potential data, suppliers must be willing to share information on field, plants and management practices; in turn, the centralized manager must be transparent in its measurement and evaluation techniques. This process can be divided into three steps as explained in the two following sections.

In this step, a characterization of regionally used technological packages has to be evaluated, this can be based or made at the same time as the Yield and Cost characterization process; these packages may consist of agrochemicals used, soil treatment, physical manipulation (e.g., hedging, pruning, shaping, etc.), and machinery used, among other things. During this phase, experts are needed not only to characterize the production systems at field, but also to help classify them by level of sophistication. The result should be a manageable set of categories in which an average is used as a typical example per category that describes how production systems work (e.g., "organic" production, average production and intensive production system). Although it is important to have a well-developed approach in order to classify, this falls out of the scope of this research. It is assumed that this step is performed through expert opinion; for the case study, information from a regional government-funded agricultural research center is used.

Once the categories are developed, a systematic evaluation is achieved to define the basic indicators that will be used in the modelling and optimization process. These indicators may consist of operational functions such as the average yield obtained per area, plant or tree. It can also include economic functions such as average cost per unit of product given a technological package used, which are developed during the Yield and Cost characterization; in addition, environmental impact indicators such as CO_2 emissions or eutrophication are calculated per area, plant or tree in a given timeframe, e.g., per year. The environmental impacts are evaluated by LCA. This analysis requires different levels of information provided by field and literature research, expert collaborators and dedicated LCA software tools that are commercially available.

The environmental impact assessment is then integrated in the model as well as the other indicators. The model is useful to predict the impact that a decision alternative has not only towards operational and economic performance, but also towards environmental aspects given a set of decision variables.

2.2. Partner for Sustainability: Supplier Network Model

In order to directly improve the overall performance of the Supplier Network (SN) by incorporating a long-term partnership in which an interchange of technological knowledge and risk sharing is made by contract, a multi-objective optimization formulation is proposed. This approach allows the consideration of multiple and possibly antagonistic objectives to be concurrently optimized [23,42].

The general model is described below:

Index and Sets

- i Supplier index of a set I
- g Technology package of a set G

Variables

- b_i Binary variable used to select a supplier (i)
- s_i Production capacity to be contracted per supplier (i) as a continuous measurement of land area (ha)
- $Y_{i,g}$ Production yield estimated per technology package (g) used at each supplier (i) (ton/ha/yr)
- $CT_{i,g}$ Production cost estimated per technology package (g) used at each supplier (i) ($/ha)
- EI Environmental impact measurement of the full set of suppliers (i) (EI unit e.g., KgCO2-eq)
- EIs_i Environmental impact estimation for each supplier (i) (EI unit/supplier)
- CO_i Cost incurred for operations for each supplier (i) ($/yr)
- CE_g Environmental cost estimated per unit of production based on technology (g) (EI unit/kg)
- Cost   Cost incurred of the full set of suppliers ($/yr)
- LLC_i Lowest value of land capacity to be contracted of each supplier (i) (ha)
- P   Total raw material produced (ton/yr)
- PCap   Processing plant raw material requirement (ton/yr)

Objective functions

Z_1 = min (Costs)
Z_2 = min (Environmental Impacts)

The cost variable represents the cumulative cost of all orchards; this is to say the sum of the cost of each supplier given the technology package used and the capacity contracted represented by CO_i (see Equation (1)):

$$Cost = \sum_{i=1}^{n} CO_i. \tag{1}$$

The CO_i expression in Equation (2) is calculated considering the selection of a supplier, the capacity that is contracted multiplied by the cost per hectare given a given technological package, here represented by the (g) index, all of this calculated per supplier (i):

$$CO_i = b_i s_i \times CT_{i,g} \; from \; i = 1, 2, \ldots, n. \tag{2}$$

The global impact generated by all suppliers can be evaluated by Equation (3):

$$EI = \sum_{i=1}^{n} EIs_i. \tag{3}$$

In Expression (4), the environmental impact based on selection variable (b), capacity to be contracted (s) and technology package selected (g) is expressed by:

$$EIs_i = b_i s_i Y_{i,g} CE_g \; for \; all \; i = 1, 2, \ldots, n \tag{4}$$

subject to, raw material requirement of the processing plant (FC) expressed in Equation (5):

$$P \leq PCap. \tag{5}$$

A restriction on the contract specification of minimum land to be guaranteed in contract phase is represented by Equation (6):

$$s_i \leq LLC_i \text{ for all } i = 1,2,\ldots,n. \qquad (6)$$

Assumption 1. *Each supplier has a given physical or contractual capacity constraint.*

Assumption 2. *Each supplier is willing to accept the technological package selected for the optimal SN in the negotiation and contracting step.*

Assumption 3. *The total quantity requirement of raw material is fixed given the capacity at the processing plant.*

It is important to mention that although this approach seeks to improve the overall supply chain, the scope of this study is yet limited since some important environmental impacts and cost producing elements related to transport distance, mode and size are neglected.

2.3. Partner for Sustainability: Supplier Network Optimization and Selection

The optimization approach proposed in this work is performed by a genetic algorithm method. This choice was made according to the flexibility of this approach to tackle problems with multiple objectives, its potential to solve problems without restriction on the type of variables, either integer or continuous, in addition to its capacity to solve linear and non-linear problems [43]. Generally, the engineering design problem tends to be of a multiobjective nature with different variables and characteristics. Other methods may be considered that can also handle multiple objectives at once with variable complexity.

The final selection process is made using a multi-criteria decision-making process that takes into account the optimal alternatives found in the Pareto front. These alternatives are found to be non-dominated solutions near optimal value, and although the decision maker may use judgment to make the final selection from the alternatives, a formal method based on TOPSIS (Technique for Order of Preference by Similarity to Ideal Solution) is proposed in this paper [44,45]. This method is based on the idea of choosing the best alternative solution from a set by analysing the shortest geometric distance from the positive ideal solution and the longest distance from the negative ideal solution. It also requires weights to be assigned per criterion and normalizes the information, so that the various alternatives are ranked. Although other ranking and classification methods exist, TOPSIS has proven its efficiency in the final alternative selection process obtained through GA [46,47].

Supplier contracting is proposed within a Contract Farming (CF) framework, where a partnership is made by a contractual agreement in order to share knowledge and risk. Through this type of contracting, the possibility of incorporating centralized decision-making regarding technology used during the farming stage is allowed because of the shared risk and growth that contract farming promotes [38]. Although the use of CF may be difficult in some circumstances and regions, in the case of many developing countries, where a large part of the production systems have not yet become intensified, this type of collaborative framework can be seen as an opportunity for both parties.

3. Results

Case Study: Orange Juice Production

The orange juice industry supply chain network serves as an illustration of the proposed methodology and raises a lot of issues of GSCM. The case study is located in the Mexican Gulf Coast region of Martinez de la Torre, Veracruz, which is the most important citrus fruit producing region in Mexico. The SSP implies orange fresh fruit supplying orchards that are to be selected as suppliers for an orange juice producing company. The case study follows the steps proposed in the methodology section. They are presented in what follows.

The regional characteristics of production systems involve four basic technological packages. These packages range from organic (1), basic (2), standard (3), to intensive (4) systems, as shown in Table 2.

Table 2. Characteristics of each type of technology package.

Agricultural Practices	Technology Package			
	1 (Organic)	2 (Basic)	3 (Standard)	4 (Intensive)
Soil loosening	0	1	2	2
Weeding	2	2	2	0
Plantation	1	2	2	1
Chemical weeding	0	1	2	5
Pruning	0	0	1	1
Trunk protection	1	1	1	2
Chemical insecticide	0	0	4	6
Chemical fungicide	0	0	4	6
Urea (N)	0	0	2	4
K_2O_5	0	0	2	4
P_2O_5	0	0	2	4
Fuel *	0	0	90.3	105.35
Communication	1	1	1	1
Harvesting	1	1	1	1

Note: Unit is in number of applications during one year; * in liters of standard gasoline.

Each technological package was studied in order to determine the mean production yield and production cost as input parameters of the modelling stage; the values are presented in Table 3.

Table 3. Yield and cost matrix per technological package.

Indicator	Unit	1 (Organic)	2 (Basic)	3 (Standard)	4 (Intensive)
Production yield	ton/ha	5	8	15	25
Production cost (contract)	$(mxn)/ha	635	1275	3064	4205
Production cost (product)	$(mxn)/ton	127	159.38	204.27	168.2

An investigation related to this case study was performed by the Research Centre on Economic, Social and Technological Aspects of International Agriculture Policies (CIESTAAM), a research institution which is part of the Autonomous University of Chapingo (UACh). The study involved specialists on agricultural issues of the region [48]. The average yield of production as well as the operational cost related to these types of technological packages and practices were determined and presented. An orange production manual for the geographical region made by the National Institute for Forestry, Agricultural, and Animal Husbandry Research (INIFAP) [49] provided the information needed to perform the LCA.

Using the information gathered during the previous steps, a LCA was carried out through the use of the specialized software tool Simapro® [50].

Table 4 shows the selected environmental impact indicators from the IMPACT 2002+ method that is adopted in the optimization phase. Global Warming Potential (GWP) is one of the most widely used and understood environmental performance indicators. This is one of the reasons why it was selected for the simulation model; other indicators related to agricultural practices such as Acidification and Terrestrial Eutrophication, among others [51], can also be used when modelling environmental performance within the proposed methodology. Key environmental indicators selection should be based on requirements and goals of the study. It must be emphasized that aquatic eutrophication was selected given the effect of chemical fertilizers, regarding water nutrient contamination [51,52]. In [53], the authors present a meta-analysis of LCA applied to orange juice production and contrast with direct findings of their own LCA study applied to a Spanish production system. In their findings,

Climate Change (related to GWP) and Eutrophication (terrestrial and marine) are mainly contributed by N-fertilizers in the orange production stage. These indicators serve as an appropriate proxy for other indicators and sources (i.e., other agrochemicals) given the incremental usage behaviour described in the technological packages used in the case study (see Table 2). It is, nevertheless, important to point out that some important indicators such as human toxicity and acidization may also be appropriate depending on the crop/cultivar being produced.

Table 4. Environmental indicators output table per technology package.

Impact Category	Unit (per kg of Orange)	Technology Package			
		1 (Organic)	2 (Basic)	3 (Standard)	4 (Intensive)
Aquatic eutrophication	kg PO_4 P-lim	0	5.060×10^{-6}	1.230×10^{-5}	1.620×10^{-5}
Global warming	kg CO_2 eq	0	1.355×10^{-3}	1.149×10^{-2}	1.291×10^{-2}

The supplier network model was then developed using the information developed in the previous steps and by integrating it to the set of systems parameters gathered by the procurement analyst or model developer. This includes the following information:

- In the case of suppliers, a set of 20 suppliers is evaluated which are related to index (i) of set (I). Four technological packages are considered and integrated as the (g) index of the set (G).
- The minimum requirement for the processing plant (see Equation (5)) is set at a value of 60,000 tons of oranges, which is the capacity of a medium to large orange juice processing plant in Mexico.

Three objective functions are considered, one related to cost and two to environmental impacts:

Z1 = minimization of operational cost = min (Cost)
Z2 = minimization of CO_2 emissions equivalent = min (GWP)
Z3 = minimization of aquatic eutrophication equivalent = min (Eutro)

This cost criterion was selected since it can be viewed as useful to convince the participating potential supplier that this strategy is mutually beneficial in contrast to supplier sale prices or other forms of economic criteria such as initial investment, net present value, etc.

The input for the restriction related to minimum and maximum land area to be contracted (see Equation (6)), is presented in Table 5. A minimum value within the contract scheme stipulates that at least half of the capacity is contracted in the case of a selected supplier. This is done to promote the supplier participation in the type of proposed collaboration, as well as to reduce the number of suppliers that the optimization process can select.

Table 5. Relation between suppliers and respective field size that can be contracted for use.

Supplier	Contractible Hectares (t_i)	Supplier	Contractible Hectares (t_i)
1	226.48	11	50
2	101.38	12	298.82
3	190.73	13	107.57
4	650.81	14	115.81
5	43.12	15	69.11
6	512.61	16	258.94
7	43.05	17	273.76
8	560.81	18	250.52
9	26.29	19	221.52
10	22.97	20	17.75

Optimization is then carried out using the MULTIGEN®genetic algorithm extension library [54]. The optimization simulations parameters used are shown in Table 6. The population size and number of generations was empirically evaluated by a preliminary analysis showing that a ratio roughly set at 20 individuals per variable, and doubling of population size is effective within the MULTIGEN environment. The crossover and mutation rates were set at default values as suggested in [54]. The use of the NSGA II optimization algorithm is selected given its capacity to find a non-dominated set of alternatives to develop the Pareto fronts needed [55].

Table 6. Optimization run parameters.

Parameters	Values
Population size	400
Number of generations	800
Algorithm	NSGA II
Crossover rate	0.9
Mutation rate	0.5

The optimization was run five times in order to validate and search a wider area of the feasible space, and this generated a set of 192 alternatives to be analyzed. Figure 3 shows a three-dimensional scatter plot where the vertical axis is Cost (in Mexican pesos or $MXN), the right axis is GWP (in kg CO_2 eq) and the depth axis is Eutrophication (in kg PO_4 P-lim).

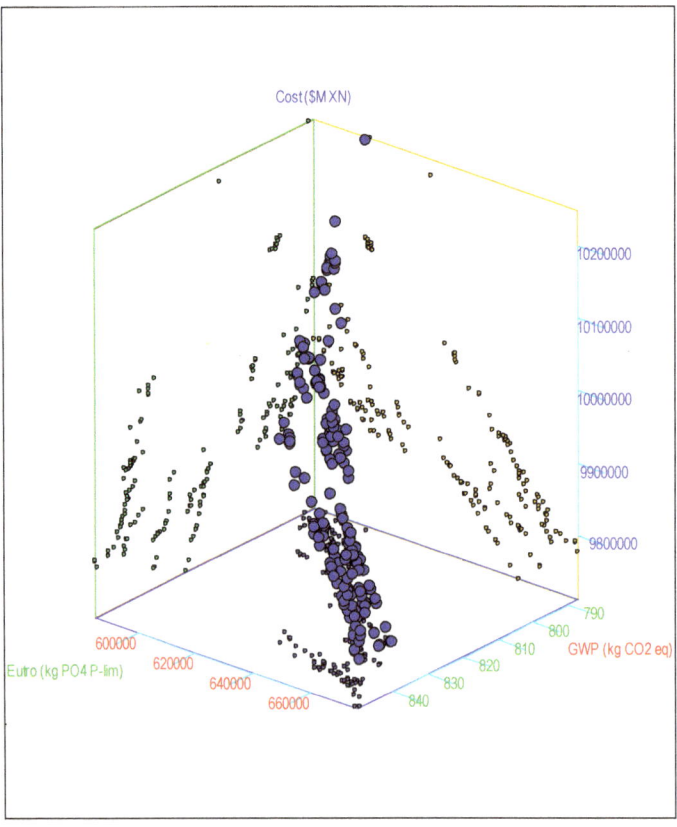

Figure 3. Three-dimensional scatter plot of all five Pareto Fronts.

Figure 3 exhibits the set of the different series of Pareto front runs. As this is a set of many Pareto fronts that were not evaluated concurrently, there are dominated points within the data set, which must be eliminated prior to applying the final selection.

Given that the resulting set of five Pareto fronts does not show a clear optimal decision alternative or region, the use of a decision-making tool becomes even more necessary. This final selection process is carried out through the use of a modified TOPSIS method proposed by Ren et al. [45] that consists of ranking the alternatives through a comparison with the values of the "ideal" curve. Before applying the TOPSIS method, a selection is made in order to keep only non-dominated alternatives from the five Pareto fronts. This series of steps is described below:

$$U\{\text{Pareto Front runs } i = 1, \ldots, 5\} \rightarrow \text{Pareto } \{U\} \rightarrow \text{TOPSIS } \{\text{Pareto}\}. \tag{7}$$

By applying steps 1 and 2 of the abovementioned procedure, leading to the Pareto of the Paretos (i.e. Pareto {U}), a lower number of 46 non-dominated alternative solutions is obtained. From them, the TOPSIS method [45] is applied in order to find the best ranked values. Figure 4 shows the resulting values with the location of special TOPSIS values called 1, 13 and 45, that correspond to the overall top-ranked one, the best TOPSIS value in relation to GWP and the best TOPSIS value in relation to Eutrophication, respectively.

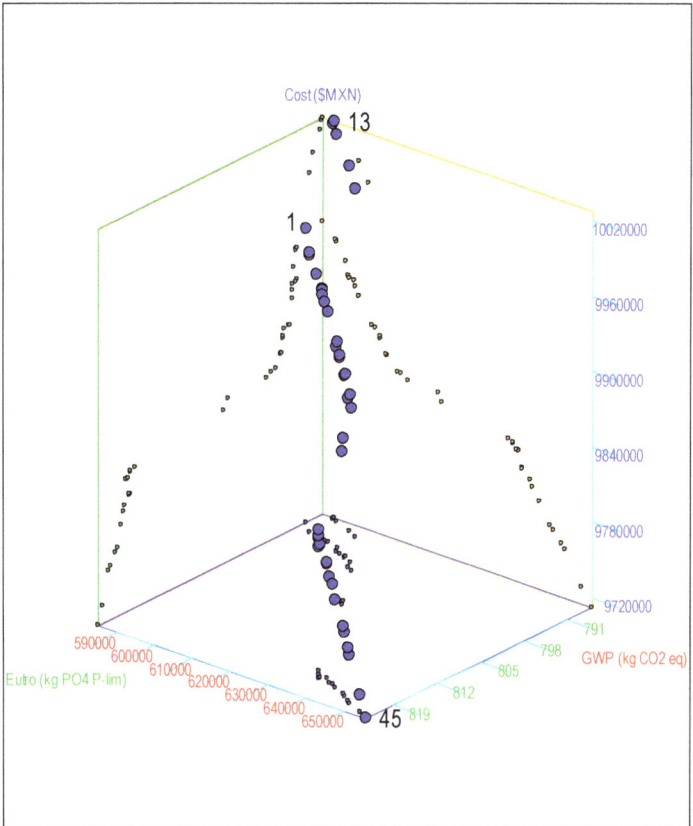

Figure 4. Three-dimensional scatter plot of the Final Set with TOPSIS ranked points.

Figure 4 shows that the top-ranked TOPSIS value is in the higher cost range. This solution is yet selected due to the trade-off against the environmental criteria. The improvement of cost is low, relative to the gains in environmental performance.

Table 7 presents the values for each criterion for some significant TOPSIS ranked alternative: number 1 (TOPSIS 1), the 13th (TOPSIS 13) and 45th (TOPSIS 45) values. TOPSIS 1 has the best compromise since it provides the best value for GWP and only differs from the best value for Eutrophication criteria by 0.33% (TOPSIS 13). Its cost is slightly higher of 2.35% of the best cost criterion performing alternative (TOPSIS 45). The three points can also be visualized in Figure 4 where TOPSIS 13 and 45 are at the extremes, whereas TOPSIS 1 is located at the upper part of the scatter plot.

Table 7. TOPSIS evaluation per criterion.

Criterion	TOPSIS 1	TOPSIS 13	TOPSIS 45	Discrepancy 1 vs. 13	Discrepancy 1 vs. 45
Cost ($MXN)	9,945,457	10,027,688	9,712,002	0.83%	2.35%
GWP (kg CO_2 eq)	585,024	587,981.24	655,538	0.51%	12.05%
Eutro (kg PO4P-lim)	789	787.09	823	0.33%	4.26%

Table 8 presents a comparison between the TOPSIS 1 solution and the TOPSIS 1 of a sample taken from the first Pareto front run at the 10th generation. The 10th generation was selected because it was the first generation in which all the individuals (solutions) are in the feasible space.

Table 8. Comparison between TOPSIS 1 and samples from 10th generation optimization value of the first Pareto front.

Criterion	TOPSIS 1	TOPSIS 1 of Sample PF		Average of Sample PF	
		Value	Discrepancy (%)	Value	Discrepancy (%)
Cost ($MXN)	9,945,457	10,348,080	4.05%	11,358,729	14.21%
GWP (kg CO2 eq)	585,024	669,783	14.49%	790,669	35.15%
Eutro (kg PO4P-lim)	789	849	7.57%	973	23.32%

The differences that can be observed are significant for all criteria, which justifies the application of the optimization procedure. A comparison between the average values of the sample is used in order to visualize the improvement that can be achieved through Partnership for Sustainability method, exhibiting a significant performance between 14% to 35% for the different criterion.

Supplier contracting is then the final stage in the selection process. For the case study, the optimization results then allow to select the suppliers, the type of technological package to use, and the area of land to be guaranteed in the final contract. Table 9 displays the final set of values for the decision variables for TOPSIS 1 alternative. It can be first observed that it implies all suppliers, since the optimization search leads to select different types of technologies of low yield but with a better overall performance. The second interesting point is that the mix of technologies used does not include technology package 3, which is the most commonly used technological package in the region. It must be also highlighted that there was only one field of a small area with a technology package type 1. This is most probably due to the fact that technology package 2 yields more products for a similar environmental impact performance.

The application of the methodology results in a set of alternatives given by rank. Although other external factors such as agricultural, economic and environmental policies have to be considered in the final judgment, the decision aid provides the insight needed for an objective and efficient supplier selection tool.

Table 9. Decision variable results for TOPSIS 1.

Supplier	Technology	Selection *	Land Area
1	2	1	217.630
2	2	1	100.407
3	4	1	188.867
4	2	1	649.984
5	4	1	42.206
6	2	1	506.885
7	4	1	42.982
8	4	1	490.285
9	4	1	25.740
10	4	1	20.785
11	4	1	48.021
12	2	1	297.277
13	4	1	107.083
14	4	1	113.144
15	4	1	68.255
16	4	1	245.091
17	2	1	271.078
18	4	1	235.275
19	4	1	204.422
20	1	1	14.454

* 1 = supplier selected; 0 = not selected.

4. Discussion

Partnership for Sustainability can be a useful tool in tackling the supplier selection problem by providing a paradigm shift regarding collaboration as a means to improve overall performance of sustainable food supply chains. The steps laid out in this paper provide a roadmap to improve and incorporate different ways on how GSCM may be used and structured to overcome weaknesses in conventional management practices when confronting strategic long-term decisions such as the Supplier Selection Problem. The case study that is presented may prove useful in understanding how to incorporate modelling and optimization within the Guidance provided by ISO20400 on Sustainable procurement. Reviewing Table 1 where a 7-step supplier selection process has been developed. Pre-selection and Short listing stages, knowing that the long-term goal is to develop collaborators and procurement staff can help develop useful knowledge for better judgment in favor of organizational attributes that will be highly valued in later stages (e.g., organizational culture, reputation, supplier local socio-economic indicators, etc.). This can feed the Supplier characterization stage that takes environmental performance as an attractive attribute. Some indicators could be used if they comply with norms and standards (e.g., organic production, ISO14000 Environmental management, ISO26000 Guidance on social responsibility). This will allow the FC to create a supplier network model that has many design alternatives, creating a wider space to optimize and choose from in the final stages (i.e., supplier network optimization and selection). Finally, this paper presented a complete and comprehensive case study in order to illustrate the capabilities and limitations of the proposed approach, and provided insight on what type of information is required and how it could be used in order to develop feasible and non-dominated optimal solutions.

The case study also illustrates some of the patterns one could confront when developing other specific green supplier selection models based on the methodology. Although it may have some areas to improve, it provides the foundation for continued work to analyze and integrate important factors in the decision-making process in the context of collaboration with suppliers that may become partners.

Some of the main perspectives to this paper are firstly, the further development and integration of contract policies into the supplier selection process methodology. This is a difficult task since information flow and synchronization among parties are not always functional. Yet exploring the potential of Contract Farming and collaborative farming in the context of the SSP may be valuable in some organizations. A second opportunity is to target the information collection, in the sense of better establishing which type of information a buyer requires from its potential and current suppliers, in order to make the best decision and to continuously maintain an effective supplier network.

Finally, an important perspective is to integrate the results of this model into a global green supply chain optimization model that may use a similar approach for visualizing and solving the other large-scale strategic challenges of GSCM.

Author Contributions: Conceptualization, C.A.-P. and M.A.M.-A.; investigation, A.A.A.-L.; data curation, K.C.A.-S.; writing—original draft preparation, M.A.M.-A.; writing—review and editing, C.A.-P.; visualization, A.B.-S.

Funding: This research received no external funding.

Conflicts of Interest: The authors declare no conflicts of interest.

Abbreviations

ASC	Agro-food Supply Chain
CF	Contract Farming
EI	Focal Company
GSCM	Green Supply Chain Management
LCA	Life Cycle Assessment
LCI	Life Cycle Inventory
MCDM	Multiple Criteria Decision Making
MOO	Multi-Objective Optimization
MS	Multiple Strength
M-TOPSIS	Modified Technique for Order of Preference by Similarity To Ideal Solution
NSGA	Non-Dominated Sorting Genetic Algorithms
PfS	Partnership for Sustainability
SC	Supply Chain
SCM	Supply Chain Management
SCND	Supply Chain Network Design
SN	Supplier Network
SS	Single Strength
SSP	Supplier Selection Problem
TOPSIS	Technique for Order of Preference by Similarity to Ideal Solution

References

1. Bougherara, D.; Combris, P. Eco-labelled food products: What are consumers paying for? *Eur. Rev. Agric. Econ.* **2009**, *36*, 321–341. [CrossRef]
2. Roheim, C.A.; Asche, F.; Santos, J.I. The Elusive Price Premium for Ecolabelled Products: Evidence from Seafood in the UK Market. *J. Agric. Econ.* **2011**, *62*, 655–668. [CrossRef]
3. Marsden, T.; Murdoch, J.; Morgan, K. Sustainable agriculture, food supply chains and regional development: Editorial introduction. *Int. Plan. Stud.* **1999**, *4*, 295–301. [CrossRef]
4. Van der Werf, H.M.; Petit, J. Evaluation of the environmental impact of agriculture at the farm level: A comparison and analysis of 12 indicator-based methods. *Agric. Ecosyst. Environ.* **2002**, *93*, 131–145. [CrossRef]
5. García-Alcaraz, J.L.; Alvarado-Iniesta, A.; Blanco-Fernández, J.; Maldonado-Macías, A.A.; Jiménez-Macías, E.; Saenz-Díez Muro, J.C. The Impact of Demand and Supplier Relationship. *J. Food Process Eng.* **2016**, *39*, 645–658. [CrossRef]
6. De Boer, L.; Labro, E.; Morlacchi, P. A review of methods supporting supplier selection. *Eur. J. Purch. Supply Manag.* **2001**, *7*, 75–89. [CrossRef]
7. Abdallah, T.; Farhat, A.; Diabat, A.; Kennedy, S. Green Supply Chains with Carbon Trading and Environmental Sourcing: Formulation and Life Cycle Assessment. *Appl. Math. Model.* **2012**, *36*, 4271–4285. [CrossRef]

8. Hugo, A.; Pistikopoulos, E.N. Environmentally conscious long-range planning and design of supply chain networks. *J. Clean. Prod.* **2005**, *13*, 1471–1491. [CrossRef]
9. Jolliet, O.; Saadé, M.; Crettaz, P. *Analyse Du Cycle De Vie: Comprendre Et Réaliser Un Écobilan*; Presses Polytechniques Et Universitaires Romandes: Lausanne, Switzerland, 2010.
10. Seuring, S.; Muller, M. From a literature review to a conceptual framework for sustainable supply chain management. *J. Clean. Prod.* **2008**, *16*, 1699–1710. [CrossRef]
11. Srivastava, S.K. Green supply-chain management: A state-of-the-art literature review. *Int. J. Manag. Rev.* **2007**, *9*, 53–80. [CrossRef]
12. Mittal, A.; Krejci, C.C.; Craven, T.J. Logistics Best Practices for Regional Food Systems: A Review. *Sustainability* **2018**, *10*, 168. [CrossRef]
13. Bertramsen, S.; Nquyen, G.; Dobbs, T. *Quality and Eco-Labeling of Food Products in France and the United States*; Department of Economics, South Dakota State University: Brookings, SD, USA, 2002.
14. Amin, S.H.; Zhang, G. An integrated model for closed-loop supply chain configuration and supplier selection: Multi-objective approach. *Expert Syst. Appl.* **2012**, *39*, 6782–6791. [CrossRef]
15. Ramudhin, A.; Chaabane, A.; Kharoune, M.; Paquet, M. Carbon Market Sensitive Green Supply Chain Network Design. In Proceedings of the 2008 IEEE International Conference on Industrial Engineering and Engineering Management, Singapore, 8–11 December 2008; pp. 1093–1097.
16. Corsano, G.; Vecchietti, A.R.; Montagna, J.M. Optimal design for sustainable bioethanol supply chain considering detailed plant performance model. *Comput. Chem. Eng.* **2011**, *35*, 1384–1398. [CrossRef]
17. Guillén-Gosálbez, G.; Grossmann, I.E. Optimal design and planning of sustainable chemical supply chains under uncertainty. *AIChE J.* **2009**, *55*, 99–121. [CrossRef]
18. Bouzembrak, Y.; Allaoui, H.; Goncalves, G.; Bouchriha, H. A multi-objective green supply chain network design. In Proceedings of the 2011 4th International Conference on Logistics (LOGISTIQUA), Hammamet, Tunisia, 20 May 2011; pp. 357–361.
19. Cao, S.; Zhang, K. Optimization of the flow distribution of e-waste reverse logistics network based on NSGA II and TOPSIS. In Proceedings of the 2011 International Conference on E-Business and E-Government (ICEE), Potomac, MD, USA, 21–26 August 2011; pp. 1–5.
20. Lin, S.-S.; Juang, Y.-S. Selecting green suppliers with analytic hierarchy process for biotechnology industry. *J. Oper. Supply Chain Manag.* **2008**, *1*, 115–129. [CrossRef]
21. Collette, Y.; Siarry, P. *Multiobjective Optimization: Principles and Case Studies*; Springer Science & Business Media: New York, NY, USA, 2003. Available online: https://books.google.fr/books?hl=en&lr=&id=XNYF4hltoF0C&oi=fnd&pg=PA1&dq=optimisation+multi+objective+yann+siarry&ots=K1C-say5RE&sig=5OJ6Dg6yuBaG2Z1tC7eSk1nB6Nw (accessed on 20 March 2015).
22. Freitas, A.A. A Critical Review of Multi-objective Optimization in Data Mining: A Position Paper. *SIGKDD Explor. Newsl.* **2004**, *6*, 77–86. [CrossRef]
23. Miranda-Ackerman, M.A.; Azzaro-Pantel, C.; Aguilar-Lasserre, A.A. A green supply chain network design framework for the processed food industry: Application to the orange juice agrofood cluster. *Comput. Ind. Eng.* **2017**, *109*, 369–389. [CrossRef]
24. Ware, N.R.; Singh, S.P.; Banwet, D.K. Supplier selection problem: A state-of-the-art review. *Manag. Sci. Lett.* **2012**, *2*, 1465–1490. [CrossRef]
25. Zsidisin, G.A.; Siferd, S.P. Environmental purchasing: A framework for theory development. *Eur. J. Purch. Supply Manag.* **2001**, *7*, 61–73. [CrossRef]
26. Walton, S.V.; Handfield, R.B.; Melnyk, S.A. The Green Supply Chain: Integrating Suppliers into Environmental Management Processes. *Int. J. Purch. Mater. Manag.* **1998**, *34*, 2–11. [CrossRef]
27. Tate, W.L.; Ellram, L.M.; Dooley, K.J. Environmental purchasing and supplier management (EPSM): Theory and practice. *J. Purch. Supply Manag.* **2012**, *18*, 173–188. [CrossRef]
28. Humphreys, P.K.; Wong, Y.K.; Chan, F.T.S. Integrating environmental criteria into the supplier selection process. *J. Mater. Process. Technol.* **2003**, *138*, 349–356. [CrossRef]
29. Lee, A.H.; Kang, H.Y.; Hsu, C.F.; Hung, H.C. A green supplier selection model for high-tech industry. *Expert Syst. Appl.* **2009**, *36*, 7917–7927. [CrossRef]
30. Van Hoof, B.; Lyon, T.P. Cleaner Production in Small Firms taking part in Mexico's Sustainable Supplier Program. *J. Clean. Prod.* **2013**, *41*, 270–282. Available online: http://www.sciencedirect.com/science/article/pii/S095965261200488X (accessed on 29 October 2013). [CrossRef]

31. Rehber, E. Vertical Coordination in the Agro-Food Industry And Contract Farming: A Comparative Study of Turkey and the USA. In *Food Marketing Policy Center Research Reports 052*; Department of Agricultural and Resource Economics, Charles J. Zwick Center for Food and Resource Policy, University of Connecticut: Mansfield, CT, USA, 2000. Available online: http://ideas.repec.org/p/zwi/fpcrep/052.html (accessed on 15 October 2013).
32. Richard, L.; Kohls, J.N.U. *Marketing of Agricultural Products*; Prentice Hall: Upper Saddle River, NJ, USA, 1998.
33. Minot, N. *Contract Farming and Its Effect on Small Farmers in Less Developed Countries*; Michigan State University, Department of Agricultural, Food, and Resource Economics: East Lansing, MI, USA, 1986. Available online: http://ideas.repec.org/p/ags/midiwp/54740.html (accessed on 23 October 2013).
34. Bai, C.; Sarkis, J. Green supplier development: Analytical evaluation using rough set theory. *J. Clean. Prod.* **2010**, *18*, 1200–1210. [CrossRef]
35. Leppelt, T.; Foerstl, K.; Reuter, C.; Hartmann, E. Sustainability management beyond organizational boundaries–sustainable supplier relationship management in the chemical industry. *J. Clean. Prod.* **2013**, *56*, 94–102. [CrossRef]
36. Rao, P.; Holt, D. Do green supply chains lead to competitiveness and economic performance? *Int. J. Oper. Prod. Manag.* **2005**, *25*, 898–916. [CrossRef]
37. Zhu, Q.; Sarkis, J. Relationships between operational practices and performance among early adopters of green supply chain management practices in Chinese manufacturing enterprises. *J. Oper. Manag.* **2004**, *22*, 265–289. [CrossRef]
38. Eaton, C.; Shepherd, A.W. *Contract Farming—Partnerships for Growt*; Food & Agriculture Organization: Rome, Italy, 2001.
39. Vachon, S.; Klassen, R.D. Green project partnership in the supply chain: The case of the package printing industry. *J. Clean. Prod.* **2006**, *14*, 661–671. [CrossRef]
40. Albino, V.; Dangelico, R.M.; Pontrandolfo, P. Do inter-organizational collaborations enhance a firm's environmental performance? A study of the largest US companies. *J. Clean. Prod.* **2012**, *37*, 304–315. Available online: http://www.sciencedirect.com/science/article/pii/S095965261200371X (accessed on 15 October 2013). [CrossRef]
41. Bowen, F.E.; Cousins, P.D.; Lamming, R.C.; Farukt, A.C. The Role of Supply Management Capabilities in Green Supply. *Prod. Oper. Manag.* **2001**, *10*, 174–189. [CrossRef]
42. Azzaro-Pantel, C.; Ouattara, A.; Pibouleau, L. Ecodesign of Chemical Processes with Multi-Objective Genetic Algorithms. In *Multi-Objective Optimization in Chemical Engineering*; Rangaiah, G.P., Bonilla-Petriciolet, A., Eds.; John Wiley & Sons Ltd.: Hoboken, NJ, USA, 2013; pp. 335–367.
43. Dietz, A.; Azzaro-Pantel, C.; Pibouleau, L.; Domenech, S. Multiobjective optimization for multiproduct batch plant design under economic and environmental considerations. *Comput. Chem. Eng.* **2006**, *30*, 599–613. [CrossRef]
44. Lai, Y.-J.; Liu, T.-Y.; Hwang, C.-L. TOPSIS for MODM. *Eur. J. Oper. Res.* **1994**, *76*, 486–500. [CrossRef]
45. Ren, L.; Zhang, Y.; Wang, Y.; Sun, Z. Comparative Analysis of a Novel M-TOPSIS Method and TOPSIS. *Appl. Math. Res. Express* **2007**, *2007*, abm005. [CrossRef]
46. Gen, M.; Cheng, R. *Genetic Algorithms and Engineering Optimization*; John Wiley & Sons, Inc.: Hoboken, NJ, USA, 1999. Available online: https://www.google.fr/search?q=Genetic+Algorithms+and+Engineering+Optimization&oq=Genetic+Algorithms+and+Engineering+Optimization&aqs=chrome.0.57j61j62l2.265&sugexp=chrome,mod=8&sourceid=chrome&ie=UTF-8 (accessed on 17 December 2012).
47. Gen, M.; Kumar, A.; Ryul Kim, J. Recent network design techniques using evolutionary algorithms. *Int. J. Prod. Econ.* **2005**, *98*, 251–261. [CrossRef]
48. Gomez Cruz, M.; Schwentesius Rindermann, R. *La Agroindustria De Naranja En Mexico*; Ciestaam: Estado de México, México, 1997.
49. Curti-Díaz, S.A.; Diaz-Zorrilla, U.; Loredo-Salazar, X. Manual De Produccion De Naranja Para Veracruz Y TABASCO. Available online: http://www.concitver.com/archivosenpdf/MANUAL%20DE%20PRODUCCION%20DE%20NARANJA%20PARA%20VERACRUZ%20Y%20TABASCO.pdf (accessed on 10 January 2015).
50. Goedkoop, M.; De Schryver, A.; Oele, M. *Introduction to LCA with SimaPro 7*; PRé Consultants Report 4; PRé Consultants: Amersfoort, The Netherlands, 2008.

51. Brentrup, F. Life Cycle assessment to evaluate the environmental impact of arable crop production. *Int. J. LCA* **2003**, *8*, 156. [CrossRef]
52. Huijbregts, M.A.; Seppälä, J. Life Cycle Impact assessment of pollutants causing aquatic eutrophication. *Int. J. Life Cycle Assess.* **2001**, *6*, 339–343. [CrossRef]
53. Doublet, G.; Jungbluth, N.; Flury, K.; Stucki, M.; Schori, S. Life cycle assessment of orange juice. In *SENSE-Harmonised Environmental Sustainability in the European Food and Drink Chain, Seventh Framework Programme: Project No. 288974*; Funded by EC; Deliverable D 2.1 ESU-services Ltd.: Zürich, Switzerland, 2013.
54. Gomez, A.; Pibouleau, L.; Azzaro-Pantel, C.; Domenech, S.; Latgé, C.; Haubensack, D. Multiobjective genetic algorithm strategies for electricity production from generation IV nuclear technology. *Energy Convers. Manag.* **2010**, *51*, 859–871. [CrossRef]
55. Deb, K.; Pratap, A.; Agarwal, S.; Meyarivan, T.A. A fast and elitist multiobjective genetic algorithm: NSGA-II. *IEEE Trans. Evol. Comput.* **2002**, *6*, 182–197. [CrossRef]

© 2019 by the authors. Licensee MDPI, Basel, Switzerland. This article is an open access article distributed under the terms and conditions of the Creative Commons Attribution (CC BY) license (http://creativecommons.org/licenses/by/4.0/).

Article

A Linear Programming Model with Fuzzy Arc for Route Optimization in the Urban Road Network

Elías Escobar-Gómez *, J.L. Camas-Anzueto *, Sabino Velázquez-Trujillo,
Héctor Hernández-de-León, Rubén Grajales-Coutiño, Eduardo Chandomí-Castellanos and
Héctor Guerra-Crespo

Tecnológico Nacional de México/I.T. Tuxtla Gutiérrez, Tuxtla Gutiérrez 29050, Chiapas, Mexico;
svelazquez@ittg.edu.mx (S.V.-T.); hhernandezd@ittg.edu.mx (H.H.-d.-L.); rgrajales@ittg.edu.mx (R.G.-C.); eduardo.chandomi@hotmail.com (E.C.-C.); hguerra@ittg.edu.mx (H.G.-C.)
* Correspondence: eescobar@ittg.edu.mx (E.E.-G.); jcamas@ittg.edu.mx (J.L.C.-A.)

Received: 1 November 2019; Accepted: 20 November 2019; Published: 25 November 2019

Abstract: In the transport system, it is necessary to optimize routes to ensure that the distance, the amount of fuel used, and travel times are minimized. A classical problem in network optimization is the shortest path problem (SPP), which is used widely in many optimization problems. However, the uncertainty that exists regarding real network problems makes it difficult to determine the exact arc lengths. In this study, we analyzed the problem of route optimization when delivering urban road network products while using fuzzy logic to include factors which are difficult to consider in classical models (e.g., traffic). Our approach consisted of two phases. In the first phase, we calculated a fuzzy coefficient to consider the uncertainty, and in the second phase, we used fuzzy linear programming to compute the optimal route. This approach was applied to a real network problem (a portion of the distribution area of a delivery company in the city of Tuxtla Gutierrez, Chiapas, Mexico) by comparing the travel times between the proposed model and a classical model. The proposed model was shown to predict travel time better than the classical model in this study, reducing the mean absolute percentage error (MAPE) by 25.60%.

Keywords: delivering products; travel time uncertainty; fuzzy logic; fuzzy linear programming; urban road network; reliability

---

1. Introduction

In the literature, optimization of routes in the urban road network involves the determination of the shortest distance between locations while considering traffic, capacity, delivery windows, and the number of turns or intersections along the route, among others. Route optimization is necessary to ensure that distance, the amount of fuel used, and travel times are minimized.

The shortest path problem (SPP) is one of the oldest key network optimization problems [1–18]. SPP is important for many real-world problems, such as communications, computer networks, logistics management, routing, and transportation systems (route optimization). In a network, the weights of arcs may represent the travel time or cost rather than representing geographical distances. In practical applications, the weights of arcs have parameters that are not usually precise, such as weather conditions, traffic flow, type of street corner (intersections), road conditions, or speed [19–25]. Therefore, it is not practical to consider the arc weight as a deterministic value.

To minimize the uncertainty regarding transportation activities, researchers have previously included as many factors as possible by using probability theory [26–29] or fuzzy logic [16,21,22,30–36], among other methods [17,18,24]. Either a predictable behavior or an occurrence frequency is required to find probability distributions. As an alternative, the fuzzy set theory provides the capability of handling uncertain information and is widely used in these situations.

Fuzzy logic is applied in many fields, such as aggregate planning and prediction [37–41], pattern recognition [42–49], and decision making [50–52]. In the search for the shortest path in an uncertain environment, many researchers considered the fuzzy shortest path problem (FSPP) as their chosen method [20–22,30–35,53–57].

Linear programming (LP) is one of the most widely used operations research tools and is a decision-making aid in almost all organizations. LP refers to a planning process that allocates resources optimally by minimizing costs or maximizing profits. One of the most important and successful methods of quantitative analysis used to solve business problems is the physical distribution of products [58–61].

The shortest path, which is computed via the application of deterministic models, is often not optimal if one considers the uncertainty of the environment regarding arc travel times, which occurs as one travels through the network, particularly around peak time.

Fuzzy linear programming (FLP) has been previously studied [37,62–67], usually under the consideration that the coefficients of the objective function or the constraints have fuzzy components; these constraints are modeled by assigning a fuzzy set with triangular, trapezoidal, left–shoulder, or right–shoulder membership functions.

The urban road network is a complex problem compounded by ever-increasing levels of traffic, which is affected by the popularity of cars, the traffic density of the area, weather, driver behavior, company policies, and so on. This problem further generates traffic accidents, environmental pollution, and energy waste. Urban road networks have been previously studied [36,66,68,69].

In this study, we propose an approach for route optimization in an urban zone through the implementation of a fuzzy adjustment coefficient, taking into consideration the uncertainty due to traffic alongside the number of intersections along the route, thereby increasing the reliability of the results.

The traffic factor was determined from the fuzzy variables, namely, the time of day and traffic density. The time of day was defined as the time at which the delivery was made, whereas the traffic density was defined as the level of congestion that was present in the area of distribution.

In this approach, we computed the optimal route using FLP. Using the objective function of FLP, the arc lengths were adjusted using the fuzzy coefficient, and the type of intersection (street corner) was considered.

In this study, we analyzed a portion of the distribution area of a delivery company in the city of Tuxtla Gutierrez, Chiapas, Mexico. We obtained travel times close to the real travel times with a mean absolute percentage error of 9.60%; the variance of the absolute percentage error of the fuzzy model was 46.57%.

This paper is organized as follows: In Section 1, we consider the preliminaries; in Section 2, we introduce the proposed method; in Section 3, we include an application to a real network to demonstrate our approach (four blocks and nine street corners); in Section 4, we present the results of our approach considering a portion of the distribution area of a delivery company (78 blocks and 98 street corners) and also compare the results between our approach and a classic model (LP); in Section 5, we present our conclusions.

2. Materials and Methods

Let $G(N, E)$ be an acyclic directed graph defined by a nonempty and finite set, $N = \{1, 2, \ldots, n\}$, of n nodes and a set, $E = \{(i, j)\ i, j\ N\ \text{and}\ i \neq j\}$, of m directed arcs/edges. Each arc, (i, j), had an associated cost, c_{ij}, that denotes the cost per unit flow on that edge. We assumed that the flow cost varied linearly with the amount of flow on the arc.

With each node, $i \in N$, we associated an integer number, $b(i)$, representing its source or target. If $b(i) < 0$, then node i was a source node, if $b(i) > 0$, then node i was a target node, and if $b(i) = 0$, then node i was a trans-shipment node.

A path was a walk in which the nodes, and therefore the arcs, were distinct. We defined a path, x_{ij}, as a sequence, $x_{ij} = \{i, (i, i_1), i_1, \ldots, i_k, (i_k, j), j\}$, of alternating nodes and arcs.

Definition 1. *Let X be the universe of discourse, with a range of interest variables (e.g., the time of day), where x is an element of X. A fuzzy subset, $\widetilde{S}$, in the universe of discourse, X, with discrete and finite variables, could be represented as a set of ordered pairs of an element, x, and its membership grade in $\mu_{\widetilde{S}}(x)$. The membership function mapped each element of X to a membership grade (or membership value) between 0 and 1.*

$$\widetilde{S} = \{(x, \mu_{\widetilde{S}}(x)) \setminus x \in X\} \tag{1}$$

where

$$0 \leq \mu_{\widetilde{S}}(x) \leq 1 \tag{2}$$

Definition 2. *In the literature, despite being a simplistic description of a population, triangular distribution is typically used if a variable requires subjective estimation. A triangular fuzzy number was represented by a triplet, $\widetilde{A} = (a_1, a_2, a_3)$, with the membership function, $\mu_{\widetilde{A}}(x)$, defined by Equation (3) (Figure 1). The triangular membership functions were simple and therefore facilitated calculation (Equation (3)).*

$$\mu_{\widetilde{A}}(x) = \begin{cases} 1 - \frac{a_2 - x}{a_2 - a_1} & a_1 < x < a_2 \\ 1 - \frac{x - a_2}{a_3 - a_2} & a_2 \leq x < a_3 \\ 0 & \text{otherwise} \end{cases} \tag{3}$$

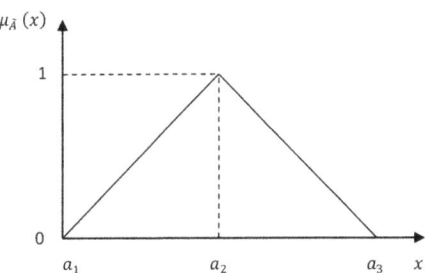

Figure 1. Triangular fuzzy number.

Definition 3. *In the literature, the trapezoidal membership functions are a generalization of the triangular membership functions in fuzzy modeling; applications of trapezoidal distribution include applied physics and risk analysis problem-solving. A trapezoidal fuzzy number was defined as $\widetilde{B} = (b_1, b_2, b_3, b_4)$, with the membership, $\mu_{\widetilde{B}}(x)$, defined by Equation (4) (Figure 2). The trapezoidal membership functions were simple and therefore facilitated easy computation (Equation (4)).*

$$\mu_{\widetilde{B}}(x) = \begin{cases} 1 - \frac{b_2 - x}{b_2 - b_1} & b_1 < x < b_2 \\ 1 & b_2 \leq x \leq b_3 \\ 1 - \frac{x - b_3}{b_4 - b_3} & b_3 < x < b_4 \\ 0 & \text{otherwise} \end{cases} \tag{4}$$

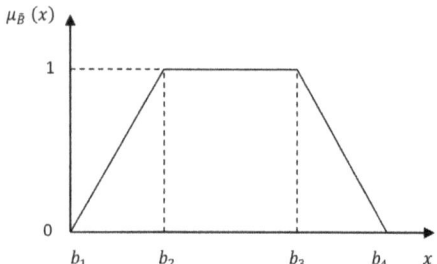

Figure 2. Trapezoidal fuzzy number.

2.1. Linear Programming Formulation of the Shortest Path Problem

For SPP, we wanted to find a direct path regarding the minimum cost from a specified source, node s, to another specified target, node t, in a directed graph (digraph), in which each arc, $(i,j) \in E$, had an associated cost, c_{ij}.

The linear programming formulation of the SPP is shown below. We represented the arc flow $(i,j) \in E$ by $X_{i,j}$ (decision variables) and the arc length by $c_{i,j}$ (cost). Figure 3 shows the arc flow and the arc length of a general network. The constraints of the linear programming model corresponded to the inputs and the outputs of the network nodes (Equation (5)).

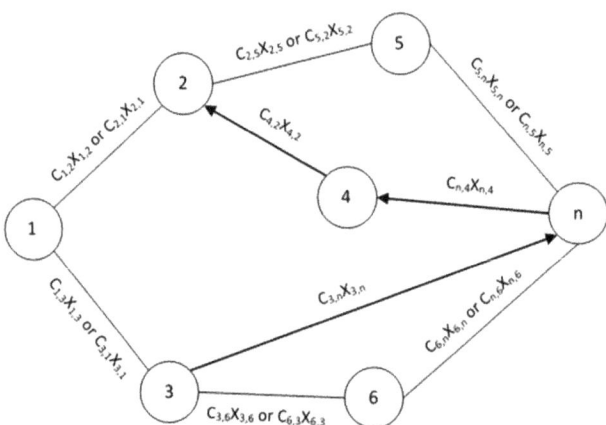

Figure 3. General network.

$$\text{Minimize} \quad Z = \sum_{i=1}^{n} \sum_{\substack{j=1 \\ j \neq i}}^{n} c_{i,j} X_{i,j}$$

$$\text{subject to} \quad \sum_{\substack{j \neq i}}^{n} X_{j,i} - \sum_{\substack{j \neq i}}^{n} X_{i,j} = b_i, \quad \forall\ i \in N \quad (5)$$

$$X_{i,j} \in \{0,1\}, \quad \forall\ (i,j) \in E$$

If we consider that the source was denoted by s and the target by t, then (Equation (6))

$$b_i = \begin{cases} -1; & \text{if } i = s \\ 1; & \text{if } j = t \\ 0; & \text{in other case} \end{cases} \quad (6)$$

The data for this model satisfied the feasibility condition (Equation (7))

$$\sum_{i=1}^{n} b_i = 0 \qquad (7)$$

that is, that the total origin must equal the total destination.

2.2. Proposed Model.

In practical applications (e.g., route optimization) there is uncertainty that makes it difficult to determine exact travel times (arc lengths). Therefore, our approach included two phases, i.e., a first phase to calculate the fuzzy coefficient and a second phase to use FLP to compute the optimal route. The fuzzy coefficient was integrated into the objective function of the model.

2.2.1. Fuzzy Coefficient Calculation

To calculate the fuzzy coefficient, we used rule-based fuzzy logic. We identified the fuzzy variables and a fuzzy rule base, then, using the inference system, we computed the fuzzy outputs. Following this, we used defuzzification [41] to obtain the fuzzy coefficient, $(\tilde{d}_{i,j})$.

To identify the fuzzy input variables, we considered expert knowledge of logistics and the transportation practices of various Mexican companies. Then, we identified a list of 49 elements that affected the distribution of products in the urban road network. These elements were analyzed and classified into seven types, namely, time of the day, traffic density, weather, driver behavior, company policies, social factors, and unnatural behaviors.

Afterward, we developed a survey for transportation workers. Through analyzing the results of the survey, we determined that the time of the day and traffic density were the main factors that modified the transportation time.

The time of day fuzzy variable was defined as the time at which the delivery took place. Six linguistic labels (Figure 4) were then formed from this definition, including:

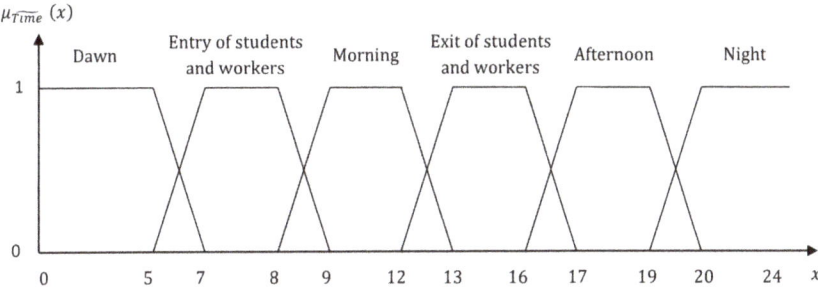

Figure 4. Time of day fuzzy variable.

Dawn—the period of the day after midnight but before to the light of the sun appeared in the sky, which was considered to be before 05:00. During this period, total vehicle transit was light.

Entry of students and workers—the period of the day when students arrived at school and employees arrived at work, which was considered to be between 07:00 and 08:00. During this period, vehicle transit was very heavy.

Morning—the time of day extending from entry of students and workers to noon, which was considered to be between 09:00 and 12:00. During this period, the traffic level was moderate.

Exit of students and workers—the period of the day when students exited school and employees left work, which was considered to be between 13:00 and 16:00. Throughout this period, the traffic level was very heavy.

Afternoon—the time of day between the exit of students and workers and evening, which was considered to be between 17:00 and 19:00. During this period, the traffic level was heavy.

Night—the period between afternoon and the midnight, which was considered to be after 20:00. During this period, the traffic level was light.

According to the definitions mentioned above, and taking into account that the trapezoidal membership functions were simple and facilitated the calculation, the membership functions of the x fuzzy variable, i.e., time of day, were trapezoidal fuzzy numbers and were obtained using Definition 3. These were substituted into Equations (8)–(13).

$$\mu_{\widetilde{D}}(x) = \begin{cases} 1 & 0 \leq x \leq 5 \\ 1 - \frac{x-5}{2} & 5 < x < 7 \\ 0 & otherwise \end{cases} \tag{8}$$

$$\mu_{\widetilde{ENSW}}(x) = \begin{cases} 1 - \frac{7-x}{2} & 5 < x < 7 \\ 1 & 7 \leq x \leq 8 \\ 9 - x & 8 < x < 9 \\ 0 & otherwise \end{cases} \tag{9}$$

$$\mu_{\widetilde{M}}(x) = \begin{cases} x - 8 & 8 < x < 9 \\ 1 & 9 \leq x \leq 12 \\ 13 - x & 12 < x < 13 \\ 0 & otherwise \end{cases} \tag{10}$$

$$\mu_{\widetilde{EXEW}}(x) = \begin{cases} x - 12 & 12 < x < 13 \\ 1 & 13 \leq x \leq 16 \\ 17 - x & 16 < x < 17 \\ 0 & otherwise \end{cases} \tag{11}$$

$$\mu_{\widetilde{A}}(x) = \begin{cases} x - 16 & 16 < x < 17 \\ 1 & 17 \leq x \leq 19 \\ 20 - x & 19 < x < 20 \\ 0 & otherwise \end{cases} \tag{12}$$

$$\mu_{\widetilde{N}}(x) = \begin{cases} x - 19 & 19 < x < 20 \\ 1 & 20 \leq x \leq 24 \\ 0 & otherwise \end{cases} \tag{13}$$

The fuzzy variable of traffic density was defined as the level of traffic congestion that was observed in the area during distribution due to proximity to school zones, industrial areas, or shopping malls. This variable was divided into three linguistic labels (Figure 5).

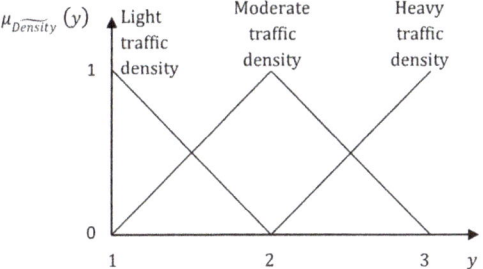

Figure 5. Traffic density fuzzy variable.

Light traffic density—vehicular flow conditions that were free-flowing due to a quiet road, which was away from any school zones, industrial areas, or shopping malls, thereby reducing travel times. This was represented by fuzzy number 1.

Moderate traffic density—vehicular flow conditions that were suitable owing to the measured use of the roads, which were located around school zones, industrial areas, or shopping malls. This was represented by fuzzy number 2.

Heavy traffic density—vehicular flow conditions that consisted of a lot of traffic due to the excess use of roads because of proximity to school zones, industrial areas, or shopping malls. This was characterized by slower speeds, longer travel times, and increased vehicular queuing and was represented by fuzzy number 3.

According to the definitions mentioned previously, triangular distribution was typically used if a variable required subjective estimation; the triangular membership functions were simple and facilitated calculation, therefore, the membership functions of the y fuzzy variable, i.e., the traffic density, were triangular fuzzy numbers and were obtained through Definition 2. These were substituted into Equations (14)–(16).

$$\mu_{\widetilde{LT}}(y) = \begin{cases} 1 & y(14) < 1 \\ 2 - y & 1 \leq y \leq 2 \\ 0 & y > 2 \end{cases} \tag{14}$$

$$\mu_{\widetilde{MT}}(y) = \begin{cases} y - 1 & 1 < y < 2 \\ 3 - y & 2 \leq y \leq 3 \\ 0 & \text{otherwise} \end{cases} \tag{15}$$

$$\mu_{\widetilde{HT}}(y) = \begin{cases} 0 & y < 2 \\ y - 2 & 2 \leq y \leq 3 \\ 1 & y > 3 \end{cases} \tag{16}$$

The z output fuzzy variable, the adjusted time, was used to modify travel times for the delivery of goods and materials. Six linguistic labels were formed for this, namely, very quickly, quickly, normal, slowly, very slowly, and extremely slowly (Figure 6). The membership functions of the fuzzy variable were triangular fuzzy numbers and trapezoidal fuzzy numbers and were obtained through Definitions 2 and 3.

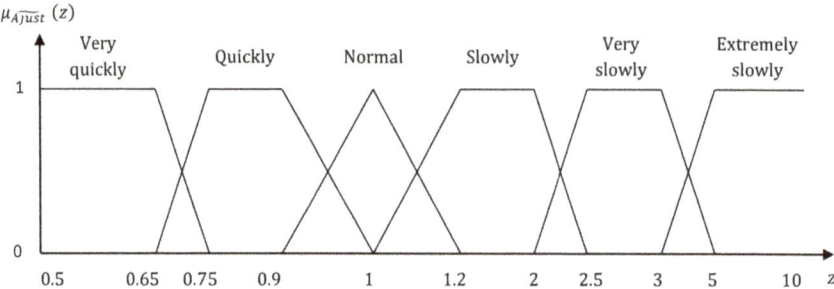

Figure 6. Adjusted time output fuzzy variable.

To consider every possible situation, the 14 fuzzy rules were used, as listed in Table 1. For the membership functions of the output fuzzy variable, VQ represents very quickly, Q represents quickly, N represents normal, S represents slowly, VS represents very slowly, and ES represents extremely slowly.

Table 1. Fuzzy rules.

		Time of Day					
		Dawn	Entry of Students and Workers	Morning	Exit of Students and Workers	Afternoon	Night
Traffic density	Light		N	N	S	Q	
	Moderate	VQ	S	S	VS	N	VQ
	Heavy		ES	VS	ES	VS	

To perform the inference system, fuzzy inputs were required to define the fuzzy rule base. The inference system considered the input values and the fuzzy rule base to determine the set of rules that was activated and the related conclusions, which were the fuzzy sets of the fuzzy output variable. To calculate the fuzzy value of the activated rule, we used the operations and intersections between the fuzzy sets in order to calculate the fuzzy value of an activated rule; unions were used to determine the fuzzy value of a set of rules which was activated with the same conclusion. We used triangular Norma and triangular Conorma to generalize the functions that defined the intersection and the union of a fuzzy set, respectively [41]. The membership grade of the output fuzzy variable was calculated using Equation (17).

$$\mu_{\widetilde{Ajust}}(z) = \vee \left(\mu_{\widetilde{Time}}(x) \wedge \mu_{\widetilde{Density}}(y) \right) \tag{17}$$

Defuzzification consisted of transforming the fuzzy output values into values that had a practical meaning using the geometrical center method. This method consisted of the following four steps [41]:

1) In regular figures, the area formed by the values of the fuzzy sets of the output variable was decomposed. To achieve this, the relationship between the membership grade of two adjacent fuzzy sets was analyzed, thereby defining two cases: The first case implied that the membership grade of the first fuzzy set was less than or equal to the value of belonging to the second set. For the second case, the membership grade of the first fuzzy set was greater than the membership grade of the second set. Four regular figures regarding the relationship between two adjacent fuzzy sets were formed.
2) The surface of each figure obtained in step 1 was calculated.
3) The centroid of each figure obtained in step 1 was determined.
4) The total centroid was calculated, with the result being the value of the defuzzification of the response variable (fuzzy coefficient, $\widetilde{d}_{i,j}$).

2.2.2. Computing the Optimal Route

The general distribution network is illustrated in Figure 7, including the source node (1), the sink node (n), the average time taken to travel from node i to node j ($c_{i,j}$), and the path from node i to node j ($X_{i,j}$).

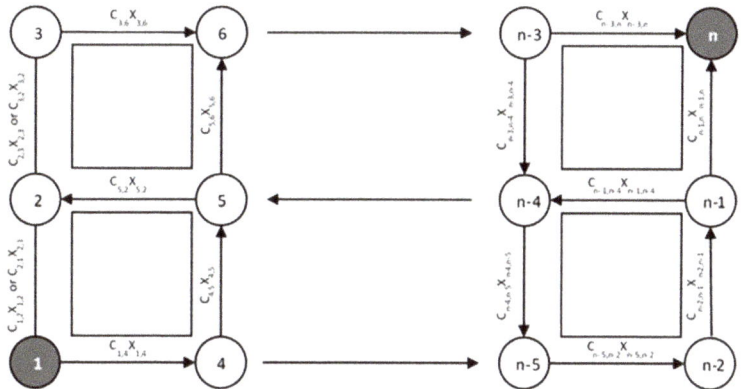

Figure 7. Diagram of the distribution network.

To select the optimal route, we took into account travel time, travel time uncertainty, and type of street corner. We assumed that the arcs were a road set in the urban road network and the nodes were a set of street corners (Figure 7). Then, we considered the fuzzy coefficient, $\tilde{d}_{i,j}$, for each arc $(i, j) \in E$, $c_{i,j}$, the average time taken to travel from node i to node j, and $v_{i,j}$, the average time taken to cross street corners. Therefore, the fuzzy linear programming model was Equation (18).

$$\min Z = \sum_{i=1}^{n} \sum_{j \neq i}^{n} [\tilde{d}_{i,j} c_{i,j} + v_{i,j}] X_{i,j}$$

$$s.t. \sum_{j \neq i}^{n} X_{j,i} - \sum_{j \neq i}^{n} X_{i,j} = b_i, \quad i = 1, 2, 3, \ldots, n \quad (18)$$

$$X_{i,j} \in \{0, 1\} \quad \forall \ i, j$$

3. Results

3.1. Application to a Real Network

Figure 8 shows an application to a portion of the distribution area of a delivery company in the city of Tuxtla Gutierrez, Chiapas, Mexico. To demonstrate our approach, Figure 9 shows a portion of the distribution area with four blocks and nine street corners, where node 15 is the source, and node 43 is the sink.

Considering Chiapas's traffic regulations, we used an average speed of 35 km/h for the avenues ($as_1 = 9.72$ m/s), and an average speed of 20 km/h for the streets ($as_2 = 5.55$ m/s). As shown in Figure 8, the distance of the arc was $a_{i,j}$ meters, therefore, the average travel time ($c_{i,j}$) in seconds was calculated using Equation (19).

$$c_{i,j} = \frac{a_{i,j}}{as_k}, \quad (19)$$

where k = 1 for the avenues, k = 2 for the streets, and $a_{i,j}$ represents the distance of the arc (i, j).

The total time (in seconds) was calculated using Equation (20).

$$TT_{i,j} = \tilde{d}_{i,j} c_{i,j} + v_{i,j} \quad (20)$$

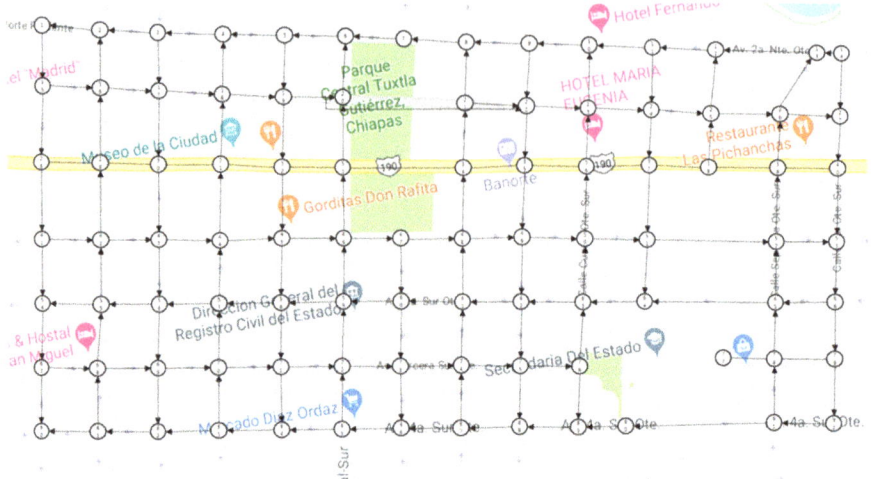

Figure 8. Diagram of the application zone.

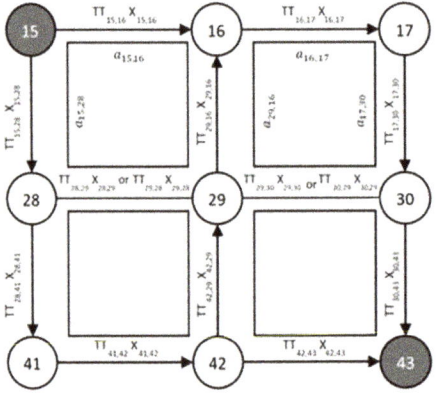

Figure 9. Diagram of a portion of the distribution area.

Table 2 illustrates the time in seconds used for each type of street corner. We sampled the distribution area to determine the average times for each type of street corner, as shown in Figure 8. For each intersection, we sampled 14 street corners at different times and noted 1519 observations. For the preferred street corner, we sampled four street corners at different times and noted 225 observations. For the non-preferred street corner, we sampled one street corner at different times and noted 50 observations. For traffic light intersections, we sampled four street corners at different times and noted 525 observations.

We analyzed the conditions of the fuzzy variables of the road and street corners of the real network by obtaining model parameters. Table 3 shows the time of day and traffic density fuzzy variables and calculated the parameters $a_{i,j}$, $c_{i,j}$, $\tilde{d}_{i,j}$, $\tilde{d}_{i,j} c_{i,j}$, $v_{i,j}$, and total time ($TT_{i,j}$).

Table 2. Time taken to cross street corners ($v_{i,j}$).

Type of Street Corner	Time in Seconds
Traffic light	50
Traffic light turn right	53
Traffic light turn left	57
No preference	8
Preference	3
Preference turn right	6
Preference turn left	7
One for one	4
One for one turn right	6
One for one turn left	6

Table 3. Model parameters.

i	j	k	Time of Day	Traffic Density	$a_{i,j}$	$c_{i,j}$	$\tilde{d}_{i,j}$	$\tilde{d}_{i,j} c_{i,j}$	$v_{i,j}$	Total Time
15	16	2	13:36:57	2	80	14	3.21	44.94	4	48.94
15	28	2	13:36:57	1.1	98	18	2.06	37.08	6	43.08
16	17	2	13:36:57	1.1	83	15	2.06	30.90	4	34.90
17	30	2	13:36:57	2	94	17	3.21	54.57	6	60.57
28	29	1	13:36:57	2.1	83	9	4.21	37.89	50	87.89
28	41	2	13:36:57	2	102	18	3.21	57.78	53	110.78
29	16	2	13:36:57	2	99	18	3.21	57.78	53	110.78
29	28	1	13:36:57	2.1	83	9	4.21	37.89	50	87.89
29	30	1	13:36:57	2.1	79	8	4.21	33.68	50	83.68
30	29	1	13:36:57	2.1	79	8	4.21	33.68	50	83.68
30	43	2	13:36:57	2	100	18	3.21	57.78	53	110.78
41	42	2	13:36:57	1.1	81	15	2.06	30.90	4	34.90
42	29	2	13:36:57	2	101	18	3.21	57.78	6	63.78
42	43	2	13:36:57	1.8	81	15	3.05	45.75	4	49.75

As shown in Table 3, node i was the initial node of arc (i, j) and node j was the final node. For arc (15,16), node 15 was the initial node and node 16 was the final node. For arc (15,16), the type of road was a street, therefore $k = 2$ and $as_2 = 5.55$ m/s. For arc (15,16), the street length was $a_{15,16} = 80$ meters. Equation (19) was therefore used.

$$c_{15,16} = \frac{a_{15,16}}{as_2} = \frac{80}{5.55} \approx 14 \; seconds$$

The time of day variable determined the scheduled goods delivery. In this study, the scheduled time of day was 13:36:57.

The traffic density variable was obtained via the following steps: Firstly, we considered the definition of the linguistic labels, as described in Section 2.1, to assign a real value ($1 \leq y \leq 3$) to the y fuzzy variable, i.e., traffic density, for each arc/street. Then, we designed a travel time study for each arc/street, with the purpose of adjusting the values. If necessary, we adjusted the value assigned to the fuzzy variable.

For arc (15,16), the roads next to the vehicular flow were located around shopping malls, therefore the fuzzy variable traffic density was 2. The fuzzy coefficient, ($\tilde{d}_{i,j}$), was computed using the time of day and traffic density fuzzy variables, as described in Section 2.1, therefore $\tilde{d}_{15,16} = 3.21$.

For arc (15,16), the type of street corner was one for one, therefore $v_{15,16} = 4$ seconds. The total time was computed using Equation (20).

$$TT_{15,16} = \tilde{d}_{15,16} c_{15,16} + v_{15,16} = 44.94 + 4 = 48.94 \; seconds$$

The fuzzy linear programming problem for the application to the network was applied to Equations (6) and (18), therefore,

$$minZ = 48.94X_{15,16} + 43.08X_{15,28} + 34.90X_{16,17} + 60.57X_{17,30} + 87.89X_{28,29} + 110.78X_{28,41} + 110.78X_{29,16}$$
$$+ 87.89X_{29,28} + 83.68X_{29,30} + 83.68X_{30,29} + 110.78X_{30,43} + 34.90X_{41,42} + 63.78X_{42,29} + 49.75X_{42,43}$$

$$\begin{aligned}
s.t. \quad -X_{15,16} - X_{15,28} &= -1 \\
X_{15,16} - X_{16,17} + X_{29,16} &= 0 \\
X_{16,17} - X_{17,30} &= 0 \\
X_{15,28} - X_{28,29} - X_{28,41} + X_{29,28} &= 0 \\
X_{28,29} - X_{29,16} - X_{29,28} + X_{42,29} &= 0 \\
X_{17,30} + X_{29,30} - X_{30,29} - X_{30,43} &= 0 \\
X_{28,41} - X_{41,42} &= 0 \\
X_{41,42} - X_{42,29} - X_{42,43} &= 0 \\
X_{30,43} + X_{42,43} &= 1 \\
X_{i,j} \in \{0,1\} \quad \forall \; i,j
\end{aligned}$$
(21)

We calculated the optimal solution of the FLP problem using a version of the LP simplex method. Table 4 and Figure 10 show the optimal path {(15,28), (28,41), (41,42), (42,43)} and time in seconds. The computed optimal travel time was 00:03:58.51 (238.51 seconds).

Table 4. Fuzzy optimal solution.

Variable	Value	Cost (Time)
$X_{15,16}$	0	48.94
$X_{15,28}$	1	43.08
$X_{16,17}$	0	34.90
$X_{17,30}$	0	60.57
$X_{28,29}$	0	87.89
$X_{28,41}$	1	110.78
$X_{29,16}$	0	110.78
$X_{29,28}$	0	87.89
$X_{29,30}$	0	83.68
$X_{30,29}$	0	83.68
$X_{30,43}$	0	110.78
$X_{41,42}$	1	34.90
$X_{42,29}$	0	63.78
$X_{42,43}$	1	49.75
Z		238.51

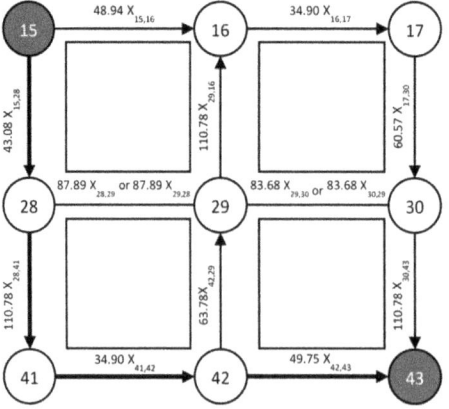

Figure 10. Diagram of the optimal solution.

We used Equation (21) to pose the fuzzy linear programming model for the network. Then, we calculated the optimal solution of the FLP problem. The computed optimal time of the FLP (00:03:58) was different to the time computed for the LP (00:02:13). Moreover, the time computed for the FLP was closer to the observed travel time (00:04:17).

3.2. Results Analysis

To analyze the reliability of the model, we designed a studio with 21 real network observations from a portion of a distribution area for a delivery company in the city of Tuxtla Gutierrez, Chiapas, Mexico, consisting of 68 blocks and 92 street corners (Figure 8). The source node was node 15 and the sink node was scheduled. Table 5 shows the source node and the sink node in ascending order.

Table 5. Results obtained.

Route	Schedule	Real Travel Time	Calculated Time of the Classical Model (LP)	Absolute Percentage Error (LP) (%)	Calculated Time of the Fuzzy Model (FLP)	Absolute Percentage Error (FLP) (%)
15–13	6:22:17	0:05:29	0:03:37	34.04	0:03:53	29.18
15–13	14:28:37	0:07:35	0:03:37	52.31	0:06:37	12.75
15–13	19:03:55	0:02:51	0:03:37	26.90	0:03:11	11.70
15–40	10:15:26	0:04:17	0:03:47	11.67	0:04:34	6.61
15–40	13:49:21	0:07:35	0:03:47	50.11	0:07:05	6.59
15–40	21:47:33	0:02:50	0:03:47	33.53	0:02:34	9.41
15–50	7:15:30	0:04:25	0:03:42	16.23	0:04:41	6.04
15–50	11:45:17	0:04:11	0:03:42	11.55	0:04:41	11.95
15–50	23:08:05	0:02:35	0:03:42	43.23	0:02:46	7.10
15–59	7:26:02	0:05:37	0:04:03	27.89	0:05:56	5.64
15–59	17:25:32	0:04:34	0:04:03	11.31	0:04:21	4.74
15–59	20:21:59	0:03:19	0:04:03	22.11	0:03:03	8.04
15–64	8:01:19	0:06:32	0:04:23	32.91	0:06:02	7.65
15–64	13:54:58	0:08:58	0:04:23	51.12	0:08:39	3.53
15–64	20:31:44	0:02:58	0:04:23	47.75	0:03:15	9.55
15–82	10:06:48	0:04:44	0:03:15	31.34	0:04:22	7.75
15–82	14:03:51	0:07:02	0:03:15	53.79	0:06:34	6.64
15–82	22:47:33	0:01:58	0:03:15	65.25	0:02:31	27.97
15–88	5:32:28	0:06:38	0:04:27	32.91	0:06:51	3.27
15–88	13:36:57	0:10:49	0:04:27	58.86	0:10:15	5.24
15–88	18:04:30	0:05:53	0:04:27	24.36	0:05:17	10.20

We computed the optimal solution with fuzzy linear programming for each day and executed the route following the optimal path found in the FLP. We also calculated the optimal solution using linear programming. The absolute percentage error (APE) was calculated using Equation (22).

$$APE = \frac{|RT - CT|}{RT} * 100 \qquad (22)$$

where RT is the real time of travel, CT is the calculated time using the classical model or the calculated time using the fuzzy model.

Table 5 shows the route, the schedule, the real travel time, the calculated time of the classical model (LP), the absolute percentage error using LP, the calculated time of the fuzzy model (FLP), and the absolute percentage error using FLP.

We observed that the mean absolute percentage error of the classical model (LP) was 35.20%, while the mean absolute percentage error of the fuzzy model (FLP) was 9.60%. Moreover, the maximum absolute percentage error of the classical model was 65.25%, whereas the maximum absolute percentage error of the fuzzy model was 29.18%. Furthermore, the minimum absolute percentage error of the classical model was 11.31%; meanwhile, the minimum absolute percentage error of the fuzzy model was 3.27%. Finally, the variance of the absolute percentage error of the classical model was 264.14%, whilst the variance of the absolute percentage error of the fuzzy model was 46.57%.

We also attempted to determine whether the calculated time of the fuzzy model was adjusted for the real travel time. For the statistical analysis between the mean real travel time and the mean time calculated for FLP, we used paired t-tests, where $\alpha = 5\%$. We proposed a null hypothesis (H_0) and an alternative hypothesis (H_1).

$$H_0 : \mu_{RT} - \mu_{CT} = 0$$
$$H_1 : \mu_{RT} - \mu_{CT} \neq 0 \ ,$$

where μ_{RT} is the mean real travel time and μ_{CT} is the mean time calculated from the FLP.

Next, we evaluated the paired t-tests. Table 6 shows the paired t-tests used to ascertain whether the null hypothesis (i.e., that the mean real travel time and mean time calculated from FLP were equal) was accepted or rejected. The p-value determined statistical significance.

Table 6. Paired t-test for the mean real travel time vs. the calculated time from fuzzy linear programming (FLP).

	Real Travel Time	Calculated Time from FLP
Mean	5.277777777	5.101587301
Variance	5.247425925	4.260914021
Observations	21	21
Hypothesized Mean Difference		0
Df		20
t-stat		1.506943922
p-value		0.147458176
t-critical		2.085963447

Table 6 shows that the p-value (0.1475) was of a higher significance level (0.05) and that the t-stat (calculated statistic) was lower than the t-critical (t distribution table). Therefore, there was not sufficient statistical evidence to reject the null hypothesis. Hence, it was concluded that the mean real travel time was statistically equal to the mean calculated time in the proposed model. Therefore, the proposed model (FLP) adjusted to the travel time of the urban road network in this study.

Next, we aimed to determine whether the absolute percentage error of the classical model (LP) was higher than the absolute percentage error of the proposed model (FLP). For the statistical analysis, we used paired t-tests where $\alpha = 5\%$. We proposed a null hypothesis (H_0) and an alternative hypothesis (H_1).

$$H_0 : \mu_{LP} - \mu_{FLP} = 0$$
$$H_1 : \mu_{LP} - \mu_{FLP} > 0 \ ,$$

where μ_{LP} was the mean absolute percentage error of LP and μ_{FLP} was the mean absolute percentage error of FLP.

Table 7 shows the paired t-tests used to ascertain whether the null hypothesis (i.e., that the mean absolute percentage error of the classical model and mean absolute percentage error of fuzzy model were equal) was accepted or rejected.

Table 7 shows that the p-value (2.4035×10^{-7}) was less than the significance level (0.05), and that the t-stat (calculated statistic) was higher than the t-critical (t distribution table), therefore the null hypothesis was rejected. Accordingly, it was concluded that the mean absolute percentage error of the LP was higher than the means absolute percentage error of the FLP, i.e., the proposed model. Therefore, the proposed model (FLP) predicted real travel times better than the classical model (LP) in this study, thereby providing more reliable travel times in the urban road network.

In this approach, we considered fuzzy variables that influence the travel times of routes within an urban area. These fuzzy variables were not considered in the classic models, thus, we obtained values closer to the real route times with this proposal.

Table 7. Paired *t*-test for mean absolute percentage error of linear programming (LP) vs. the absolute percentage error from fuzzy linear programming.

	Absolute Percentage Error (LP)	Absolute Percentage Error (FLP)
Mean	35.199228509	9.597056757
Variance	264.135940142	46.574056590
Observations	21	21
Hypothesized Mean Difference	0	
Df	20	
t-stat	7.284123131	
p-value	$2.403528976 \times 10^{-7}$	
t-critical	1.724718243	

4. Discussion

In this paper, the implementation of fuzzy logic allowed us to model the uncertainty of the route optimization problem that occurs when delivering products in an urban road network through two linguistic variables, namely, time of day and traffic density. By adjusting the travel times and applying fuzzy linear programming, we computed the shortest path, thereby providing a more reliable result than the results of the classical model (Table 5), with a mean absolute percentage error (MAPE) of 9.60% in this study. However, if conditions change due to the fuzzy variables, it is possible that the optimal route will change.

This approach provides a good tool for route selection when delivering products in urban road networks, increasing the reliability of the transportation system through route optimization. In this study, the maximum absolute percentage error of the classical model (LP) was reduced by 36.07%. By applying the proposed method to a larger proportion of the distribution area of a delivery company, we expect even better results.

5. Conclusions

The shortest path problem is important for practical applications, e.g., in route optimization for a delivery company. The use of fuzzy logic to consider arc weights as a non-deterministic value further improves the obtained results.

With the classic models, the optimal route of product delivery is determined. However, this optimal route is unable to change during the analysis period, regardless of whether the delivery conditions are changing. On the other hand, with the approach raised in this investigation, the determined optimal route of goods delivery is able to adjust to different conditions, i.e., the time of day and traffic density fuzzy variables. With this approach, we were able to calculate travel times that were closer to the real travel times of the routes.

In this study, we observed that the optimal route sometimes changed according to the changing fuzzy variables. Therefore, with this approach, we determined the optimal route of goods delivery for the different fuzzy variable conditions, which guaranteed the shortest travel time for each delivery. Minimizing delivery route travel times is expected to reduce pollution and the level of noise generated by delivery units and increase the transfer capacity, thereby contributing to sustainable transport.

Furthermore, it is worth bearing in mind that when the conditions of the urban road network change, it is necessary to adapt the membership function's fuzzy variable parameters to the new conditions.

This paper puts forward elements to help consider a larger number of fuzzy variables in future projects. Different uncertain elements, such as rain and the condition of the streets, among others, could be analyzed. We suggest amplifying the street corner crossing time used in this study in order to consider the uncertain factors that affect street corner crossing time.

This study empowers workers to make better decisions in the area of distribution regarding their travel routes.

Author Contributions: Conceptualization, E.E.-G., S.V.-T., and R.G.-C; data curation, H.H.-d.-L., R.G.-C., and E.C.-C.; formal analysis, J.L.C.-A., S.V.-T., R.G.-C., E.C.-C., and H.G.-C.; investigation, E.E.-G. and J.L.C.-A.; methodology, E.E.-G. and S.V.-T.; project administration, E.E.-G.; Software, H.H.-d.-L., and H.G.-C.; supervision, E.E.-G.; validation, J.L.C.-A., H.H.-d.-L., E.C.-C., and H.G.-C.; writing—original draft, E.E.-G. and E.C.-C.; writing—review and editing, E.E.-G.

Funding: This research received no external funding.

Conflicts of Interest: The authors declare no conflict of interest.

References

1. Bellman, E. On a routing problem. *Q. Appl. Math.* **1958**, *16*, 87–90. [CrossRef]
2. Dijkstra, E.W. A note on two problems in connection with graphs. *Numer. Math.* **1959**, *1*, 269–271. [CrossRef]
3. Dantzig, G.B. On the shortest route through a network. *Manag. Sci.* **1960**, *6*, 187–190. [CrossRef]
4. Floyd, R.W. Algorithm 97: Shortest path. *Commun. ACM* **1962**, *5*, 345. [CrossRef]
5. Butas, L.F. A directionally oriented shortest path algorithm. *Transp. Res.* **1968**, *2*, 253–268. [CrossRef]
6. Johnson, D.B. Efficient algorithms for shortest paths in sparse networks. *J. Assoc. Comput. Mach.* **1977**, *24*, 1–13. [CrossRef]
7. Zhan, F.B.; Noon, C.E. Shortest path algorithms: An evaluation using real road networks. *Transp. Sci.* **1998**, *32*, 65–73. [CrossRef]
8. Zamirian, M.; Farahi, M.; Nazemi, H.A.R. An applicable method for solving the shortest path problems. *Appl. Math. Comput.* **2007**, *190*, 1479–1486. [CrossRef]
9. An, P.T.; Hai, N.N.; Hoai, T.V. Direct multiple shooting method for solving approximate shortest path problems. *J. Comput. Appl. Math.* **2013**, *244*, 67–76. [CrossRef]
10. Bode, C.; Irnich, S. The shortest-path problem with resource constraints with (k,2)-loop elimination and its application to the capacitated arc-routing problem. *Eur. J. Oper. Res.* **2014**, *238*, 415–426. [CrossRef]
11. Duque, D.; Lozano, L.; Medaglia, A.L. An exact method for the biobjective shortest path problem for large-scale road networks. *Eur. J. Oper. Res.* **2015**, *242*, 788–797. [CrossRef]
12. Marinakis, Y.; Migdalas, A.; Sifaleras, A. A hybrid particle swarm optimization-variable neighborhood search algorithm for constrained shortest path problems. *Eur. J. Oper. Res.* **2017**, *261*, 819–834. [CrossRef]
13. Rostami, B.; Chassein, A.; Hopf, M.; Frey, D.; Buchheim, C.; Malucelli, F.; Goerigk, M. The quadratic shortest path problem: Complexity, approximability, and solutions methods. *Eur. J. Oper. Res.* **2018**, *268*, 473–485. [CrossRef]
14. Arun Prakash, A. Pruning algorithm for the least expected travel time path on stochastic and time-dependent networks. *Transp. Res. Part B* **2018**, *108*, 127–147. [CrossRef]
15. Chen, B.Y.; Li, Q.; Lam, W.H.K. Finding the k reliable shortest paths under travel time uncertainty. *Transp. Res. Part B* **2016**, *94*, 111–135. [CrossRef]
16. Strehler, M.; Merting, S.; Schwan, C. Energy-efficient shortest routes for electric and hybrid vehicles. *Transp. Res. Part B* **2017**, *103*, 189–203. [CrossRef]
17. Shi, N.; Zhou, S.; Wang, F.; Tao, Y.; Liu, L. The multi-criteria constrained shortest path problem. *Transp. Res. Part E* **2017**, *101*, 13–29. [CrossRef]
18. Zhang, Y.; Shen, Z.M.; Song, S. Lagrangian relaxation for the reliable shortest path problem with correlated link travel times. *Transp. Res. Part B* **2017**, *104*, 501–521. [CrossRef]
19. Cooke, K.L.; Halsey, E. The Shortest Route Through a Network with Time-Dependent Internodal Transit Times. *J. Math. Anal. Appl.* **1966**, *14*, 493–498. [CrossRef]
20. Hernandes, F.; Lamata, M.T.; Verdegay, J.L.; Yamakami, A. The shortest path problem on networks with fuzzy parameters. *Fuzzy Sets Syst.* **2007**, *158*, 1561–1570. [CrossRef]
21. Deng, Y.; Chen, Y.; Zhang, Y.; Mahadevan, S. Fuzzy Dijkstra algorithm for shortest path problem under uncertain environment. *Appl. Soft Comput.* **2012**, *12*, 1231–1237. [CrossRef]
22. Dou, Y.; Zhu, L.; Wang, H.S. Solving the fuzzy shortest path problem using multi-criteria decision method based on vague similarity measure. *Appl. Soft Comput.* **2012**, *12*, 1621–1631. [CrossRef]
23. Kok, A.L.; Hans, E.W.; Schutten, J.M.J. Vehicle routing under time-dependent travel times: The impact of congestion avoidance. *Comput. Oper. Res.* **2012**, *39*, 910–918. [CrossRef]

24. Farhanchi, M.; Hassanzadeh, R.; Mahdavi, I.; Mahdavi-Amiri, N. A modified ant colony system for finding the expected shortest path in networks with variable arc lengths and probabilistic nodes. *Appl. Soft Comput.* **2014**, *21*, 491–500. [CrossRef]
25. Lakouari, N.; Ez-Zahraouy, H.; Benyoussef, A. Traffic flow behavior at a single lane roundabout as compared to traffic circle. *Phys. Lett. A* **2014**, *378*, 3169–3176. [CrossRef]
26. Frank, H. Shortest paths in probabilistic graphs. *Oper. Res.* **1969**, *17*, 583–599. [CrossRef]
27. Mirchandani, P.B. Shortest distance and reliability of probabilistic networks. *Comput. Oper. Res.* **1976**, *3*, 347–355. [CrossRef]
28. Sigal, C.E.; Pritsker, A.A.B.; Solberg, J.J. The stochastic shortest route problem. *Oper. Res.* **1980**, *28*, 1122–1129. [CrossRef]
29. Noorizadegan, M.; Chen, B. Vehicle routing with probabilistic capacity constraints. *Eur. J. Oper. Res.* **2018**, *270*, 544–555. [CrossRef]
30. Dubois, D.; Prade, H. Algorithmes de plus courts chemins pour traiter des donnees floues. *RAIRO Oper. Res.* **1978**, *12*, 213–227. [CrossRef]
31. Chanas, S.; Kamburowski, J. The fuzzy shortest route problem. In *Interval and Fuzzy Mathematics*; Albrycht, J., Wisniewski, H., Eds.; Technology University of Poznan: Poznan, Poland, 1983; pp. 35–41.
32. Chanas, S.; Delgado, M.; Verdegay, J.L.; Vila, M.A. Fuzzy optimal flow on imprecise structures. *Eur. J. Oper. Res.* **1995**, *83*, 568–580. [CrossRef]
33. Klein, C.M. Fuzzy shortest paths. *Fuzzy Sets Syst.* **1991**, *39*, 27–41. [CrossRef]
34. Dey, A.; Pal, A.; Pal, T. Interval type 2 Fuzzy Set in Fuzzy Shortest Path Problem. *Mathematics* **2016**, *4*, 62. [CrossRef]
35. Tajdin, A.; Mahdavi, I.; Amiri, N.M.; Sadeghpour-Gildeh, B. Computing a fuzzy shortest path in a network with mixed fuzzy arc lengths using α-cuts. *Comput. Math. Appl.* **2010**, *60*, 989–1002. [CrossRef]
36. Ramazani, H.; Shafahi, Y.; Seyedabrishami, S.E. A Shortest Path Problem in an Urban Transportation Network Based on Driver Perceived Travel Time. *Trans. A Civ. Eng.* **2010**, *17*, 285–296.
37. Dai, L.; Fan, L.; Sun, L. Aggregate production planning utilizing a fuzzy linear programming. *J. Integr. Des. Process Sci.* **2003**, *7*, 81–95.
38. Fung, R.; Tang, J.; Wang, D. Multiproduct Aggregate Production Planning with Fuzzy Demands and Fuzzy Capacities. *IEEE Trans. Syst. Man Cybern. Part A* **2003**, *33*, 302–313. [CrossRef]
39. Sheng-Tun, L.; Yi-Chung, C. Deterministic Fuzzy Time Series Model for Forecasting Enrollments. *Comput. Math. Appl.* **2007**, *53*, 1904–1920.
40. Castillo, O.; Melin, P. Automated mathematical modelling for financial time series prediction combining fuzzy logic and fractal theory. In *Soft Computing for Financial Engineering*; Kacprzyk, J., Ed.; Springer: Berlin/Heidelberg, Germany, 1999; pp. 93–106.
41. Escobar-Gómez, E.N.; Díaz-Núñez, J.J.; Taracena-Sanz, F.L. Model for Adjustment of Aggregate Forecasts using Fuzzy Logic. *Ingeniería Investigación y Tecnología.* **2010**, *11*, 289–302. [CrossRef]
42. Hosseini, R.; Dehmeshki, J.; Barman, S.; Mazinani, M.; Qanadi, S. A genetic type-2 fuzzy logic system for pattern recognition in computer aided detection system. In Proceedings of the 2010 IEEE World Congress on Computational Intelligence, Barcelona, Spain, 18–23 July 2010.
43. Deng, Y.; Shi, W.K.; Du, F.; Liu, Q. A new similarity measure of generalized fuzzy numbers and its application to pattern recognition. *Pattern Recognit. Lett.* **2004**, *25*, 875–883.
44. Bellman, R.E.; Kalaba, R.E.; Zadeh, L.A. Abstraction and pattern classification. *J. Math. Anal. Appl.* **1966**, *13*, 1–7. [CrossRef]
45. Mitra, S.; Pal, S.K. Fuzzy sets in pattern recognition and machine intelligence. *Fuzzy Sets Syst.* **2005**, *156*, 381–386. [CrossRef]
46. Pedrycz, W. Fuzzy sets in pattern recognition: Accomplishments and challenges. *Fuzzy Sets Syst.* **1997**, *90*, 171–176. [CrossRef]
47. Mitchell, H.B. Pattern recognition using type-II fuzzy sets. *Inf. Sci.* **2005**, *170*, 409–418. [CrossRef]
48. Tizhoosh, H.R. Image thresholding using type II fuzzy sets. *Pattern Recognit.* **2005**, *38*, 2363–2372. [CrossRef]
49. Das, S. Pattern Recognition using the Fuzzy c-means Technique. *Int. J. Energy Inf. Commun.* **2013**, *4*, 1–14.
50. Zadeh, L.A. Outline of a New Approach to the Analysis of Complex Systems and Decision Processes. *IEEE Trans. Syst. Man Cybern.* **1973**, *3*, 28–44. [CrossRef]

51. Bellman, R.E.; Zadeh, L.A. Decision-making in a fuzzy environment. *Manag. Sci.* **1970**, *17*, 141–164. [CrossRef]
52. Mardani, A.; Jusoh, A.; Zavadskas, E.K. Fuzzy multiple criteria decision-making techniques and applications-two decades review from 1994 to 2014. *Expert Syst. Appl.* **2015**, *42*, 4126–4148. [CrossRef]
53. Boulmakoul, A. Generalized path-finding algorithms on semirings and the fuzzy shortest path problem. *J. Comput. Appl. Math.* **2004**, *162*, 263–272. [CrossRef]
54. Mahdavi, I.; Nourifar, R.; Heidarzade, A.; Amiri, N.M. A dynamic programming approach for finding shortest chains in a fuzzy network. *Appl. Soft Comput.* **2009**, *9*, 503–511. [CrossRef]
55. Ghatee, M.; Hashemi, S.M.; Zarepisheh, M.; Khorram, E. Preemptive priority based algorithms for fuzzy minimal cost flow problem: An application in hazardous materials transportation. *Comput. Ind. Eng.* **2009**, *57*, 341–354. [CrossRef]
56. Ghatee, M.; Hashemi, S.M. Application of fuzzy minimum cost flow problems to network design under uncertainty. *Fuzzy Sets Syst.* **2009**, *160*, 3263–3289. [CrossRef]
57. Keshavarz, E.; Khorram, E. A fuzzy shortest path with the highest reliability. *J. Comput. Appl. Math.* **2009**, *230*, 204–212. [CrossRef]
58. Marien, E.J. The application of linear programming to a distribution system orientated toward service. *Int. J. Phys. Distrib.* **1972**, *3*, 191–204. [CrossRef]
59. Ali, M.A.M.; Sik, Y.H. Transportation problem: A special case for linear programming problems in mining engineering. *Int. J. Min. Sci. Technol.* **2012**, *22*, 371–377. [CrossRef]
60. García, J.; Florez, J.E.; Torralba, A.; Borrajo, D.; López, C.L.; García-Olaya, A.; Sáenz, J. Combining linear programming and automated planning to solve intermodal transportation problems. *Eur. J. Oper. Res.* **2013**, *227*, 216–226. [CrossRef]
61. Luathep, P.; Sumalee, A.; Lam, W.H.K.; Li, Z.; Lo, H.K. Global optimization method for mixed transportation network design problem: A mixed-integer linear programming approach. *Transp. Rese. Part B Methodol.* **2011**, *45*, 808–827. [CrossRef]
62. Faddel, S.; Aldeek, A.; Al-Awami, A.T.; Sortomme, E. ZAl-Hamouz, Ancillary Services Bidding for Uncertain Bidirectional V2G Using Fuzzy Linear Programming. *Energy* **2018**, *160*, 986–995. [CrossRef]
63. Zimmermann, H.J. Fuzzy programming and linear programming with several objective functions. *Fuzzy Sets Syst.* **1978**, *1*, 45–56. [CrossRef]
64. Okada, S.; Soper, T. A shortest path problem on a network with fuzzy arc length. *Fuzzy Sets Syst.* **2000**, *109*, 129–140. [CrossRef]
65. Buckley, J.J. Possibilistic linear programming with triangular fuzzy numbers. *Fuzzy Sets Syst.* **1988**, *26*, 135–138. [CrossRef]
66. Foulds, L.R.; Nascimento, H.A.D.D.; Calixto, I.C.A.C.; Hall, B.R.; Longo, H. A fuzzy set-based approach to origin–destination matrix estimation in urban traffic networks with imprecise data. *Eur. J. Oper. Res.* **2013**, *231*, 190–201. [CrossRef]
67. Ebrahimnejad, A.; Tavana, M. A novel method for solving linear programming problems with symmetric trapezoidal fuzzy numbers. *Appl. Math. Model.* **2014**, *38*, 4388–4395. [CrossRef]
68. Zhu, G.; Song, K.; Zhang, P.; Wang, L. A traffic flow state transition model for urban road network based on Hidden Markov Model. *Neurocomputing* **2016**, *214*, 567–574. [CrossRef]
69. Tian, Z.; Jia, L.; Dong, H.; Su, F.; Zhang, Z. Analysis of Urban Road Traffic Network Based on Complex Network. *Procedia Eng.* **2016**, *137*, 537–546. [CrossRef]

© 2019 by the authors. Licensee MDPI, Basel, Switzerland. This article is an open access article distributed under the terms and conditions of the Creative Commons Attribution (CC BY) license (http://creativecommons.org/licenses/by/4.0/).

Article
Supply Chain Optimization for Energy Cogeneration Using Sugarcane Crop Residues (SCR)

Leonardo Rivera-Cadavid, Pablo Cesar Manyoma-Velásquez and Diego F. Manotas-Duque *

School of Industrial Engineering, Universidad del Valle, Cali 13 No 100-00, Colombia;
leonardo.rivera.c@correounivalle.edu.co (L.R.-C.); pablo.manyoma@correounivalle.edu.co (P.C.M.-V.)
* Correspondence: diego.manotas@correounivalle.edu.co; Tel.: +57-2-321-21-67

Received: 31 October 2019; Accepted: 19 November 2019; Published: 21 November 2019

Abstract: Access to clean and non-polluting energy has been defined as a Sustainable Development Goal (SDG). In this context, countries such as Colombia have promoted policies and incentives for the implementation of energy projects with non-conventional sources of energy. One of the main energy alternatives available is related to the use of residual biomass left by agribusiness supply chains, such as sugarcane. In Colombia, sugar cane is grown and harvested all year round, due to the local tropical climate. The model we propose addresses the question of the selection of the plots whose crop residue will be transported for energy production on a given day. We built a Mixed-Integer Programming model to decide which plots to harvest on a given day. Although no additional energy is generated in the model, the results show that it is feasible to replace all coal used in the boilers with sugarcane crop residues (SCRs) for power cogeneration.

Keywords: Sugarcane Crop Residues (SCRs); cogeneration; optimization of bioenergy supply chain decisions

1. Introduction

1.1. Background and Motivation

Countries all over the world are focusing on renewable energy resources as an attractive option for achieving future energy security. Access to clean and non-polluting energy has been defined as a Sustainable Development Goal (SDG). In this context, many countries are promoting strategies to diversify the power generation mix considering clean energy projects to mitigate the climate change originated by greenhouse gas emissions from fossil fuel consumption. To achieve this purpose, one of the main energy resources available is the biomass. Biomass is defined as "a biological material derived from living or recently living organisms," There are three main biomass types—(1) first-generation biomass, which includes edible crops, such as corn and sugarcane; (2) second-generation biomass, which includes wood, organic waste, food waste, and specific biomass crops residues; and (3) third-generation biomass, mainly based on algae [1]. In this paper, we present a model of supply-chain design for sugarcane crop residues in the sugar industry. Currently, the sugar industry has been participating in power cogeneration projects. Usually, in these projects, sugar factories use a mix of fuels, mainly bagasse—a by–product of the production process—and coal. There is an opportunity to include part of the sugarcane crop residues (SCRs) originated in crop activities in the mix. The use of SCRs arises as a possible substitute for the coal used in boilers. Authors like those of reference [2] have made a complete analysis of the state of research and trends in biomass for renewable energy from 1978 to 2018, focusing on the current situation and future trends. This information is very useful for making decisions about the future of scientific policy in the field of renewable energy projects.

The use of agricultural crop residues in electrical co-generation processes has been widely discussed in the literature. Some authors, such as those of reference [3], systematically describe energy needs and targets, biofuel feedstocks, conversion processes, and provide a comprehensive review of biomass supply chain (BSC) design and modeling. In reference [4], the authors describe the key challenges and opportunities in the modeling and optimization of biomass-to-bioenergy supply chains. The optimization process of a biomass supply chain is essential to overcome barriers and uncertainties that may inhibit the development of a sustainable and competitive bioenergy market. In reference [5], the authors give an overview of the optimization techniques and models, focusing on decisions regarding the design and management of the upstream segment of the biomass-for-bioenergy supply chain. Another research approach is related to the integration of different optimization objectives in biomass supply chain models. There is a recent tendency to integrate economic, environmental, and social aspects in the evaluation and optimization of biomass supply chains. Some authors (for example, those of reference [6]) have reviewed the studies that assessed or optimized economic, social, and environmental aspects of forest biomass supply chains for the production of bioenergy and bio-products.

The structure of this paper is as follows: In Section 1.2, we present a literature review of bioenergy supply chain optimization decisions. In Section 2, the optimization model is explained, taking into account the parameters, variables, objective functions, and constraints. Sections 3 and 4 present the results and discussion. Section 5 presents conclusions and future research options. The Julia code of the model is included as an appendix.

1.2. Literature Review—Bioenergy Supply Chain Optimization

For tropical and subtropical countries, sugarcane is one of the most important crops, from economic, social, and environmental points of view. The area and productivity of sugarcane differ from one country to another. It is estimated that this crop is grown in 200 countries, with Asia being the region with the greatest contribution to world production (44%), while South America contributes 34%, including Brazil, the largest sugarcane producer in the world, accounting for 25% of total world production [7]. Ninety-two percent of the sugarcane harvested in the world is directed toward sugar production. Sugar production and trade flows are concentrated in large producers, exporters, and importers, such as Brazil, India, Thailand, the European Union, Australia, and China [8].

There are different studies about the use of sugarcane crop residues in the bioenergy production. According to reference [9], sugarcane crop residues can be used as raw material in the production of bioenergy or biofuels. The use of these resources can avoid environmental problems and concerns. Each country has its unique set of crop residues that can be used for the generation of biofuels or bioenergy. In reference [10], the authors assess the bioelectricity potential from ecologically available sugarcane straw in the state of Sao Paulo (Brazil) considering the spatial distribution of sugarcane fields, the spatial variation of sugarcane yield, the location, and the milling data of each mill. The bioelectricity potential from ecologically available sugarcane straw is estimated to be between 18.7 and 45.8 TWh in Sao Paulo, equal to 22–37% of the electricity demand. Reference [11] presents a techno-economic analysis of upgraded sugarcane bio-refineries in Brazil, aiming at utilizing surplus bagasse and cane trash for electricity and/or ethanol production. This study investigates the trade-off on sugarcane biomass use for energy production—bioelectricity versus 2G ethanol production. In references [12–14], the authors value the cogeneration potential of the sugar industry as a new business model for sugar mills.

The main activities in the sugarcane industry are crop, harvest, and factory (sugar mill plant and distilleries). SCR is a by-product of the crop activities (Figure 1). SCRs are not used in the amount required by current circumstances. It has been estimated that, for each ton of cane, between 70 and 80% are clean stems, and the rest is what is called crop residues (green leaves, dried leaves, buds, etc.). The SCR can be used for different purposes, such as soil improvement, animal feed, energy generation, and other applications (boards, binders for construction, handicraft products, etc.). These residues, obviously, constitute a renewable energy source, but their use depends largely on aspects such as collection, transport, storage, and technological processes to transform them.

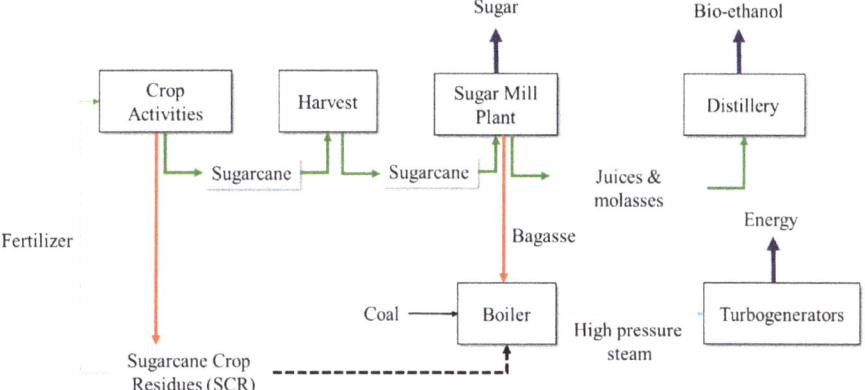

Figure 1. Cogeneration process and main activities in the sugar industry.

The collection process, traditionally, can be done in two places—in collection centers (cleaning stations) or in the field itself. Among the aspects related to residue handling, one of the most important is transport due to its technical-economic impact [15]. Agricultural residue is a low density material, and the amount that can be moved is limited by volume and not by weight. Therefore, the use of all available space is essential for transport economics. For this, some mechanical treatments have been carried out that can eliminate the geometric heterogeneity of the residue and make compaction possible for later use. Finally, some authors warn of the need to keep 50% of the residual in the field to preserve appropriate soil conditions [16].

In Colombia, it is estimated that the cultivation of sugarcane has a production potential of SCR per ton of cane close to 25%. This availability can be affected by factors such as: variety of cane planted, crop agro climatic conditions, agronomic practices and the harvest types, among others [17].

Table 1 presents the potential production of SCR in the state of Valle del Cauca, depending on the crop variety.

Table 1. Percent of sugarcane residue (SCR) by sugarcane variety.

Variety	Total (t SCR/t cane)
CC 01–1940	0.2657
CC 85–92	0.2479
CC 93–4418	0.2624
Others	0.2587

Source: [18].

Under this structure and according to the amount of cane milled by the sector (25 million tons in 2018), approximately 6.25 million tons of SCR could be generated. However, total SCR that would actually be available is reduced to approximately 1.8 million tons per year due to crop type and the material that should be left in the field for soil conservation.

The SCR is an important energy resource for the sugar sector, but its use requires overcoming some technical challenges. One of these is its low energy content compared to traditional fossil fuels (higher calorific value: coal 25,983.7 KJ/kg; SCR 16,963.52 KJ/kg). Its low density (100 kg/m^3), added to its energy characteristics, makes the biomass volumes necessary to meet the energy need for a possible replacement of the coal used in the boilers of the sector significantly higher. In that sense, 1.53 tons of SCR are required to replace the calorific output of a ton of coal. This substitution has the clear benefit of reducing the carbon footprint, since 1 ton of coal would generate 2254 kg of equivalent CO_2, while 1.53 tons of SCR generate 53.7 kg of equivalent CO_2. [18]

The use of mathematical models to analyze the performance of the bioenergy supply chain has been widely studied in the literature. [19] Some authors, such as those of references [4,20,21], describe the key challenges and opportunities in modeling and optimization of biomass-to-bioenergy supply chains. According to references [5,22], one of the most important barriers to the development of a strong bioenergy sector is the cost of the biomass supply chain—specifically, the cost of SCR preparation, transport, and storage. In this context, the SCR supply chain cost competes directly with fossil fuel cost (coal). To overcome all these barriers and uncertainties, biomass supply chain optimization is essential.

Optimization decisions usually refer to the choice of highly productive non-food crops; the coordination of transportation, pre-treatment and storage at operational, tactical and strategic level; and the use of advanced efficient biomass-to-bioenergy conversion technologies to enable relevant reductions in environmental and biomass production costs [5].

A first step in the study of bioenergy supply chain optimization models is to build a taxonomy of the main optimization decisions for operational, tactical and strategic level [23]. Decisions related to biomass supply chain to energy conversion usually consider activities such as preparing and conditioning, transport, storage, and processing to energy generation. Normally, these decisions are taken at different levels. Some decisions can be strategic, such as technology type, kind of biomass, contracting, selection of transportation mode, supply network design, inventory levels, etc. At the tactical level the set of decisions includes storage conditions, vehicle routing, and harvest scheduling. The main biomass supply chain optimization decisions considered are presented in Figure 2 [3].

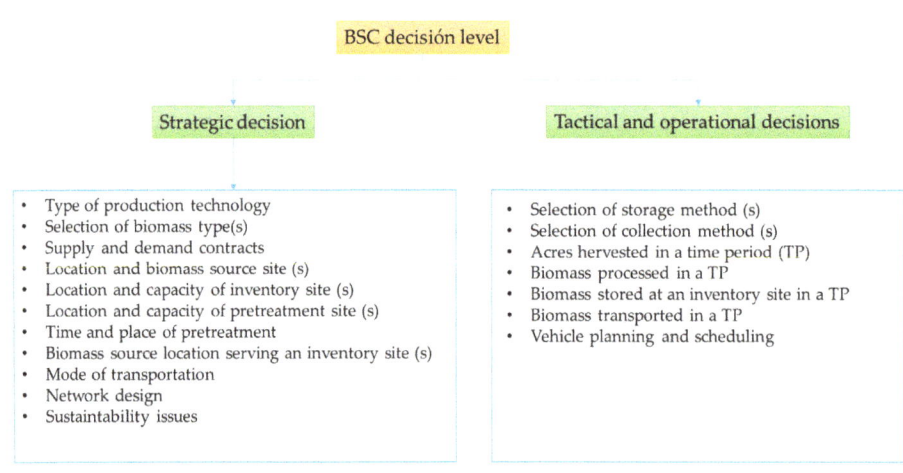

Figure 2. Biomass supply chain decision taxonomy [3].

On the other hand, when we consider the strategies to model biomass supply chain, usually we find stochastic, deterministic, hybrid, and IT-driven models. In the field of deterministic models, we can develop single and multiple objective models. The techniques used to solve the deterministic models include linear, mixed, integer linear and non-linear programming.

Specifically, in the sugar agro-industrial production chain, different strategies have been developed for the use of the biomass produced in the cultivation process. An important part of the total fiber produced by the sugarcane can be used as a renewable energy resource. The use of sugarcane crop residues as a substitute fuel in electric cogeneration systems begins to become a topic with great research potential in an effort to seek power generation systems with less environmental impact. Reference [24] presents an analysis of the potential of electricity generation from sugarcane crop residues in South Africa.

Some authors, like those of reference [25], propose a multi-objective integer linear programming optimization model to choose sugarcane varieties to minimize costs in the use of crop residue and simultaneously maximize the energy balance in such a process. In that paper, the authors considered different logistics costs. First, the cost to collect, compact, and load the truck with residual biomass of different sugarcane variety. In addition to these costs, the model considered included the fixed and variable costs associated with the movement of the cargo vehicles. The inclusion of sugarcane variety in the model introduces another dimension related with different quantities of SCR production.

Other papers in which information regarding decision optimization models associated with the sugarcane supply chain can be found include reference [26], which proposed a stochastic optimization model for sugarcane supply chain planning integrating sowing, growing, and harvesting operations. Reference [27] presented a model to improve the integration between harvest decisions and transportation decisions in sugar industry supply chain in Australia. Finally, in references [28,29], the authors proposed a decision support systems for the integration of supply chain decision in the sugar industry. The model developed was applied in the Cuban sugar industry.

2. Model

2.1. Problem Description

Harvesting crops leaves biomass on the ground (leaves, stems. and other plant materials). Sometimes, this residue is left in the plot to decompose and fertilize the soil [30]. In other cases, it is necessary to remove it for the process of soil preparation. In other instances, there are productive uses for the residue, such as producing biofuel or other agro-industrial products. In this case, we will consider the use of crop residue as combustible material for energy cogeneration.

In this paper, we use sugar cane as a hypothetical case study. In Colombia, sugar cane is grown and harvested all year round, due to the local tropical climate. Sugar mills harvest sugar cane according to a schedule that reflects the appropriate moment to do so for every plot. These companies have the logistics for harvesting sugar cane and transporting it to the mill. In addition to the resources devoted to this activity, it will be necessary to assign vehicles and operators to the logistics of collecting and transporting the crop residue. The model we propose addresses the question of the selection of the plots whose crop residue will be transported for energy production on a given day.

2.2. The Process

Sugar cane producers plant the plots staggered throughout the year. The crop takes 13 months on average to grow to maturity. The sugar cane mills build a daily schedule of the harvest of the plots in order to maximize the amount of sugar they will produce. To build this schedule, they must consider several factors, such as

- Distance from the plot to the factory;
- Sucrose contents in the crop;
- Availability of vehicles, harvesting equipment, and crews;
- In-factory waiting time for raw materials (the sugar contents in the canes start decreasing the moment they are cut, a process called inversion).

The harvest is executed using large mechanized harvesters or by hand by crews of human operators. The harvest leaves behind a substantial amount of biomass—the crop residue. This residue has been left in the field to decompose and fertilize the soil, but there is a growing interest in collecting and transporting it to the factory to burn it and generate electricity.

2.3. Using Crop Residue to Generate Electricity

Sugar cane mills are working to become independent of the interconnected electricity grid; they can generate excess electricity and sell it back to the network at some times during the day. They are currently using coal in their boilers to generate steam (for the production process) and electricity for general usage [31,32]. The use of crop residue has some apparent advantages over coal, including

- Crop residue generates less CO2 equivalent than coal;
- Sugar cane mills have to purchase coal or exchange it for post-production byproducts. Crop residue is basically free.

2.4. Mathematical Model

We built a Linear Programming model to decide which plots to harvest on a given day. We will explain the elements of the model in the following subsections [32].

Sets: Plots scheduled to be harvested on a given day. Index: i.

2.4.1. Parameters

- TCR_i Tons of crop residue available on plot i.
- CP_i: Energy value of the crop residue from plot i (kw*h/ton).
- P_i: Pollution caused by burning the crop residue from plot i (CO_2 tons/ton of crop residue burned).
- D_i: Rectilinear distance (km) from plot i to the factory.
- $FCOST$: Fixed cost of collecting the crop residue from plot i ($).
- $COST$: Transportation cost ($/(ton*km)).
- $COAL$: Baseline level of coal consumption (tons).
- CCL: Cost of coal ($/ton).
- CPC: Energy value of coal (kw*h/ton).
- POL: Pollution caused by burning coal (CO_2 tons/ton of coal burned).
- $CBOND$: Value of carbon bonds ($/tons of CO_2 removed) [33].
- SPE: Sales price of energy.
- VAV: Vehicles and crews available today for collection and transportation of crop residue.

2.4.2. Detailed Information about the Parameters

- TCR_i: These are the tons of sugar cane crop residue that can be obtained from lot i. On average, it is possible to harvest 110 tons of sugar cane from a hectare of crop. From that, it has been estimated that the crop residue lies between 19% and 37% of that total weight [18]. The TCR from a given plot depends on its area, the variety of sugar cane grown, and the harvesting tools that are in use.
- CP_i: This represents the energy that can be obtained from a ton of crop residue. Reference [18] shows that sugar cane residue has a calorific potential of 16 kilojoules per ton. Converting this potential into kilowatts-hour, we get that on average sugar cane crop residue may generate 0.00444 kwh/ton.
- P_i: Pollution caused by burning the crop residue from plot i. This parameter has been estimated as an average of 35.1 kg of CO_2 equivalent per ton of crop residue burned [34].
- D_i: Rectilinear distance from a plot to the sugar cane mill. The structure of plots and roadways is predominantly rectilinear. For a factory located at (A, B) the distance to a plot with coordinates (x_i, y_i) would be $D_i = |A - x_i| + |B - y_i|$ (Figure 3).

Figure 3. Sugar cane plots. Source: Google Maps.

- *FCOST*: Fixed cost of collecting the crop residue from plot i ($). This is the cost related to fixed salary of the driver of the vehicle and the cost of ownership of said vehicle. It depends neither on the tons transported nor on the distance. We have calculated this cost at $32 USD per trip, assuming three trips per day per crew. We calculated this cost considering the fixed salaries of the crews, assuming full utilization of their work, and also considering the cost of ownership of the vehicles required for the transportation of the sugarcane crop residue. The cost of operation of the vehicles is included in the parameter *COST*.
- *COST*: This is the variable transportation cost. We have calculated its value at $0.115 USD per ton-km.
- *COAL*: This is a user-supplied value. For our exercise we will use a value of 100 ton/day as a baseline (obtained from an actual sugar cane mill).
- *CCL*: Cost of acquisition of coal in $/ton USD. This cost may vary according to the sources the individual company uses to acquire its coal. There are different acquisition strategies, including long-term contracts, open market contracts and the exchange of bagasse for coal with the paper mills. For this model, we used $75/ton USD.
- *CPC*: Energy value of coal (kwh/ton). The coal used by sugar cane mills has a calorific power of 0.00666 kwh/ton on average.
- *POL*: Pollution caused by burning coal (CO_2 tons/ton of coal burned). This value has been estimated as 2.66 tons of CO_2 per ton of coal burned [34].
- *CBOND*: Value of carbon bonds ($/tons of CO_2 removed). We will use $25 USD per ton of equivalent CO_2, as this is close to the average in the first part of 2019 [33].
- *SPE*: Sales price of energy. In Colombia, this price is currently $0.028 USD/kwh [35].
- *VAV*: Trips for residue collection and transportation available today. In our example, VAV ranges between 15 and 25 for any given day.

2.4.3. Variables

- y_i: Binary: 1 if the residue from plot i is collected, 0 if not.
- x_i: Amount of tons of crop residue collected from plot i.
- *cpur*: Tons of coal burned today.
- *bond_income* : Income obtained through the sale of green bonds.
- *elec_sold* : Electricity sold.
- *coal_savings* : Savings realized if the coal purchased is less than the baseline.
- *fixed_transp* : Total fixed transportation costs.
- *var_transp* : Total variable transportation costs.

2.4.4. Objective Function: *Maximize Z = Total Profit*

The total profit is the result of the sum of the income from the sale of carbon credits, income from the sale of energy, and savings from the replacement of coal, less the fixed and variable costs of transport and logistics (Equation (1)).

$$Total\ Profit = bond\_income + elec\_sold + coal\_savings - fixed\_transt - var\_transp \quad (1)$$

Revenue from carbon credits is estimated from the reduction of tons of carbon dioxide by replacing coal as fuel in boilers (Equation (2)).

$$bond\_income = (COAL * POL - \sum_i (x_i * P_i) - cpur * POL) * CBOND \quad (2)$$

The marginal revenues from the sale of energy due to the replacement of coal by SCR. This substitution reduces the variable cost of generation and increases the probability of plant use (Equation (3)).

$$elec\_sold = (\sum_i (x_i * CP_i) + (cpur - COAL) * CPC) * SPE \quad (3)$$

Savings derived from lower coal consumption as a boiler fuel (Equation (4)).

$$coal\_savings = (COAL - cpur) * CCL \quad (4)$$

The fixed and variable transport cost are presented in Equations (5) and (6).

$$fixed\_transp = \sum_i (y_i * FCOST) \quad (5)$$

$$var\_transp = \sum_i (x_i * D_i * COST) \quad (6)$$

2.4.5. Constraints

(1) The new level of pollution must not be larger than the baseline (Equation (7)).

$$cpur * POL + \sum_i (x_i * P_i) \leq COAL * POL \quad (7)$$

(2) The electricity generated today must not be less than the baseline (Equation (8)).

$$\sum_i (x_i * CP_i) + (cpur * CPC) \geq COAL * CPC \quad (8)$$

(3) The number of plots in which to collect the crop residue is limited by the number of crews available (Equation (9)).

$$\sum_i y_i \leq VAV \quad (9)$$

(4) If a plot is not selected, no crop residue will be collected from it (Equation (10)).

$$x_i \leq TCR * y_i;\ \forall i \quad (10)$$

3. Results

The main results obtained are showed in the Table 2. The total profit is $13,708 USD. The plots collected are plot 2, plot 5, plot 9, plot 11, and plot 16 (Table 2). The model exactly replaces the energy generated by coal with energy from the crop residue. There is an interesting opportunity to obtain additional financial resources associated with the participation in the carbon bond market from the replacement of coal in the boiler. The model does not recommend the collection of more crop residue to generate electricity and sell it back to the grid. This is due to the fact that in a country like Colombia, the marginal cost of electricity generation is determined through optimal economic dispatch. Given the predominant character of hydro power generation sources in the electrical system, thermal power plants have less opportunity to be used.

Table 2. Model outputs.

Total Profit	\$13,708.04 USD				
Plots collected	P2	P5	P9	P11	P16
Tons collected	30.0	26.0	21.0	30.8	36.0
Coal to buy	0.0				
Bond Income	\$6523.53 USD				
Electricity Sold	0.0				
Coal Savings	\$7500 (USD				
Fixed Transp.	160.0				
Variable Transp.	155.50				
Residue Collected	143.823				

We have used a hypothetical instance with 20 candidate plots for the collection and transportation of their SCR on a given day. Out of them, only five were attractive enough to be collected. This means that the decision to collect the crop residue is not automatic, it should include the plots that are beneficial to the objective function, and otherwise a model would not be required to collect all plots.

For this instance, the sale of electricity to the grid is not attractive. In an electricity market such as the Colombian, the price of the kwh varies greatly depending on the rains, given that most of the electricity comes from hydropower. This fact makes the use of the model on a daily basis a requirement, since variations on the electricity price might alter the results.

The revenues of the model in this instance come from the green bonds (reductions in pollution levels) and the substitution of coal for crop residue (savings in the purchase of coal). Out of the three possible revenue components of the model (green bonds, coal savings, and sales of electricity), only two are attractive.

In this model, we have been conservative in the estimation of the transportation costs under current forms of operation. In discussions with engineers who work in the harvesting department of sugar cane mills, different forms of operation have been suggested to save on transportation costs, such as hitching additional wagons to sugar cane trains, resulting in smaller marginal transportation costs for the crop residue.

Since the model includes binary variables, it is not possible to use linear programming sensitivity analysis, so building scenarios and observing their results should be the procedure to follow. For this task, a general programming language with an optimization module such as Julia [36] (Appendix A) is a better choice than specialized optimization languages because generating the values of the parameters, running the model multiple times, storing the results, and analyzing the output data are easier in such a language. The instance data is presented in Appendix B.

4. Discussion

The authors share the perception that the use of sugarcane crop residue (SCR) has an economic potential that is just beginning to be studied and discovered. The first version of this model is a conservative one, trying not to overestimate the attractiveness of energy cogeneration using the SCR. However, there is a variety of strategies in the preliminary stages of discussion, such as modifying the current harvesting practices to obtain a more compact SCR, in order to transport more weight in less trips. The current practices include manual and mechanized harvesting, so additional treatment of the SCR in the field or the addition of specialized compacting equipment to the harvesters are recommended to reduce the cost of collecting and transporting the SCR.

Highway infrastructure on the region is in a continuous development process. This fact, added to the previous point, could potentially reduce variable transportation costs in the future and make more plots attractive to harvest their SCR.

The price of electricity that the Colombian grid would pay the sugar mills fluctuates during the day, following the peaks and valleys of electricity consumption in the country. In this sense, the optimal moments to generate electricity and dispatch it to the grid should also be studied to maximize revenues. When sugarcane is harvested, every minute that passes before grinding to extract sugar-bearing juices destroys value, because the usable sugar content decreases over time. This behavior is not the same with SCR, because sugar content is not a variable of interest, and some holding time could be used to let SCR dry and make combustion easier and more productive.

The price of coal is another parameter of interest. For future research, the authors propose to conduct Monte Carlo simulation experiments considering inputs, such as the price of electricity, the price of coal, and the state of the plots (dry or wet if it rains), to design robust decision strategies to maximize the economic and environmental benefit of SCR collection.

Models that can be found in the technical literature sought to maximize profits of the sugarcane mills through the utilization of sugarcane for the production of sugar or ethanol. This model proposes a contribution to the theory of the field by including an alternate use of the sugarcane crop residue, not for the reduction of fertilizers (as it has been studied) but for the substitution of coal to be burned in the boilers and the resulting reduction in pollution and possible revenue through the sale of electricity to the interconnected electricity network of a country.

5. Conclusions

For the sugarcane sector, finding alternatives that provide notions of sustainability is fundamental, since its environmental role has always been questioned because of the amount of resources used, from the land itself to the water required for production. The use of their own agricultural residue in the generation of their own energy generates an agribusiness concept based on the conservation of the environment and the efficient use of available resources.

The results obtained look very promising for the sugar industry to the extent that the replacement of fossil fuel (coal) with agricultural crop residues generate a double effect as can be seen. On the one hand, the savings from less coal consumed, taking into account that the mills buy this coal in national and international markets. The price of coal is very volatile and depends on world supply and demand. Savings for substituted coal in this case reach $7500 USD. Additionally, the replacement of coal generates another very significant effect and is the income obtained from the sale of emission reduction certificates (more than $6500 USD).

In Colombia, Law 1715 of 2014 has become a strong incentive for the promotion of projects with non-conventional energy sources. In this context, the sugar sector power cogeneration has become a key element of the national electricity system. Power co-generation plants are backup plants in a predominantly hydropower system. This country has a non-typical behavior, since sugarcane can be sown and harvested all year round, due to the local tropical climate. This characteristic enables the use of SCR as an energy source as a part of day-to-day operations of the sugarcane mills and, thus, as part of the energy generation portfolio for this companies. Sugarcane mills in Colombia are already independent of the energy distributed through the national interconnected network, and the use of SCR would allow them to profit from selling electricity to the public network.

The co–generation of electricity using sugarcane crop residue (SCR) is going to be studied with interest in the near future. This activity shows promise in some specific aspects.

- SCR releases less equivalent CO_2 than coal when burned for electricity generation;
- SCR is basically free. It is a byproduct of the harvesting of the sugarcane. CO_2 is a raw material that has to be acquired through private contracts or the commodities market;
- SCR can be collected using crews, equipment, and vehicles that are similar to those used for the harvesting and transportation of sugarcanes.

This model is limited by the information it requires to choose a plot over others. Since the Colombian environment allows the harvest of canes throughout the year, there are professionals gathering information and inspecting the growth of the plots to gather the parameters required by the model. This might not be the case in environments with a limited agricultural season.

Also, this model might require adaptations for use in areas with seasonal weather. The crop residue would be collected in a shorter time frame and some storage costs would be accrued to use the residue for electricity generation spread throughout more months of the year.

It would also be interesting to use this model to study the impact of the price of coal, electricity, and green bonds on the decision to collect and transport residue. A time-series analysis or econometric study of the data of these prices would be required as part of the analysis.

Author Contributions: All authors contributed in this paper. P.C.M.-V. collected the data and developed the conceptual model. D.F.M.-D. developed the economic model. L.R.-C. developed the methodology and mathematical model. All authors contributed in the results analysis and writing.

Funding: This research received no external funding

Acknowledgments: All authors acknowledge the School of Industrial Engineering at the Universidad del Valle for the support to develop this paper.

Conflicts of Interest: All authors declare no conflict of interest.

Appendix A

Julia Code

```
using JuMP
using GLPK
using MathOptInterface
using CSV

# Read data from file
data = CSV.read("Model-Data.csv")

# Create indexed parameters
TCR = data.TCR
CP = data.CP
P = data.P
D = data.D

# Create constants
FCOST = 32; COST = 0.115; COAL = 100
CCL = 75; CPC = 0.00666; POL = 2.66
CBOND = 25; SPE = 0.028; VAV = 18

# Count the number of plots
plots = size(TCR)[1]

# Create optimization model
m = Model(with_optimizer(GLPK.Optimizer))

# Declare decision variables
@variable(m, y[1:plots], Bin)
@variable(m, x[1:plots] >= 0)
@variable(m, cpur >= 0)
```

```
# Declare cost variables for the objective function
@variable(m, bond_income >= 0)
@variable(m, elec_sold >= 0)
@variable(m, coal_savings >= 0)
@variable(m, fixed_transp >= 0)
@variable(m, var_transp >= 0)

# Declare the objective function
@objective(m, Max, bond_income + elec_sold + coal_savings - fixed_transp -
var_transp)

# The new level of pollution must not exceed the baseline
@constraint(m, pol_base, cpur*POL + sum(x[i]*P[i] for i in 1:plots) <= COAL*POL)

# The electricity generated must not be less than the baseline
@constraint(m, elec_gen, sum(x[i]*CP[i] for i in 1:plots) + (cpur*CPC) >= COAL*CPC)

# The number of plots to collect crop residue is limited by
# the number of trips available per day
@constraint(m, collect_plots, sum(y[i] for i in 1:plots) <= VAV)

# If a plot is not selected, no crop residue can be collected from it
@constraint(m, residue[i=1:plots], x[i] <= TCR[i]*y[i])

# Assign value to the cost variables of the objective function
@constraint(m, bondincome, bond_income == (COAL*POL - sum(x[i]*P[i] for i in
1:plots) - cpur*POL) * CBOND)
@constraint(m, elecsold, elec_sold == (sum(x[i]*CP[i] for i in 1:plots) +
((cpur-COAL)*CPC))*SPE)
@constraint(m, coalsavings, coal_savings == (COAL - cpur)*CCL)
@constraint(m, fixedtransp, fixed_transp == sum(y[i]*FCOST for i in 1:plots))
@constraint(m, vartransp, var_transp == sum(x[i]*D[i]*COST for i in 1:plots))

print(m)

status = optimize!(m)

println("Profit: ", JuMP.objective_value(m))
println("Plots to collect: ", JuMP.value.(y))
println("Crop residue: ", JuMP.value.(x))
println("Coal to buy: ", JuMP.value(cpur))
println("Bond Income: ", JuMP.value(bond_income))
println("Electricity Sold: ", JuMP.value(elec_sold))
println("Coal Savings: ", JuMP.value(coal_savings))
println("Fixed Transp: ", JuMP.value(fixed_transp))
println("Variable Transp: ", JuMP.value(var_transp))
println("Residue Collected: ", sum(JuMP.value.(x)))
```

Appendix B

Instance Data

Table A1. Instance data.

Plot.	TCR	CP	P	D
P1	37	0.0030336	0.0352	17.14
P2	30	0.0032734	0.0334	3.42
P3	26	0.0058655	0.0363	21.16
P4	39	0.0041573	0.0331	15.65
P5	26	0.0032552	0.0361	3.24
P6	35	0.0032061	0.0338	15.31
P7	37	0.0057260	0.0347	22.56
P8	33	0.0040761	0.0345	21.03
P9	21	0.0059792	0.0336	5.55
P10	33	0.0037543	0.036	24.73
P11	33	0.0052091	0.0354	17.78
P12	32	0.0044785	0.0333	22.54
P13	27	0.0030945	0.0343	11.12
P14	27	0.0044857	0.0363	16.47
P15	30	0.0046803	0.0353	22.67
P16	36	0.0054730	0.0367	13.91
P17	24	0.0042343	0.034	22.60
P18	23	0.0039274	0.0349	14.76
P19	30	0.0044093	0.0341	22.49
P20	26	0.0055547	0.0353	16.93

References

1. Biomass Energy Resource Center Home Page. Available online: https://www.biomasscenter.org/ (accessed on 12 November 2019).
2. Perea-Moreno, M.-A.; Samerón-Manzano, E.; Perea-Moreno, A.-J. Biomass as renewable energy: Worldwide research trends. *Sustainability* **2019**, *11*, 863. [CrossRef]
3. Sharma, B.; Ingalls, R.G.; Jones, C.L.; Khanchi, A. Biomass supply chain design and analysis: Basis, overview, modeling, challenges, and future. *Renew. Sustain. Energy Rev.* **2013**, *24*, 608–627. [CrossRef]
4. Yue, D.; You, F.; Snyder, S.W. Biomass-to-bioenergy and biofuel supply chain optimization: Overview, key issues and challenges. *Comput. Chem. Eng.* **2014**, *66*, 36–56. [CrossRef]
5. De Meyer, A.; Cattrysse, D.; Rasinmäki, J.; Van Orshoven, J. Methods to optimise the design and management of biomass-for-bioenergy supply chains: A review. *Renew. Sustain. Energy Rev.* **2014**, *31*, 657–670. [CrossRef]
6. Cambero, C.; Sowlati, T. Assessment and optimization of forest biomass supply chains from economic, social and environmental perspectives—A review of literature. *Renew. Sustain. Energy Rev.* **2014**, *36*, 62–73. [CrossRef]
7. Sindhu, R.; Gnansounou, E.; Binod, P.; Pandey, A. Bioconversion of sugarcane crop residue for value added products—An overview. *Renew. Energy* **2016**, *98*, 203–215. [CrossRef]
8. Sector Agroindustrial de la Caña. Informe Anual 2018–2019. Available online: https://www.asocana.org/modules/documentos/3/362.aspx (accessed on 29 October 2019).
9. Go, A.W.; Conag, A.T.; Igdon, R.M.B.; Toledo, A.S.; Malila, J.S. Potentials of agricultural and agro-industrial crop residues for the displacement of fossil fuels: A Philippine context. *Energy Strategy Rev.* **2019**, *23*, 100–113. [CrossRef]
10. Cervi, W.R.; Lamparelli, R.A.C.; Seabra, J.E.A.; Junginger, M.; van der Hilst, F. Bioelectricity potential from ecologically available sugarcane straw in Brazil: A spatially explicit assessment. *Biomass Bioenergy* **2019**, *122*, 391–399. [CrossRef]
11. Khatiwada, D.; Leduc, S.; Silveira, S.; McCallum, I. Optimizing ethanol and bioelectricity production in sugarcane biorefineries in Brazil. *Renew. Energy* **2016**, *85*, 371–386. [CrossRef]

12. Olivério, J.L.; Ferreira, F.M. Cogeneration—A new source of income for sugar and ethanol mills or bioelectricity—A new business. *Proc. Int. Soc. Sugar Cane Technol.* **2010**, *27*.
13. Bocci, E.; Di Carlo, A.; Marcelo, D. Power plant perspectives for sugarcane mills. *Energy* **2009**, *34*, 689–698. [CrossRef]
14. Hiloidhari, M.; Araújo, K.; Kumari, S.; Baruah, D.C.; Ramachandra, T.V.; Kataki, R.; Thakur, I.S. Bioelectricity from sugarcane bagasse co-generation in India—An assessment of resource potential, policies and market mobilization opportunities for the case of Uttar Pradesh. *J. Clean. Prod.* **2018**, *182*, 1012–1023. [CrossRef]
15. Pierossi, M.A.; Bertolani, F.C. Sugarcane trash as feedstock for biorefineries: Agricultural and logistics issues. In *Advances in Sugarcane Biorefinery: Technologies, Commercialization, Policy Issues and Paradigm Shift for Bioethanol and By-Products*; Chandel, A.K., Silveira, M.H.L., Eds.; Elsevier: Amsterdam, The Netherlands, 2017; pp. 17–39.
16. León-Martinez, T.; Dopíco-Ramírez, D.; Triana-Hernández, O.; Medina-Estevez, M. *Sobre los Derivados de la Caña de Azúcar*; ICIDCA: Habana, Cuba, 2013; Volume 47, pp. 13–22.
17. Cobo-Barrera, D.F. Pirólisis de Residuos de Cosecha de Caña de Azúcar (RAC) como Alternativa de Aprovechamiento en Procesos de Cogeneración. Master's Thesis, Universidad del Valle, Cali, Colombia, May 2012.
18. Ojeda, W.; Saltaren, J.; Lucuara, J.; Gómez, A.; Gil, N. Valorización de residuos agrícolas de cosecha (RAC) de caña de azúcar mediante densificación mecánica. In Proceedings of the XI Congreso Atalac-Tecnicaña, Cali, Colombia, 24–28 September 2018; pp. 488–496.
19. Zandi Atashbar, N.; Labadie, N.; Prins, C. Modelling and optimisation of biomass supply chains: A review. *Int. J. Prod. Res.* **2018**, *56*, 3482–3506. [CrossRef]
20. Flores Hernández, U.; Jaeger, D.; Islas Samperio, J. Bioenergy potential and utilization costs for the supply of forest woody biomass for energetic use at a regional scale in Mexico. *Energies* **2017**, *10*, 1192. [CrossRef]
21. Hofsetz, K.; Silva, M.A. Brazilian sugarcane bagasse: Energy and non-energy consumption. *Biomass Bioenergy* **2012**, *46*, 564–573. [CrossRef]
22. Rentizelas, A.A.; Tolis, A.J.; Tatsiopoulos, I.P. Logistics issues of biomass: The storage problem and the multi-biomass supply chain. *Renew. Sustain. Energy Rev.* **2009**, *13*, 887–894. [CrossRef]
23. Agustina, F.; Vanany, I.; Siswanto, N. Biomass Supply Chain Design, Planning and Management: A Review of Literature. In Proceedings of the 2018 IEEE International Conference on Industrial Engineering and Engineering Management, Bangkok, Thailand, 16–19 December 2018; pp. 884–888.
24. Smithers, J. Review of sugarcane trash recovery systems for energy cogeneration in South Africa. *Renew. Sustain. Energy Rev.* **2014**, *32*, 915–925. [CrossRef]
25. Florentino, H.D.O.; De Lima, A.D.; De Carvalho, L.R.; Balbo, A.R.; Homem, T.P.D. Multiobjective 0-1 integer programming for the use of sugarcane residual biomass in energy cogeneration. *Int. Trans. Oper. Res.* **2011**, *18*, 605–615. [CrossRef]
26. Carvajal, J.; Sarache, W.; Costa, Y. Addressing a robust decision in the sugarcane supply chain: Introduction of a new agricultural investment project in Colombia. *Comput. Electron. Agric.* **2019**, *157*, 77–89. [CrossRef]
27. Higgins A., J.; Laredo, L.A. Improving harvesting and transport planning within a sugar value chain. *J. Oper. Res. Soc.* **2006**, *57*, 367–376. [CrossRef]
28. López-Milán, E.; Plà-Aragonés, L.M. A decision support system to manage the supply chain of sugar cane. *Ann. Oper. Res.* **2014**, *219*, 285–297. [CrossRef]
29. López-Milán, E.; Plà-Aragonés, L.M. Optimization of the supply chain management of sugarcane in Cuba. *Int. Ser. Oper. Res. Manag. Sci.* **2015**, *224*, 107–127.
30. Go, A.W.; Conag, A.T. Utilizing sugarcane leaves/straws as source of bioenergy in the Philippines: A case in the Visayas Region. *Renew. Energy* **2019**, *132*, 1230–1237. [CrossRef]
31. Alonso-Pippo, W.; Luengo, C.A.; Felfli, F.F.; Garzone, P.; Cornacchia, G. Energy recovery from sugarcane biomass residues: Challenges and opportunities of bio-oil production in the light of second generation biofuels. *J. Renew. Sustain. Energy* **2009**, *1*, 063102. [CrossRef]
32. Portugal-Pereira, J.; Soria, R.; Rathmann, R.; Schaeffer, R.; Szklo, A. Agricultural and agro-industrial residues-to-energy: Techno-economic and environmental assessment in Brazil. *Biomass Bioenergy* **2015**, *81*, 521–533. [CrossRef]

33. Roca, R. El precio del CO_2, imparable: Marca un nuevo récord por encima de €27 por tonelada, el doble que hace un año. *El Periodico de la Energía*. 12 April 2019. Available online: https://elperiodicodelaenergia.com/el-precio-del-co2-imparable-marca-un-nuevo-record-por-encima-de-27-e-por-tonelada-el-doble-que-hace-un-ano/ (accessed on 30 October 2019).
34. Castillo, J.L.; Lucuara, J.E. Uso de residuos agrícolas de cosecha (RAC) como combustible no convencional para generar energía eléctrica en calderas de potencia. In Proceedings of the 3er. Encuentro de Usuarios de Calderas Colombia, Bogota, Colombia, 12–13 May 2016. Available online: https://www.ceaca.com/wp-content/uploads/2016/06/7-USO-DE-RESIDUOS-AGRICOLAS-DE-COSECHA-COMO-COMBUSTIBLE.pdf (accessed on 17 November 2019).
35. Fuentes renovables tendrían participación del 8% en la torta energética nacional. *La República*. 23 October 2019. Available online: https://www.larepublica.co/economia/fuentes-renovables-tendran-participacion-del-8-en-la-torta-energetica-nacional-2923630 (accessed on 29 October 2019).
36. Bezanson, J.; Edelman, A.; Karpinski, S.; Shah, V.B. Julia: A fresh approach to numerical computing. *Siam Rev.* **2017**, *59*, 65–98. [CrossRef]

© 2019 by the authors. Licensee MDPI, Basel, Switzerland. This article is an open access article distributed under the terms and conditions of the Creative Commons Attribution (CC BY) license (http://creativecommons.org/licenses/by/4.0/).

Article

Collection of Solid Waste in Municipal Areas: Urban Logistics

Jania Astrid Saucedo Martinez [1,]* Abraham Mendoza [2,]* and
Maria del Rosario Alvarado Vazquez [1]

[1] Facultad de Ingenieria Mecanica y Electrica, Universidad Autonoma de Nuevo Leon, Ciudad Universitaria, San Nicolas de los Garza, Nuevo Leon 66455, Mexico; maria.alvaradovzqz@uanl.edu.mx
[2] Universidad Panamericana. Facultad de Ingenieria. Alvaro del Portillo 49, Zapopan, Jalisco 45010, Mexico
* Correspondence: jania.saucedomrt@uanl.edu.mx (J.A.S.M.); amendoza@up.edu.mx (A.M.)

Received: 29 August 2019; Accepted: 23 September 2019; Published: 1 October 2019

Abstract: A sustainable process satisfies the current needs without compromising the ability of future generations to satisfy their own needs; that is, it must have a triple impact (sustainability): social, economic, and environmental. In México, there are several services that the government must provide to society for its proper development, for example, the collection of solid waste. Urban logistics include all the processes and operations that provide a service to the community, such as water, safety, health, waste collection, etc., providing the service with the lowest possible cost (economic, social, and environmental) that contributes to the sustainability of the city. Due to the accelerated growth of the world population, several environmental problems have arisen, among them, the generation of solid waste in important quantities; their proper management is relevant for adequate development of the population. The collection of solid waste in municipal areas aims to grant green spaces and recreation areas for the citizens. Although an outstanding effort has been made by the government to provide an adequate service, there are still gaps in the application of correct tools that guarantee efficiency in operations and continuity in services. This article presents a proposal to improve the planning of the design of territories for the cleaning, weeding, and collection of solid waste in municipal areas, using two MILP (Mixed Integer Linear Programming) models. The main contribution of the adaptation of this model is the application to the weeding and waste collection service municipality of the Monterrey Metropolitan Area, which considers important factors among which are the amount of waste, frequency, and service coverage.

Keywords: waste management; residue collection; urban logistics; design of territories

1. Introduction

In Mexico, the government is responsible for urban solid waste management (USWM) and the cleanup of green and public spaces. This activity is very important for cities; however, it has become a difficult task to perform as this service has very limited funds.

In the last few years, the term *Urban Logistics* have come to play [1], referring to the activities that provide a service to the community, such as water, safety, health and waste collection, and green spaces weeding, at the lowest possible cost (having economical, social, and environment impact) in order to contribute to a sustainable city. Also Taniguchi et al. [2], Winkenbach et al. [3], Macário et al. [4], and Villamizar et al. [5] mention that it covers multiple objectives. These types of activities also have a positive impact on the city's industrial supply chains.

The increasing traffic congestion and the need for preserving the environment (CO_2 reduction), make USWM an area opportunity to achieve an economic, social, and environmental impact by reducing urban transport using adequate collection techniques [6–8]. The need to control the damage to the population and the environment transforms waste collection in a public and private matter.

Therefore, this leads to the need for developing solutions, mainly focused on the administration, collection, and disposal of waste.

According to official data from 2014 INEGI [9], in Mexico, 102,887,315 kg of garbage is collected daily throughout the country, that is, 0.86 kg of waste per person per day; thus, everyday, a person generates approximately one kilogram of garbage. Figure 1 shows the total generation of solid waste per state in Mexico. Nuevo Leon represents the 3.6% that is equivalent to 4,037,198 kg.

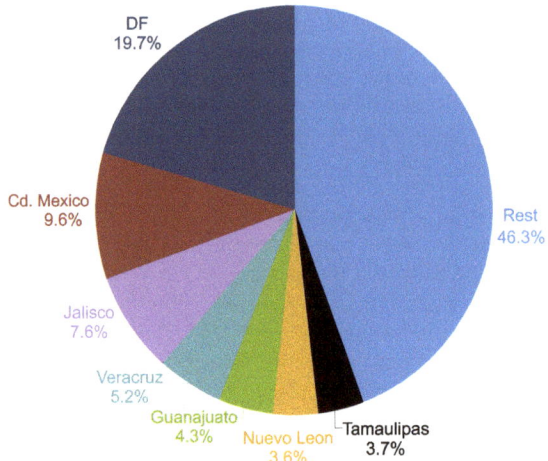

Figure 1. Total generation of solid waste. Source: INEGI [9].

The total generation of solid urban waste is a relevant variable that allows us to determine the public policies of an adequate tool to manage and dispose of that waste [10].

The purpose of this work is focused on solid-waste collection in green spaces of the Monterrey Metropolitan Area. Currently, this activity is the municipality government's responsibility and they have done it through empirical knowledge [11], usually outsourcing these services; therefore, it is important to propose tools to support decision-making that helps to design and plan waste collection.

Two classifications of vehicle routing design exist: macro-routing and micro-routing [12]. Macro-routing is the allocation of collection vehicles to various areas of the city to perform the collection, and micro-routing is the specific route that must be met by daily collection vehicles in the areas of population where they have been assigned.

However, the basic objective is to create preferably balanced territories, that is, similar in size to one or more of these criteria. In the case of space constraints, there is a basic set of conditions present in most applications that force the creation of contiguous, connected, and compact areas.

The present work develops two macro-routing mathematical models based on the design of territories for the collection of solid waste in the Monterrey Metropolitan Area.

In the next section, a brief description of the concepts related to this activity can be found: reverse and urban logistics, and solid waste management. In Section 3, we present a case study and the main contribution of this work: two mathematical models. We finish with the results, analysis, and conclusions.

2. Literature Review

This section presents the literature analysis focusing on two concepts very much related to solid waste collection and management: solid waste management and reverse logistics.

2.1. Reverse Logistics

Reverse logistics is a recent concept that refers to the most effective and economic upstream material flow and merchandise return to the supply chain and to the retrieval of products that were already used for their main purpose [13]. This concept emerged at the end of the 20th century in developed countries by doing recycling processes [14] as a key activity for waste management. In countries with a higher income, one-third of the waste could be retrieved, whereas those with a lower income only retrieved 4% [15]. In most cases, it is expected to generate income or to manage waste.

Such is the case of Uriarte-Miranda et al. [16], where strategies for waste tire management become important due to its economic, social, and environmental impacts. This article shows a reverse logistics model for the waste tire management systems in México and Russia (two specific processes: remanufacturing and diversification) so that it is affordable and sustainable. Hage et al. [17] presented a study about waste recycling (plastic, glass, paper, and metal) that demonstrates that, if a container is located adequately, the probability for recycling increases by 28%. Something similar is proposed by Castillo et al. [18] and applied in Spain.

In the cases of Chang and Wei [19], Gautam and Kumar [20], and Alonso [21], they use optimization models focused on waste or container retrieval for their later recycling.

As could be observed in reverse logistics, it is a sustainable process, as it considers the triple helix: society, economy, and environment. In the presented application as the contribution of this work, whilst it does not consider an economical remuneration as such, it takes it into account due to the cost reduction that it implies, as well as the social impact, as it is a service that improves the quality of life for citizens and the environmental impact related to the solid waste management in green spaces.

Below, we present the works of which the main focus is on urban logistics, that is, those where the main application is on social good.

2.2. Municipal Solid Waste (MSW Collection)

Generally the macro-routing problem is solved by design territories, whilst micro-routing employs mathematical models like *Vehicle Routing Problem* (VRP) and *Travel Salesman Problem* (TSP) [22–24] or uses computational tools and Geographic Information Systems (GIS) [25,26] to minimize the route collection time.

Design territories is a mixed-integer linear programming that consists of grouping small geographical areas, called areas, in larger geographic groupings, called territories, so that the latter are acceptable according to certain criteria. There is a classification according to the context of the problem proposed by Moreno [27]; these criteria may have economic reasons related to the average potential sales [28,29], areas of use of services and equipment located in a fixed location [30], zones for the provision of services at home [31], energy resource receiving areas [32], and political-electoral districts [33–35].

Territorial partitioning in different regions or zones is a problem that is presented in various disciplines related to earth and space sciences and has been treated under various denominations such as partitioning, regionalization, zoning, delineation of zones and/or districts, allocation of spatial units, etc. [36].

While most applications are related to the design of territories, there are other applications like the literature review presented by Phuntsho et al. [37] to determine the solid waste collection plan and schedule in the Bhutan municipality or the work presented by Jayakody et al. [38] that presents the findings of a municipal solid waste characterization study with a sample size of 2850 persons, estimating that the average for daily waste-generation rate for a household was 0.77 ± 0.13 kg/cap/day. Also, this paper distinguishes that waste composition and waste management systems play a key role in establishing an integrated sustainable solid waste management system in countries like Sri Lanka. In addition to this, Kaosol [39] presented a study of the composition of waste collected in Thailand (86%), which is mostly organic waste (paper, plastic, glass, and metal). The integrated MSW system has the potential to maximize the useable waste materials as well as to produce energy as a by-product.

While we can observe a qualitative approach in these articles, there are others with a quantitative approach such as ours.

For example, Ayantoyinbo and Adepoju [40] analyzed the relationship between waste management logistics and metrics for waste management performance. In this case, transport demand system analysis must come in the form of predictive and prescriptive models, while deterministic models make accurate predictions about a system of interest and probabilistic models entail some elements of uncertainty. Dotoli and Epicoco [7] presented a routing and scheduling vehicle model (i) to determine the optimal routes for HWC (Hazardous Waste Collection), which seeks to find the best trade-off between reduction of traveled distances, maximization of amount of collected waste, and maximization of commercial value of withdrawals and their transportation to the disposal site and (ii) to assign the obtained routes to the available fleet. Similar to this work, Buhrkal et al. [8] studied the Waste Collection Vehicle Routing Problem with Time Windows (WCVRPTW), where the problem consists of routing vehicles to collect customers waste within a given time window while minimizing travel costs. Laureri et al. [41] presented a mathematical model based on a Genoa Municipality case study and contemplates various kind of objectives and constraints, making the collection a complex task involving many actors (citizens, politicians, and technical personnel), technological expertise, and investments since it uses WSM concerning the protection of the environment and conservation of natural resources.

Finally, the case study of Son and Louati [42] developed an effective vehicle routing model that optimizes the total traveling distance of vehicles for MSW collection in Danang City (Vietnam) considering the environmental emissions; the investment cost; and the available node structures, vehicles, and parameters in a generalized context. Moreover, in Das and Bhattacharyya [43], MSW management systems suffer by the high collection and transportation cost, where typically, collection cost represents 80–90% and 50–80% of municipal solid waste management budget in low income and middle-income countries. Proposed heuristic solutions reduce more than 30% of the total waste collection path length (100 source points, 65 waste collection centers, and 50 transfer stations).

Considering the literature review and the lack of a defined algorithm or model for the design of macro-routing for cleaning and solid waste collection in municipal areas, the problem consists of the optimal grouping of green spaces in the Monterrey Metropolitan Area, so that territories are balanced regarding the workload of each provider and the coverage of the collection service. These two fundamental aspects are not currently in the planning service.

As a consequence of this fact, our proposal consists of improving the territorial design planning of the current collection service through the adaptation of a mathematical model of territory design for cleaning and solid waste collection in the municipal areas and considers the criteria of spatial integrity, homogeneity, contiguity, and compactness.

It is important to mention that the objects to be grouped are parks of the municipality of San Nicolas de los Garza, but due to a large number of parks, a first grouping has been done to reduce the scale of the problem and to simplify the solution. Thus, we then applied model 1 where we group the districts established by the municipality and model 2 to group the parks; both models are similar, and each one is described below.

3. Case Study: San Nicolas De Los Garza

The Monterrey Metropolitan Area is formed by nine municipalities of the state of Nuevo Leon (which work in a similar way in terms of the operation of primary services, so the proposal for this project is replicable to the others) and constitutes the third most populated conurbation of the Mexican Republic, according to the most recent count and official delimitation carried out in 2010; together with the Instituto Nacional de Estadistica y Geografia, the Consejo Nacional de Poblacion, and the Secretaria de Desarrollo Social del Gobierno Federal, there are a total of 4,080,329 people in an area of 6680 (km)^2. In the area of economic development, it holds the second place at the national level, only after Mexico City, the capital of Mexico.

We focus on the solid waste collection municipal areas. Most of the waste composition is grass; trees; and in minimum proportion, nonorganic products. The amount of waste collected is proportional to the dimension of the green spaces (see Table 1). It is worth mentioning that this process will be performed every 2 weeks in autumn or every 3 weeks in summer, depending on weather conditions. Because of the rain, its growth is faster as is the proliferation of mosquitoes. Besides that, the municipality has already standardized the carry out time, so we do not need to contemplate it in the models.

In this complex metropolitan frame stands the municipality of San Nicolas de los Garza, located geographically in the North of Monterrey, with important industrial, commercial, and housing areas, in which there is a constant evolution and physical transformation. San Nicolas is the second largest, at the national level, in terms of economic development, only behind Mexico City, the capital of the country [44].

We chose to apply this mathematical model in the municipality of San Nicolas because it plays a fundamental role in the Monterrey Metropolitan Area. It is considered the sixth municipality at the national level in terms of the Human Development Index and the tenth in the National highest income index [45]. Therefore, it is extremely important to provide citizens with an excellent level of quality of life, which involves factors ranging from health, housing, transportation, education, and wealth to natural conditions; hence, it is important to have green spaces in a good state. On the other hand, considering the physical conditions of San Nicolas, geographically, it is a regular municipality which facilitates the design of the territories.

San Nicolas has 24 districts which consist of strategic territorial delimitations that enjoy homogeneity in their land uses and enjoy regional and primary equipment for the service of their neighborhoods and, in some cases, the municipality. Figure 2 shows the districts, and Table 1 presents the number of parks in each district and the measure of activity.

Table 1. Districts of the municipality San Nicolas de los Garza.

Districts	Parks	Activity Measure m^2	% m^2
1. Casa Bella	11	71,344.02	5%
2. Balcones	3	25,901.85	1.8%
3. Centro	2	11,771.30	0.8%
4. CEDECO	18	82,428.73	5.7%
5. El Refugio	20	51,673.84	3.6%
6. Vicente Guerrero	10	74,529.20	5.2%
7. Santo Domingo	12	89,845.99	6.3%
8. Del Paseo	17	10,6051.64	7.4%
9. San Cristobal	7	27,787.56	1.9%
10. La Fe	4	27,285.80	1.9%
11. Casa Blanca	13	95,637.52	6.7%
12. Talaverna	14	130,072.60	9.1%
13. Del Vidrio	14	52,549.01	3.7%
14. Constituyentes	19	33,960.43	2.4%
15. Nogalar	8	45,471.45	3.2%
16. Pedregal	5	25,059.05	1.7%
17. Residencial Anahuac	7	51,956.72	3.6%
18. La Grange	6	8417.01	0.6%
19. Industrial	0	0	0%
20. Andalucia	19	97,414.25	6.8%
21. Cuauhtemoc	12	77,329.98	5.4%
22. Las Puentes	20	137,946.42	9.6%
23. Anahuac	11	42,991.38	3%
24. Jardines de Anahuac	13	67,664.64	4.7%

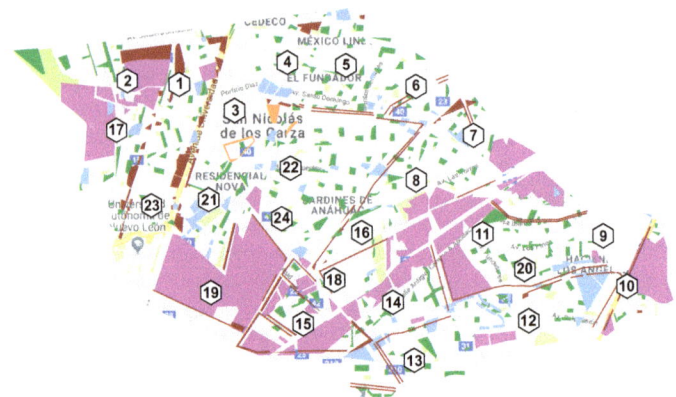

Figure 2. Districts of the municipality San Nicolas de los Garza.

3.1. Model 1

Some municipalities in the Monterrey Metropolitan Area have zoning consisting of areas that integrate and delimit a population center; their predominant uses, reserves, and destinations; as well as the delimitation of areas of conservation, improvement, and growth.

The use of this model is to determine the optimal grouping of the zones established by a municipality of the Monterrey Metropolitan Area to reduce the scale of the problem due to the great number of parks that the municipality has.

The following assumptions are considered: each basic area (districts) must be assigned to a single territory, the quantity of territories is known, and it is determined based on the number of service providers (8) employed by the municipality. Only these providers are eligible and authorized, and the measure of activity is determined by the number of square meters that makes up each of the parks since it is directly related to the amount of waste generated.

Consider model 1 displayed in Equations (1)–(10), where I is the set of basic areas (districts); J the set of centers of territories; p the number of territories; and i and j are basic areas index (where $i \in I$ y $j \in J$). The parameters of the model are as follows: w_i is the measure of the activity; $\mu = \sum w_i / p$ is the average of the size of territories; and a_{ij} is the adjacency matrix, where 1 if the districts are adjacent and 0 otherwise. The continuous variables are defined as follows: w_{max} for the greatest measure of activity of the territories formed; w_{min} for the least measure of activity of the territories formed, and $\bar{w}_j$ for the measure of activity of the territory j. Binary decision variables are defined as $x_j = 1$ if the basic area j is a center, $x_j = 0$ otherwise, $y_{ij} = 1$ if the basic area i is assigned to the center j, and $y_{ij} = 0$ otherwise.

Model 1

$$\min \quad w_{max} - w_{min} \tag{1}$$

$$\text{s.t:} \quad \bar{w}_j = \sum_i w_i y_{ij} \quad \forall j \in J \tag{2}$$

$$w_{max} \geq \bar{w}_j \quad \forall j \in J \tag{3}$$

$$w_{min} \leq \bar{w}_j \quad \forall j \in J \tag{4}$$

$$\sum_j x_j = p \tag{5}$$

$$\sum_j y_{ij} = 1 \quad \forall i \in I \tag{6}$$

$$a_{ij} x_j \geq y_{ij} \quad \forall i \in I, j \in J \tag{7}$$

$$x_j \leq y_{jj} \quad \forall j \in J \tag{8}$$

$$x_j, y_{ij} \in \{0,1\} \quad \forall i \in I, j \in J \tag{9}$$

$$w_{max}, w_{min}, \bar{w}_j \in \mathbb{R} \quad \forall j \in J \tag{10}$$

The objective function in Equation (1) minimizes the difference between the territory with the greatest measure of activity and the territory with the least measure of activity with the idea of forming balanced territories, and in this way, the workload of each service provider is balanced. The restriction in Equation (2) determines the measure of activity of each territory. The inequality of Equation (3) establishes the territory with the greatest measure of activity, while Equation (4) has the least activity. In the restriction of Equation (5), it is sought that the number of fictitious centers installed equals the number of territories; on the other hand, Equation (6) forces all districts to be assigned to a single center. The inequality of Equation (7) ensures contiguity between districts. The restriction of Equation (8) states if a district is a center of territory within the territory. Finally, in Equations (9) and (10), the nature of the variables is defined.

3.2. Model 2

The use of this model is to determine the optimal grouping of the parks of a municipality of the Monterrey Metropolitan Area in considering the criteria of planning and the frequency of cleaning and collection of the service. The objective of model 2 is to minimize the total distance of each basic area (park) to its respective center. The idea of the objective function is to obtain figures as compact as possible in the distance direction.

The following assumptions are considered: each basic area (park) must be assigned to a single territory; the number of territories is known and is determined based on the frequency of the provision of cleaning and collection services, in this case, considered 4 weeks; the measure of activity is determined by the number of square meters that makes up each of the parks since it is directly related to the amount of waste generated; the relative tolerance for capacity was considered as a percentage between 5 % and 10 % (because it is the tolerance of a vehicle's permitted surplus without risking damage to the vehicle and the allowable slack so that use is optimal); and the distances from park to park consider the meaning of the streets.

The sets considered in this model are the same as in model 1, with the difference that, in this case, the basic areas are parks. The parameters are also similar, but in model 2, we consider d_{ij} as the distance between the basic area i and the center of territory j and discard the adjacency matrix. Binary decision variables are defined as $x_j = 1$ if the basic area j is a center, $x_j = 0$ otherwise, $y_{ij} = 1$ if the basic area i is assigned to the center j, and $y_{ij} = 0$ otherwise.

Model 2

$$\min \quad \sum_i \sum_j d_{ij} y_{ij} \quad (11)$$

$$\text{s.t:} \quad \sum_j x_j = p \quad (12)$$

$$x_j \geq y_{ij} \quad \forall i \in I, j \in J \quad (13)$$

$$\sum_{j \in J} y_{ij} = 1 \quad \forall i \in I \quad (14)$$

$$(1+\tau)x_j \mu \leq \sum_{i \in I} w_i y_{ij} \leq (1-\tau)x_j \mu \quad \forall j \in J \quad (15)$$

$$x_j, y_{ij} \in \{0,1\} \quad \forall i \in I, j \in J \quad (16)$$

The objective function in Equation (11) minimizes the total distance of each basic area to its respective center. The idea of the objective function is to obtain figures as compact as possible in the distance direction. In the restriction of Equation (12), it is sought that the number of fictitious centers installed is equal to the number of territories; on the other hand, Equation (13) prohibits the assignment of a basic area to a non-installed center. Equation (14) forces all the basic areas to be assigned to a single center. The restriction in Equation (15) ensures that a load of each territory is within the lower and upper capacities established. Finally, Equation (16) defines the nature of the variables.

3.3. Other Considerations

In the literature, compliance with all planning criteria simultaneously by exact methods is difficult, so it has resorted to using heuristic and metaheuristic methods. This situation is mainly due to conflicting criteria of balance and contiguity, objectives that are in competition, since the improvement in one of them can cause the deterioration of the other.

In this case, for model 1, the territorial design planning criteria were fulfilled entirely due to the objective function that it has; however, in model 2, it did not meet the criterion of contiguity and compactness. Therefore, we used a methodology to establish this criterion; to apply it, we need the following data: the matrix of distances between the parks, the map with results obtained from GAMS (General Algebraic Modeling System) with CPLEX Optimizer. of model 2, an adjacency matrix of the parks of a territory, and the GAMS results. Table 2 describes the general steps of the proposed algorithm.

Table 2. Algorithm of the solution method for the criterion of contiguity.

Algorithm: Allocation of Non-Contiguous Parks
Input: Matrix of adjacency of the parks in each formed territory, where an element takes the value of 1 if the parks are adjacent and 0 otherwise; Adjacency matrix of all district parks; Matrix of distances between each park; Table of results of model 2 containing the measure of activity of each territory. *Output:* Contiguous parks.

1. **While** Criterion of contiguity is not met **do**
2. Identify a territory that contains at least one non-contiguous park.
3. Select a non-contiguous territory.
4. Search the adjacent parks of the park selected from the previous step.
5. **for** Adjacent parks **do**
    Comparison between distances from the parks to the center of territory.
   end for
6. Select the park with the shortest distance to the center of territory.
7. To realize interchange between the park with smaller distance to the center of territory and the non-contiguous park.
8. Calculate the activity measure of the two territory formed.
9. Evaluate criterion of contiguity.
   end while
10. **return** Compliance with the criterion of contiguity of the parks and new territories formed.

It is important to emphasize that, as a consequence of the newly formed territories, the criterion of contiguity and compactness is fulfilled; however, the criterion of balance is violated. It is important to emphasize that, in the literature, the property of contiguity is a priority spatial criterion, so that both solutions will be proposed to the service managers by letting them know that both proposals are in support of their decision-making; the zoning given by model 2 would show a balance of work whereas the proposal by the heuristic shows imbalance but contiguity.

4. Results And Analysis

4.1. Results of Model 1

For both model 1 and model 2, programming tests were performed in GAMS, which is the software used as the basis for solving these models characterized with real data obtained. GAMS was used with CPLEX version 12 and was run in a computer Intel Xeon E5-2697v2 2.7 GHz with 12 cores each, a RAM of 64 Gb, and a hard disk of 1 Tb.

For the proposed model 1, some tests were performed, considering as basic areas the 24 districts that make up the municipality of San Nicolas and considering the size of parks in square meters. Eight territories were established due to the number of service providers that count the municipality, and an adjacency matrix of the districts was constructed.

Figure 3 shows the corresponding groupings of model 1; each color indicates a territory, and each territory is assigned to a service provider, for example, the territory formed by the districts 1, 2, and 17 corresponds to provider A, the territory formed by districts 4–6 correspond to provider B, and so forth.

Table 3 shows the allocation of districts to each provider. A notable aspect is to assign the lowest capacity provider to the smallest activity measure and the highest capacity provider to the highest activity measure.

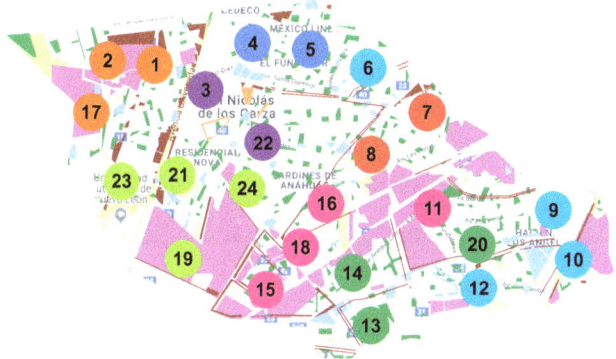

Figure 3. Proposed territory design for providers.

Table 3. Allocation of districts for each provider.

Districts	Provider	Parks	Activity Measure m^2
1, 2, 17	Provider A	21	149,202.59
4, 5, 6	Provider B	48	208,631.77
3, 22	Provider C	22	149,717.72
19, 21, 23, 24	Provider D	36	187,986
11, 15, 16, 18	Provider E	32	174,585.03
7, 8	Provider F	29	195 897.63
13, 14, 20	Provider G	52	183,923.69
9, 10, 12	Provider H	25	185,145.96

From the obtained results, the following aspects can be emphasized: in the clusters obtained, it is observed in Figure 3 that the districts are contiguous which facilitates the work to the provider in each district because their area of work would be delimited; if perfectly balanced territories were achieved, the average activity measure that would correspond to each service provider would be 179, 386.29 m^2. However, because perfectly balanced territories cannot normally be achieved, the idea of the objective function is to create territories that are almost perfectly balanced, minimizing the difference between the values of the territory with the highest activity (provider B) and the least activity (provider A); in this case, a minimum value of 59,426.18 m^2 was obtained. The main purpose of creating balanced territories is to allocate a balanced workload to each supplier because the current imbalance in workloads causes some parks to be unprocessed by provider; it is shown that not necessarily the supplier with more parks has the greater measure of activity and also is not fulfilled in a contrary way. This is because it influences the size of the parks in each district. Finally, it is very important to highlight the coverage of the service, since with this design, all the parks will be assigned to a provider which guarantees the cleanup of all the parks that make up the municipality of San Nicolas de los Garza.

4.2. Results of Model 2

Once the previous design has been obtained for each supplier, it is now necessary to carry out a second grouping because one of the most important factors of cleaning and solid waste collection in parks is the frequency; in this case, it is considered that the cleaning of each of the parks must be held every month, i.e., every 4 weeks.

There were 8 runs of model 2, one for each provider, considering the size in m² of each park and the distances between each one in meters with a 10% tolerance. The graphic solution for provider D is displayed in Figure 4, where each color represents a territory, so that the parks that make up the blue territory will be weeded on week 1; the parks that form the pink territory will be weeded on week 2; the red territory will be weeded on week 3; and finally, the yellow territory will be weeded on week 4. Table 4 shows the weeks and the number of parks to be weeded in each of them, and the weekly workload (activity measure) is shown.

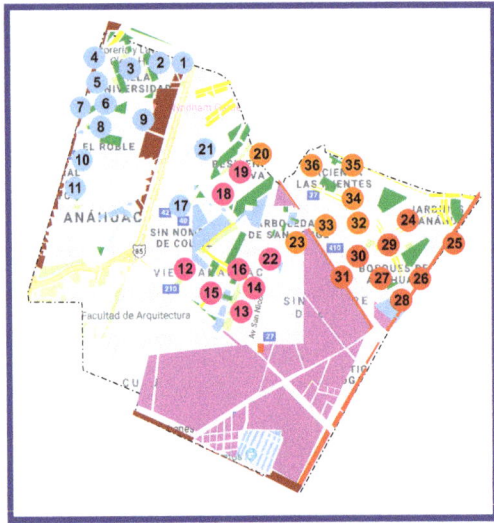

Figure 4. Design for districts 19, 21, 23, and 24.

Table 4. Planning for park maintenance for provider D.

Weeks	Parks	Activity Measure m²
Week 1	1–11, 17, 21	53,658.57
Week 2	12–16, 18, 19, 22	52,728.17
Week 3	20, 23, 32–36	52,708.06
Week 4	24–31	46,947.11

As an analysis of the previous results, we have the clusters obtained; Figure 4 shows that the parks are contiguous or connected which facilitates the work for the maintenance team, because their work area would be delimited. The workload of each week is balanced; see Table 4. This aligns with the idea that the cleaning and collection takes place in all parks and that the work is equitable every week for the staff; the coverage of the service is fulfilled, since with this design, all the parks will be weeded once a month which guarantees the maintenance of all the parks that make up the municipality of San Nicolas de los Garza. The minimum distance traveled from the parks to the center of territory (central park of the territory) is 28,150 m.

On the other hand, as mentioned previously, 8 runs were made for model 2 because the municipality has 8 providers. However, in the results that were obtained, there were cases where the criterion of contiguity is not met; for this purpose the methodology is proposed. Figure 5 shows the solution of model 2 for provider F. Also Table 5 shows the weeks and the number of parks to be weeded in each of them and the workload every week.

Figure 5. Design for districts 7 and 8—Solution 1.

Table 5. Planning for park maintenance for provider F.

Week	Parks	Activity Measure m^2
Week 1	2–5, 7, 8	52,387
Week 2	1, 6, 9, 10–16	48,546
Week 3	11, 17–23, 25	50,697
Week 4	24, 26–29	44,265

4.3. Results Algorithm: Allocation of Non-Contiguous Parks

The proposed algorithm consists of making exchanges to obtain territories where all the parks are contiguous and are next described graphically below.

Application of step 3 from Table 2, identify the non-contiguous parks: 1, 11, and 24 (see Figure 6). After selecting one of the three, changes can be done; park 24 will be selected. The next step (4) is to identify the adjacent parks: 25, 23, and 13 (see Figure 7). In order to determine the parks to be switched, we select the one closest (park 25) to the center of territory (park 27) (see Figure 8).

Figure 6. Application of step 3.

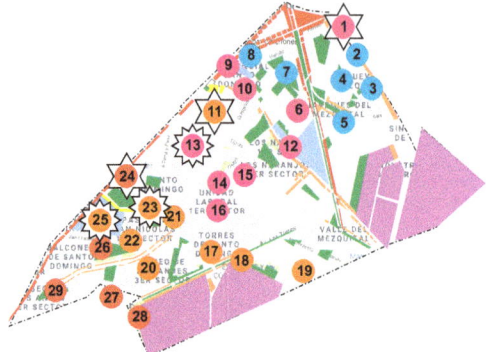

Figure 7. Application of step 4.

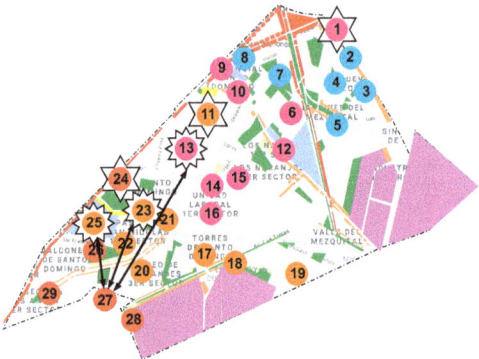

Figure 8. Application of step 5.

Similarly, exchanges are done for parks 11 and 1. Once all exchanges have been done, the activity measure is adjusted for each territory. Figure 9 shows the second solution obtained with the proposed methodology. It is important to mention that, in this new configuration, the criterion of contiguity is met but the balance criterion is unbalanced, since in each week, as shown in Figure 9, the workload is very variable, which leaves the final decision of the choice of solution to service providers (see Table 6).

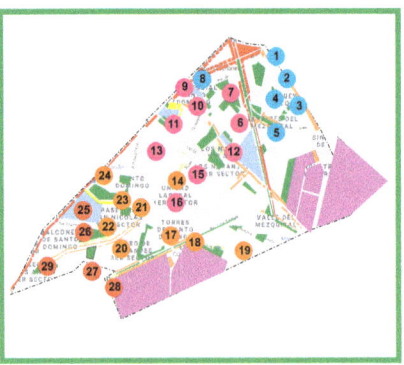

Figure 9. Design for districts 7 and 8—Solution 2.

Table 6. Planning for park maintenance for provider F, solution 2.

Week	Parks	Activity Measure m^2
Week 1	1–5, 8	43,580
Week 2	6, 7, 9, 10–13, 15, 16	56,877
Week 3	14, 17–24	48,893
Week 4	25–29	46,546

5. Conclusions

In this research, we obtain the conceptualization based on a mathematical model of territory design for the cleaning and solid waste collection service in San Nicolas de los Garza public parks, this being the main contribution, where the factors were considered important for this service, such as the amount of waste collected and the frequency of service.

As an important conclusion, the results obtained using the mathematical model proposed are satisfactory in the computational part and also offer an adequate design of territories fulfilling the criteria of planning of contiguity, balance, compactness, and spatial integrity. On the other hand, a remarkable aspect of the results is the coverage of the service, which ensures an adequate level of service, where all the parks of the municipality of San Nicolas de los Garza will be weeded once a month to generate a greater control in the allocation of parks to providers considering the payload.

The direct application of the mathematical model for the operation of public services would be determined by the service managers; however, among the benefits that would be obtained, there is greater efficiency in the use of transport equipment and service coverage.

6. Future Work

In the municipality of San Nicolas, each provider has a certain number of crews and a time available for the weeding and collection of each park; in addition, the amount of playground equipment and benches in each park differs as well as the travel times. The different capacities of provider are elements that influence this service; therefore, it would be interesting to give service providers a new proposal with these elements.

Author Contributions: Conceptualization and mathematical formulation, J.A.S.M.; Experimentation and analysis, M.d.R.A.V.; Funding acquisition, language review, model validation, A.M.; Methodology and Writing—original draft J.A.S.M. and M.d.R.A.V.

Funding: This research was funded by Becas Nacionales CONACYT grant number 416894.

Acknowledgments: Universidad Autonoma de Nuevo Leon and Universidad Panamericana

Conflicts of Interest: The authors declare that there are no conflict of interest.

References

1. Robusté, F.; Campos, J.M.; Galván, D. Nace la Logistica Urbana. In Proceedings of the Actas del IV Congreso de Ingeniería del Transporte, Valencia, Spain, 7–9 June 2000; Volume 2, pp. 683–691.
2. Taniguchi, E.; Thompson, R.G.; Yamada, T. Recent Trends and Innovations in Modelling City Logistics. *Procedia Soc. Behav. Sci.* **2014**, *125*, 4–14. [CrossRef]
3. Winkenbach, M.; Kleindorfer, P.R.; Spinler, S. Enabling Urban Logistics Services at La Poste through Multi-Echelon Location-Routing. *Transp. Sci.* **2015**, *50*, 1–21.
4. Macário, R.; Galelo, A.; Martins, P.M. Business Models in Urban Logistics. *Ing. Desarro.* **2008**, *24*, 77–96.
5. Villamizar, A.M.; Torres, J.M.; Padilla, N.H. Mathematical Programming Modeling and Resolution of the Location-Routing Problem in Urban Logistics1. *Ing. Univ.* **2014**, *18*, 271–289.
6. Álvarez, J.; Eslava, A. La logística urbana, la ciudad logística y el ordenamiento territorial logístico. *Reto* **2016**, *4*, 21–40.
7. Dotoli, M.; Epicoco, N. A Vehicle Routing Technique for Hazardous Waste Collection. *IFAC Pap. Online* **2017**, *50*, 9694–9699. [CrossRef]

8. Buhrkal, K.; Larsen, A.; Ropke, S. The Waste Collection Vehicle Routing Problem with Time Windows in a City Logistics Context. *Procedia Soc. Behav. Sci.* **2012**, *39*, 241–254. [CrossRef]
9. INEGI. *Banco de Indicadores*; Technical Report; Instituto Nacional de Estadistica y Geografia: Mexico City, Mexico, 2015. Available online: http://mapserver.inegi.org.mx/ambiental/map/indexV3FFM.html (accessed on 1 October 2019).
10. Arriaga, E.P.; Morenoz, J.R.; Caro, G.V. Los sistemas de recolección de residuos sólidos (Los métodos y sus aplicaciones). *CienciaUAT* **2007**, *1*, 58–60.
11. Galicia, F.G. Análisis del Sistema de Recolección de Residuos Sólidos Urbanos en el Centro Histórico de Morelia, aplicando Sistemas de Información Geográfica (SIG). Master's Thesis, Universidad Nacional Autónoma de México, Mexico City, Mexico, 2008.
12. SEDESOL. *Manual Para el Diseño de Rutas de Recolección de Residuos Sólidos Municipales*; Secretaría de Desarrollo Social. 1997. Available online: http://www.inapam.gob.mx/work/models/SEDESOL/Resource/1592/1/images/ManualTecnicosobreGeneracionRecoleccion.pdf (accessed on 1 October 2019).
13. Mercado, L. *Logística Inversa*; Technical Report; Revista Industria al dia; Universidad de Cordoba: Cordoba, Colombia, 2005. Available online: http://www.unicordoba.edu.co/revistas/vieja_industrialaldia/documentos/ed.1/logistica_inversa.pdf (accessed on 1 October 2019).
14. Purkayastha, D.; Majumder, M.; Chakrabarti, S. Collection and recycle bin location-allocation problem in solid waste management: A review. *Pollution* **2015**, *1*, 175–191.
15. Group, W.B. *What a Waste 2.0 A Global Snapshot of Solid Waste Management to 2050*; Technical report; World Bank: Washington, DC, USA, 2018.
16. Uriarte-Miranda, M.L.; Caballero-Morales, S.O.; Martinez-Flores, J.L.; Cano-Olivos, P.; Akulova, A.A. Reverse Logistic Strategy for the Management of Tire Waste in Mexico and Russia: Review and Conceptual Model. *Sustainability* **2018**, *10*, 3398. [CrossRef]
17. Hage, O.; Soderholm, P.; Berglund, C. Norms and economic motivation in household recycling: Empirical evidence from Sweden. *Resour. Conserv. Recycl.* **2015**, *53*, 155–165. [CrossRef]
18. Castillo, L.D.R.; Gallardo Izquierdo, A.; Pinero Guilamany, A. La distancia del domicilio al contenedor como un factor influyente en la frecuencia de separación de residuos urbanos. In *Hacia la sustentabilidad: Los Residuos Solidos Como Fuente de Energía y Materia Prima*; Red iberoamericana en gestion y aprovechamiento de residuos (REDISA). 2011; pp. 209–213. Available online: http://www.redisa.net/doc/artSim2011/GestionYPoliticaAmbiental/La%20distancia%20del%20domicilio%20al%20contenedor%20como%20un%20factor%20influyente%20en%20la%20frecuencia%20de%20separaci%C3%B3n%20de%20residuos%20urbanos.pdf (accessed on 1 October 2019).
19. Chang, N.B.; Wei, Y.L. Strategic Planning of Recycling Drop-Off Stations and Collection Network by Multiobjective Programming. *Environ. Manag.* **1999**, *24*, 247–263. [CrossRef]
20. Gautam, A.K.; Kumar, S. Strategic planning of recycling options by multi-objective programming in a GIS environment. *Clean Technol. Environ. Policy* **2005**, *7*, 306–316. [CrossRef]
21. Alonso, A.B. Localizacion optima de contenedores de residuos solidos urbanos en Alcalá de Henares. *M+A Rev. Electrón. Medioambiente* **2016**, *17*, 1–23.
22. Simon, S.; Demaldé, J.; Hernandez, J.; Carnero, M. Optimización de Recorridos para la Recolección de Residuos Infecciosos. *Inf. Tecnol.* **2012**, *23*, 125–132. [CrossRef]
23. Bing, X.; de Keizer, M.; Bloemhof-Ruwaard, J.M.; van der Vorst, J.G. Vehicle routing for the eco-efficient collection of household plastic wastes. *Waste Manag.* **2014**, *34*, 719–729. [CrossRef] [PubMed]
24. Rodríguez, A.A.; Butrón, E.G. Asignación de rutas de vehículos para un sistema de recolección de residuos sólidos en la acera. *Rev. Ing.* **2001**, *13*, 5–11. [CrossRef]
25. Aguilar, J.A.A.; Zambrano, M.E.J. Mejora del servicio de recolección de residuos sólidos urbanos empleando herramientas SIG: Un caso de estudio. *Ingeniería* **2015**, *19*, 118–1128.
26. Pastor, J.M. Optimizacion de la localizacion y recogida de residuos sólidos urbanos (RSU). Master's Thesis, Universidad Complutense de Madrid, Madrid, Spain, 2013.
27. Regidor, M.P.M.; de Lacalle, J.G.L. Estado del arte en procesos de zonificación. *Rev. GeoFocus* **2009**, *11*, 155–181.
28. Zoltners, A.; Sinha, P. Sales territory alignment: A review and model. *Manag. Sci.* **1983**, *29*, 1237–1256. [CrossRef]

29. Ríos-Mercado, R.Z.; Fernández, E. A reactive GRASP for a commercial territory design problem with multiple balancing requirements. *Comput. Oper. Res.* **2009**, *36*, 755–776. [CrossRef]
30. Armstrong, M.; Honey, R. A spatial decision support system for school redistricting. *URISA J.* **1993**, *5*, 40–52.
31. Muyldermans, L.; Cattrysse, D.; Oudheusden, D.V.; Lotan, T. Districting for salt spreading operations. *Eur. J. Oper. Res.* **2002**, *139*, 521–532. [CrossRef]
32. Tiede, D.; Strobl, J. Polygon-based regionalisation in a GIS environment. In Proceedings of the Trends in Knowledge-Based Landscape Modeling, Dessau, Germany, 18–20 May 2006; pp. 54–59.
33. Kalcsics, J.; Nickel, S.; Schröder, M. *A Generic Geometric Approach to Territory Design and Districting*; Berichte des Fraunhofer ITWM; Fraunhofer-Institut fur Techno: Kaiserslautern, Germany, 2009; pp. 1–32.
34. Benabdallah, S.; Wright, J.R. Multiple subregion allocation models. *Asce J. Urban Plan. Dev.* **1992**, *118*, 24–40. [CrossRef]
35. García, E.R.; Andrade, M.A.G. Compacidad en celdas aplicada al diseño de zonas electorales. *Supplement* **2008**, *5*, 73–95.
36. Díaz, A.; Bernábe, B.; Luna, D. Relación lagrangeana para el problema de particionamiento de áreas geográficas. *Rev. Mat. Teor. Apl.* **2012**, *19*, 169–181.
37. Phuntsho, S.; Dulal, I.; Yangden, D.; Tenzin, U.; Herat, S.; Shon, H.; Vigneswaran, S. Studying municipal solid waste generation and composition in the urban areas of Bhutan. *Waste Manag. Res.* **2010**, *28*, 545–551. [CrossRef] [PubMed]
38. Jayakody, K.P.K.; Jayakody, L.L.; Karunarathna, A.K.; Basnayake, B.F. *Municipal Solid Waste Management System and Solid Waste Characterization at Hikkaduwa Secretariat*; Technical Report; Department of Agricultural Engineering, University of Peradeniya: Peradeniya, Sri Lanka, 2008.
39. Kaosol, T. Sustainable Solutions for Municipal Solid Waste Management in Thailand. *Int. J. Environ. Ecol. Eng.* **2009**, *3*, 399–404.
40. Ayantoyinbo, B.B.; Adepoju, O.O. Analysis of Solid Waste Management Logistics and Its Attendant Challenges in Lagos Metropolis. *Logistics* **2018**, *2*, 11. [CrossRef]
41. Laureri, F.; Minciardi, R.; Robba, M. An algorithm for the optimal collection of wet waste. *Waste Manag.* **2010**, *48*, 56–63. [CrossRef]
42. Son, L.H.; Louati, A. Modeling municipal solid waste collection: A generalized vehicle routing model with multiple transfer stations, gather sites and inhomogeneous vehicles in time windows. *Waste Manag.* **2016**, *52*, 34–49. [CrossRef] [PubMed]
43. Das, S.; Bhattacharyya, B.K. Optimization of municipal solid waste collection and transportation routes. *Waste Manag.* **2015**, *43*, 9–18. [CrossRef] [PubMed]
44. Gobierno-Municipal, S. Plan de Desarrollo Urbano Sustentable 2013–2033 San Nicolás de Los Garza. 2013. Available online: http://transparencia.sanicolas.gob.mx/LTAINL/dic13/plan%20de%20desarrollo%20sustentable.pdf (accessed on 15 December 2016).
45. Calva, L.F.L.; de la Torre, R.; Reyes, A.G.; Chamussy, L.R.; Garcia, C.R.; Dominguez, F.V. El desarrollo humano de los municipios en México. Technical Report. 2015. Available online: http://www.mx.undp.org/content/mexico/es/home/library/poverty/informe-de-desarrollo-humano-municipal-2010-2015--transformando-.html (accessed on 1 October 2019).

© 2019 by the authors. Licensee MDPI, Basel, Switzerland. This article is an open access article distributed under the terms and conditions of the Creative Commons Attribution (CC BY) license (http://creativecommons.org/licenses/by/4.0/).

Article

Prices of Mexican Wholesale Electricity Market: An Application of Alpha-Stable Regression

Roman Rodriguez-Aguilar [1,*], Jose Antonio Marmolejo-Saucedo [2] and Brenda Retana-Blanco [3]

[1] Escuela de Ciencias Económicas y Empresariales, Universidad Panamericana, Augusto Rodin 498, Mexico, Mexico City 03920, Mexico
[2] Facultad de Ingeniería, Universidad Panamericana, Augusto Rodin 498, Mexico, Mexico City 03920, Mexico; jmarmolejo@up.edu.mx
[3] Facultad de Ingeniería, Universidad Anáhuac Mexico, Av. Universidad Anáhuac 46, Lomas Anáhuac, Huixquilucan, Estado de Mexico 52786, Mexico; brenda.retana@anahuac.mx
* Correspondence: rrodrigueza@up.edu.mx; Tel.: +52-5554-821-600

Received: 14 April 2019; Accepted: 27 May 2019; Published: 6 June 2019

Abstract: This paper presents a proposal to estimate prices in the Mexican Wholesale Electric Market, which began operations in February 2016, which is why it moves from a scheme with a single bidder to a competitive market. There are particularities in the case of the Mexican market, the main one being the gradual increase in the number of competitors observed until now and, on the other hand, the geographic and technical characteristics of the electric power generation. The observed prices to date show great fluctuations in the observed data due to diverse aspects; among the stems we can mention the own seasonality of the demand of electrical energy, the availability of fuel, the problems of congestion in the electrical network, as well as other risks such as natural hazards. For the above, it is relevant in a market context to have a price estimation as accurate as possible for the decision-making of supply and demand. This paper proposes a methodology for the generation of electricity price estimation through the application of stable alpha regressions, since the behavior of the electric market has shown the presence of heavy tails in its price distribution.

Keywords: electricity markets; alpha-stable distribution; alpha-stable regression; electricity prices

1. Introduction

As part of the recent reform in the Mexican electricity sector, the electricity market has been liberalized, culminating with the creation of the Wholesale Electricity Market. Before the reform, the activities of generation, transmission, distribution, and commercialization of electric power were exclusive of the State. A set of problems that affected the electricity sector were identified: (a) high generation costs; (b) low participation of clean energies in the energy matrix; (c) inadequate infrastructure; (d) transmission losses; and, (e) structural problems in the Federal Electricity Commission (State Company). The objective of the reform was to address these problems [1–3].

In a competitive market, it is necessary to have reliable information to be able to make market decisions. The recent reform in the electric sector in Mexico requires the generation of robust models to forecast the prices of electricity. Having reliable information on the electricity market will allow better decisions to be made by both the market operator and the generators and consumer participants [4].

This work presents an application of alpha-stable regression for estimation of forecasting of electricity prices in the Mexican market. There is a set of relevant proposals in the state of the art on electricity price estimation. However, the vast majority presents applications to consolidate electricity markets. The particularity of the Mexican market is that it is of recent creation and with low private participation, maintaining a practically monopolistic market structure [4]. This, together with other factors such as fuel prices, network congestion, as well as disruptive events, have generated high volatility in prices. One of the advantages of addressing the problem of estimating electricity prices through stable regressions lies in the ability to capture the cyclical behavior of demand determined by consumption peaks and, therefore, spikes in electricity prices. The modeling considering heavy tails in the data will allow capturing the behavior of the data with greater precision and taking into account the presence of peaks of demand and prices when making the expected estimates in a future horizon. The stable regression has been applied efficiently in other disciplines such as finance, machine learning, and soft computing [5–7]. Derived from the literature review, no alpha-stable regression applications were identified for the estimation of electricity prices.

The study presents the general framework of the wholesale electricity market in Mexico, as well as information regarding competitors and electricity prices. The usefulness of the use of heavy tailings distributions that allow capturing in a better way the presence of extreme values according to the behavior of the energy demand is addressed. By estimating prices using alpha-stable regressions, we seek to capture this information and generate efficient forecasts for the decision-making of market participants. Data for the Electricity Sector for Mexico of the period 2016–2017 were used. Historical information was considered from the first year of operation of the wholesale electric market for the realization of the model. The results show an efficient adjustment of the observed price trend, despite the fact that there are outliers in the distribution of prices, especially because of the recent implementation of the competitive market. This situation will be dynamic since competitors are entering the market in a constant manner, so there is an immediate impact of entry into the behavior of electricity prices.

The work is structured as follows: section one describes the Mexican Electricity Sector until 2013, before the reform. Subsequently, the main characteristics of the Wholesale Electricity Market are described, the structure of the wholesale electricity market, and the behavior of electricity prices in the first year of operation. In the second section, is analyzed the alpha-stable distributions and alpha-stable regressions framework. The third section presents the main results of the study. The final section addresses conclusions and recommendations.

2. The Mexican Electric Sector

The Mexican Electricity Sector operated with a parastatal company as a single supplier (Federal Electricity Commission (FEC)). Other generators could participate through self-generation schemes, independent producers, cogeneration, small producers, export centers, and continuous own uses with a total gross generation of 42,676 GWh (excluding independent producers) that represents 13% of the total gross generation in 2014 (302,806 GWh). The generation of independent producers was usually sold directly to FEC. By 2014, the total net electricity generation in Mexico was 302,806 GWh; the generation was concentrated in the use of fossil fuels. Of the total FEC and independent producers generation, 79% use fossil fuels and coal, 15% hydro, and the rest is clean energy. In 2011, the generators that only used fossil fuels were 73% of the total electricity generation [8]. The use of fossil fuels for power generation has been discussed due to the fact that production of CO_2 contributes to the accumulation of greenhouse gases. Therefore, there are two scenarios of capacity expansion for the period 2013–2027. The first scenario contemplates the expansion of the public power generation with a share of 32% with the use of clean technologies in 2027. The second scenario considers the expansion seeking to increase its generation with non-fossil sources to 35% in 2027 [9].

In reference to generation costs, the level cost of the different technologies for Mexico shows that turbo-gas technology with diesel and gas, and those of internal combustion, are the most expensive plants with a maximum of up to $316 per MW (diesel) (Figure 1). The levelized cost of generation is an essential factor in the determination of energy prices since it is necessary to build an energy matrix focused on reducing emissions and an electricity market that seeks to reduce energy prices.

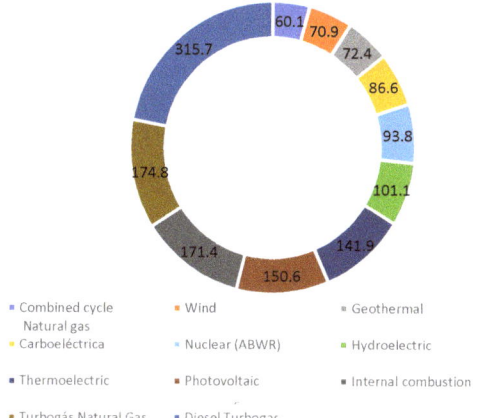

Figure 1. Level cost (dollars per net MWh), 2014.

2.1. Mexican Wholesale Electricity Market

The Wholesale Electric Market (WEM) was implemented in Mexico as a market based on variable costs, whose operation began in February 2016. Currently, there are 62 participants in the market—nine of them are from the Federal Electricity Commission and 53 are from private generators. Of the 62 market participants, 38 are generators, 15 are suppliers of qualified services, seven non-supplier marketers, an intermediation generator, and a basic services provider (Figure 2).

Figure 2. Market participants, 2015–2018.

For more details of the market participants, consult the Wholesale Electricity Market Bases published by the Ministry of Energy [10]. The WEM is integrated by a set of market instruments to operate efficiently.

- Short-term Energy Market
- Power Balance
- Clean Energy Certificates Market
- Financial Transmission Rights Market
- Medium and long-term auctions

2.2. The Locational Marginal Prices

Locational marginal pricing (LMP) reflects the value of the energy at the specific location and time it is delivered. Congestion raises the LMP in the receiving area of the congestion. Operating conditions that limit the delivery capacity of specific transmission lines also can contribute to congestion and result in LMP changes. Locational marginal prices are calculated by the market operator and published on the web. This enables market participants to factor the information into their decision-making. The calculations used to determine LMPs take into account electricity demand, generation costs, and the use of, and limits on, the transmission system. The price tells market participants the cost to serve the next megawatt of load at a specific location. LMPs give price signals that encourage new generation sources to locate in areas where they will receive higher prices [11].

LMP includes: (a) cost of the energy in the delivery node; (b) variable cost of energy generation; (c) cost per congestion; and, (d) cost of the losses. The LMP is the result of an optimization model performed by the market operator, of allocation of units in a day in advance. The model seeks to minimize generation costs subject to the generation capacity of all units and their costs, as well as to satisfy the required demand, taking into account the corresponding required reserves and the restrictions of the transmission network. The algorithm for the dispatch economic system will calculate the marginal price of energy in each node, on a time basis, and it will have three components: marginal energy component, marginal congestion component, and marginal losses component [12].

$$LMP_i = EC_i + LC_i + CC_i \qquad (1)$$

where:

EC_i corresponds to the energy component of node i
LC_i corresponds to the loss component of node i
CC_i corresponds to the congestion component of node i

The WEM started operations in February 2016; in the first year of operation, prices show volatility. The main reasons for the volatility of LMP may be due to climatic aspects, fuel availability, congestion, as well as network failures. In Mexico, there are three systems, the National Interconnected System, Baja California Sur, and Baja California Norte. In the present study, only the National Interconnected System will be addressed. The evolution of the LMP in what is carried out by the WEM in its first year of operation shows an average value of $2,144 with a standard deviation of $700 (MXN) (Figure 3).

Figure 3. Locational marginal pricing (LMP) in the National Interconnected System (NIS).

The prices in the National Interconnected System (NIS) shows high volatility, as can be seen in the dotted line. In general, the high volatility can be attributed to the fact that it is a newly created market with low private participation in the generation and with variations attributable to losses in transmission, congestion in the network, and the cost of fuels (Figure 3).

There is a correlation between energy consumption and the price of energy, so there are peaks of demand according to the time of day and the day of the week. As expected, the weeks with the highest energy demand coincide with higher prices. The highest prices observed were in weeks 25 and 42 (Figure 4).

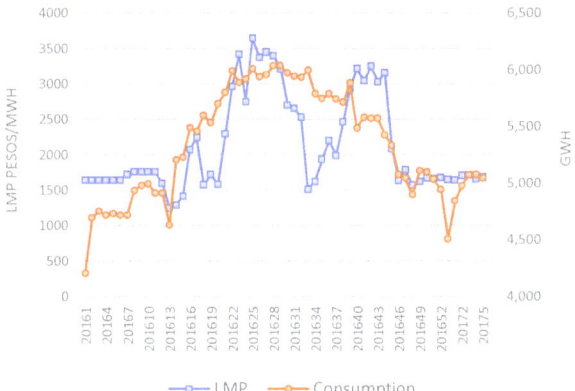

Figure 4. LMP vs. consumption by a week in the NIS, January 2016–January 2017.

The relationship between the peaks of demand and the determination of the LMP is a fundamental factor to consider for estimating electricity prices, since many methods may not correctly adjust these peaks of demand that, in turn, reflected in peaks in prices (Figure 4). The objective of the price forecasting model is that, based on the consumption and the expected demand, the price of energy can be estimated taking into account the prevalence of peaks in the prices, for which the use of alpha-stable regression is proposed.

3. Alpha-Stable Distributions and Stable Regressions

3.1. α-Stable Distributions

The Gaussian distribution has wide applications in different areas but there are variables that cannot be modeled efficiently with this distribution because of a greater degree of impulsivity, that is, events or events are possible that, described by a Gaussian distribution, would be considered as very unlikely. The α-stable distribution has been used in the literature to describe this type of phenomena. The stable distribution has been applied in many areas with efficient results.

In this section, the α-stable distribution and its main properties are described, noting that the α-stable distribution complies with the Central Limit Theorem and the stability property. On the other hand, it will also denote the complexity of working with this type of distribution, since it lacks general analytical expression. As a consequence, this distribution is not very widespread among the academic community. In spite of this, this difficulty can be overcome by making use of the property of the symmetric α-stable distribution that allows expressing a distribution of this type as a Gaussian conditional to a random variable with the α-stable distribution.

Definition. X has an α-stable distribution if it has the characteristic function [13]:

$$\varphi(w) = \begin{cases} \exp\{|\gamma w|[1 + i\, sign(w)\beta\frac{2}{\pi}\log(|w|)] + i\mu w\}, (\alpha = 1) \\ \exp\{-|\gamma w|^{\alpha}[1 - i\, sign(w)\beta \tan\left(\frac{\pi\alpha}{2}\right)] + i\mu w\}, (\alpha \neq 1) \end{cases} \quad (2)$$

$\alpha \in (0, 2]$ is the characteristic parameter; this parameter controls impulsivity of the variable X, $\beta \in [-1, +1]$ is a parameter of symmetry, $\gamma > 0$ is a scaling parameter, and μ is the position parameter. Derived from the above and by convention, the α-stable distributions will be denoted as a function of its four parameters using the following notation.

$$f(\alpha, \beta x)(.|\gamma, \mu) \quad (3)$$

3.2. Alpha-Stable Regression

Consider the standard regression model

$$y_i = \sum_{j=1}^{k} x_{ij}\theta_j + \varepsilon_i, \quad i = 1, \ldots, N \quad (4)$$

where y_i is an observed dependent variable, the x_{ij} are observed independent variables, are unknown coefficients to be estimated, and are identically and independently distributed. The standard Ordinary Least Squares (OLS) estimator in matrix form:

$$\widehat{\beta_{OLS}} = (X'X)^{-1}X'y \quad (5)$$

It can be expressed as

$$\widehat{\beta_{OLS}} - \beta = (X'X)^{-1}X'\varepsilon \quad (6)$$

Thus, in the simplest case where X is predetermined, $\widehat{\beta_{OLS}} - \beta$ is a linear sum of the elements of ε. With ε elements, i.i.d. (independent identically distributed) non-normal stable variables, and then $\widehat{\beta_{OLS}}$ has a stable distribution. The variance of ε_i cannot be estimated. Then, standard Ordinary Least Squares (OLS) inferences are invalid [14,15], proving the following properties of the asymptotic t-statistic.

a. The tails of the distribution function are normal-like at ±∞.
b. The density has infinite singularities $|1 \mp x|^{-\alpha}$ at ±1 for $0 < \alpha < 1$ and $\beta \neq \pm 1$, when $1 < \alpha < 2$ the distribution has peaks at ±1.
c. As $\alpha \to 2$, the density tends to normal and the peaks vanish

Further, the estimation of the parameters at stable distribution have the usual asymptotic properties of a Maximum Likelihood estimator [16], are asymptotically normal, asymptotically unbiased, and have an asymptotic covariance matrix $n^{-1}I(\alpha,\beta,\gamma,\delta)^{-1}$, where $I(\alpha, \beta, \gamma, \delta)$ is Fisher's Information. Assume that $\varepsilon_i = y_i - \sum_{j=1}^{k} x_{ij}\theta_j$ is alpha-stable with parameters $\{\alpha, \beta, \gamma, 0\}$. Let it be alpha-stable density function $s(x, \alpha, \beta, \gamma, \delta)$, then the density function of ε_i is [17]:

$$s(\varepsilon_i, \alpha, \beta, \gamma, \delta) = \frac{1}{\gamma} s\left(\frac{y_i - \sum_{j=1}^{k} x_{ij}\theta_j}{\gamma}, \beta, 1, 0 \right) \quad (7)$$

the likelihood function,

$$L(\varepsilon, \alpha, \beta, \gamma, \theta_1, \theta_2, \ldots) = \left(\frac{1}{\gamma}\right)^n \prod_{i=1}^{n} s\left(\frac{y_i - \sum_{j=1}^{k} x_{ij}\theta_j}{\gamma}, \beta, 1, 0 \right) \quad (8)$$

and the loglikelihood,

$$l(\varepsilon, \alpha, \beta, \gamma, \theta_1, \theta_2, \ldots) = \sum_{i=1}^{n} -n\log(\gamma) + \log\left(s\left(\frac{y_i - \sum_{j=1}^{k} x_{ij}\theta_j}{\gamma}, \beta, 1, 0\right)\right)$$
$$= \sum_{i=1}^{n} \psi(\hat{\varepsilon}_i) \quad (9)$$

The maximum likelihood estimators are the solutions of the equations:

$$\frac{\partial l}{\partial \theta_j} = \sum_{i=1}^{n} -\psi'(\hat{\varepsilon}_i)x_{ij} = 0, \quad j = 1, 2, \ldots, k \quad (10)$$

Thus,

$$\sum_{i=1}^{n} -\frac{\psi'(\tilde{\varepsilon}_i)}{\tilde{\varepsilon}_i} y_i x_{ij} = \sum_{i=1}^{n} -\frac{\psi'(\tilde{\varepsilon}_i)}{\tilde{\varepsilon}_i} \sum_{j=1}^{k} x_{ij} \theta_j \qquad (11)$$

If W is the diagonal matrix, we can write the model in the matrix format:

$$W = \begin{pmatrix} -\frac{\psi'(\tilde{\varepsilon}_1)}{\tilde{\varepsilon}_1} & 0 \cdots & 0 \\ 0 & -\frac{\psi'(\tilde{\varepsilon}_2)}{\tilde{\varepsilon}_2} \cdots & 0 \\ \vdots & \cdots & -\frac{\psi'(\tilde{\varepsilon}_n)}{\tilde{\varepsilon}_n} \end{pmatrix}$$

Then,

$$X'Wy = (X'WX)\hat{\theta} \qquad (12)$$

If $X'WX$ is not singular,

$$\hat{\theta} = (X'WX)^{-1} X'Wy \qquad (13)$$

This estimator has the format of a Generalized Least Squares estimator with heteroscedasticity, where the variance of the error term ε_i is proportional to $\frac{\psi'(\varepsilon_i)}{\varepsilon_i}$. The effect of the Generalized Least Squares adjustment is to give less weight to larger observations. The estimator for stable processes gives higher weights to the center of the distribution and smaller weights to extreme values. This effect increases as α are reduced [17]. This is consistent with what was obtained by [18].

4. Estimation of Locational Marginal Prices in National Interconnected System Using Alpha-Stable Regressions

The estimation of electricity prices has been a challenge for all markets. There is a set of methodologies used that consist of models tailored to the market that is studied. That is why, in the Mexican case, the use of alpha-stable regressions is proposed. According to the state of the art, there is a set of methodologies generally used for the estimation of energy price forecasts: multi-agent models, diffusion and Markov models, statistical models, and recently badass models in computational intelligence, as well as the development of hybrid models or ensemble modeling. Temporality is another characteristic aspect in the case of electricity prices since there are different horizons in the market. In addition, according to temporality, it is possible to apply different forecasting methodologies [19]. Figure 5 shows a classification of general methodologies applied in energy price models.

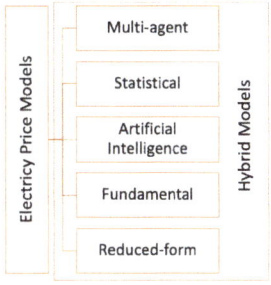

Figure 5. Electricity price modeling.

Highlights include works such as [20], where neural networks are applied for the realization of forecasts. Ref. [21] shows a general framework of statistical models for the forecast of prices of a day in advance using models ARIMA, ARMAX, GARCH and discusses the application of stochastic models for the derivation of prices as models of diffusion and Markov jumps. Ref. [22] shows the use of linear time series models and nonlinear ones. Another set of authors focuses on the use of price forecasts modeling the stochastic dynamics of prices, seeking to manage volatility risk and the valuation of derivatives, refs. [23–26]. Another approach is based on the modeling of market behavior through multiagent models such as [27–30] and [31]. However, the limitation of these methods is that they do not allow efficient forecasts to be generated for short periods of time (market in advance time and real-time market). Recently, the application of artificial intelligence models has increased, as in the case of [32–34] and [35], the application of neural network models, fuzzy neural networks are highlighted and committee machines. Refs. [36,37] perform a revision of machine-learning methods such as functional principal component analysis, support vector machine, and time series. On the other hand, ref. [38] highlights the use of time series for short-term forecasts such as ARIMA, seasonal ARIMA, and VAR models for price forecasts in one hour in advance and one day in advance.

A fundamental element in the elaboration of forecasts of electricity prices is the evaluation of its effectiveness; for this, a group of authors has dedicated to studying the efficiency of the forecasts proposing a set of generally accepted methods. It is the case of Mean Absolute Error and Mean Absolute Percentage Error, which are generally accepted methods. [39] performed a comparison of forecast performance measures highlighting the use of mean absolute scaled error. Other authors such as [40,41] use the absolute error or square scaled error. Another alternative is to use standardized error measurements as shown by [42–44], highlighting measures for short-term forecast such as daily or weekly weighted mean absolute errors and root square errors.

The present work proposes the use of alpha-stable regressions, given the presence of heavy tails in the errors, due to the impulsivity of the series. An advantage of using alpha-stable regressions in the presence of errors with heavy tails is that the inference made by OLS estimates before the presence of stable errors has no validity because of the case of infinity variance. In the case of alpha-stable regressions, both the parameters of the model and the parameters of the stable distribution of errors are estimated simultaneously, which makes it possible to generate adequate inference in such cases.

Estimation of Electricity Price in Mexico

In the case of the Mexican market that started operations in 2016, information was collected on average locational marginal prices for the National Interconnected System. The information corresponds to weekly data from January 2016 to January 2017, in addition to the real consumption per week and the weekly demand forecast estimated by the National Center for Energy Control (NCEC). The model considers the locational marginal price as a dependent variable and consumption as an independent variable. For the Mexican market, due to its recent start of operations, there is little information available as well as additional variables that could be considered in the model. However, consumption as such is a variable of great importance when determining the expected price of electricity. Based on the summary of the statistics shown in Table 1, the normality tests show that the data do not conform to a normal distribution. Similarly, the parameters of the estimated alpha-stable distribution are shown.

Table 1. Summary Statistics.

	LMP	Consumption
	(Mexican Pesos/MWh)	(GW/h)
Mean	2132.784	5340.727
St. dev	680.7264	486.5223
Skewness	0.8344004	−0.1500883
Kurtosis	2.218459	1.861669
	Goodness-of-Fit Test for Normal Distribution ($p < 0.05$)	
KS test	0.0176	0.0071
SW test	0.00000	0.00416

The series used show high variability depending on the peaks of electricity demand according to the time and day of the week. The normality tests show that the data do not adjust to a normal distribution as expected due to the presence of impulsivity in the series. The parameters of the alpha-stable distribution are estimated using the maximum likelihood method and the S_0 parameterization (Table 2) [13]. The adjustment of the parameters evidences impulsivity in the series, as well as positive asymmetry.

Table 2. Maximum Likelihood Parameters of Stable Distribution.

	LMP	Consumption
	(Mexican pesos/MWh)	(GW/h)
α	1.5340	1.7851
β	0.4121	0.2579
γ	477.179	341.044
δ	2132.70	5340.73
Goodness-of-Fit Test for alpha-stable distribution ($p < 0.05$)		
K-S test	0.82939	0.65749

Data generated by the electric power demand peaks adjust largely to the behavior of a stable distribution.

The model estimated is:

$$PML = \theta_0 + \theta_1 C + \varepsilon_i, \ i = 1, \ldots, N \tag{14}$$

The results of the alpha-stable regression are shown in Table 3. The parameters are estimated by two methods, the Maximum Likelihood method (stable model) and the Ordinary Least Squares. As expected, the parameters estimated by the alpha-stable regression have values lower than OLS, due to the penalty to larger observations, whose effect increases as a function of the value of the alpha parameter [45,46].

Table 3. Parameters of Stable Regression.

	θ_0	θ_1
Ordinary Least Squares	−3098.9843	0.9795
Maximum Likelihood Stable	−2604.1885	0.9510
Stable parameters of a stable residuals		
(Asymptotic 95% confidence intervals)		
α		1.34575 (1.19434, 2.49726)
β		0.09980 (−0.14193, 0.34152)
γ		9.56450 (8.68237, 10.44663)

Although the series used in the regression are close to the normal case, when calculating the errors, it is observed that they adjust to a stable distribution. Given the properties of the estimation of the stable regression by maximum likelihood, it is a robust method for errors with heavy tails and an efficient method with data that are not strictly stable distributions. Table 4 shows the Fisher information matrix that allows the estimation of the confidence intervals of the parameters of the model, as well as the parameters of the distribution of the errors. The dimension of the matrix is represented by (k + 3), the sum of the parameters in the regression and the coefficients of stable distribution (α, β, and γ).

Table 4. Fisher Information Matrix (k + 3) * (k + 3).

182.4849	−4.2016	−8.7818	−1.0338	−2201.844
−4.20169	79.6711	−0.5106	6.2927	13401.96
−8.7818	−0.5106	5.3781	−0.1778	−378.819
−1.0338	6.2927	−0.1778	2.8481	6065.802
−2201.844	13401.96	−378.819	6065.802	1.46E + 07

Figure 6 shows the relationship between an estimated dependent variable and an independent variable by a scatter plot, comparing the adjustment of prices with respect to consumption, showing how, in the case of the adjustment by OLS, it shows greater dispersion with respect to the observed value of the PML, this largely due to the errors that contain heavy tails. In the case of the stable model estimated by maximum likelihood, a better fit to the observed data of the LMP was observed. The alpha-stable adjustment (Maximum Likelihood Stable) is below the OLS values on average, since it weighs the values of the center of the distribution, largely and, to a lesser extent, the extreme values of the tails. This depends on the value of the alpha impulsivity parameter (Figure 6).

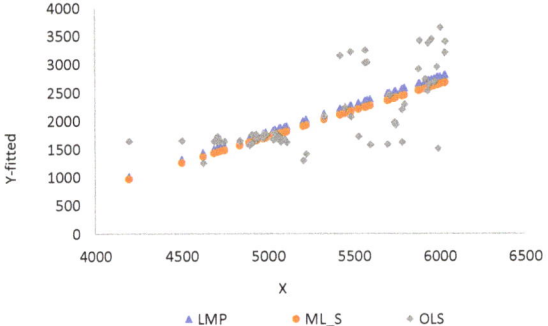

Figure 6. OLS and Stable regression.

The adjustment in two steps of the parameters of the model and the parameters of the stable distribution of the errors allows to control the impulsivity of the series and that the inference made with the estimators is robust and consistent. However, it should be noted that, in the case of the LMP for Mexico, the presence of a nonlinear behavior is observed; it is clear that the linear adjustment has limitations (Figure 7). So in another stage of the research, it is recommended to estimate other methodology as Generalized Additive Models (GAM) in combination with the stable adjustment to improve the performance in the presence of heavy tails in errors.

The estimations made by OLS and the alpha-stable regression show that the behavior of electricity prices is nonlinear, so the adjustment with both methodologies can be improved. However, in the case of OLS, it shows poor performance in the adjustment of the data when there are price peaks (peaks in demand) underestimating the values of LMP. In the case of alpha-stable regression, the adjustment to price peaks is better but has limitations in the nonlinear behavior of prices (Figure 7).

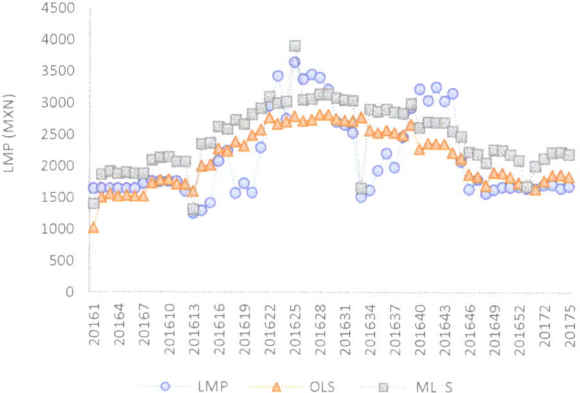

Figure 7. Estimations of Electricity Prices.

The residuals of the estimations made by the alpha-stable regression show that, on average, the stable model overestimates the LMP values, even though the adjustment to demand peaks is better. The adjustment of the stable regression (Maximum Likelihood Stable) shows a better performance in the extreme values; this is very low or very high prices of electricity (Figure 8). As inputs for making decisions about offers in the electricity market, it is necessary to have accurate estimates in this case; the Mean Absolute Percentage Error (MAPE) was estimated according to Equation (15).

$$MAPE = \frac{\sum_{t=1}^{N} \left| \frac{E_t}{Y_t} \right|}{N} \tag{15}$$

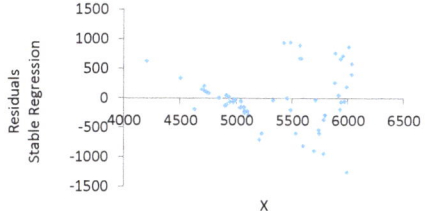

Figure 8. Stable regression residuals.

The performance of the estimations comparing both methods shows a MAPE highly influenced by the presence of extreme data that is high prices of the energy, not being the case of the stable regression. In the same way, this behavior is attributed to the nonlinear relationship observed in the variables considered in the model (Figure 9).

The MAPE for the stable estimate (Maximum Likelihood Stable) was 14% and for the OLS case of 25% (Figure 9). It is recommended to explore other adjustment methods to improve the performance of the estimates, such as the combination of the Stable Model and some nonlinear adjustment method such as Generalized Additive Models.

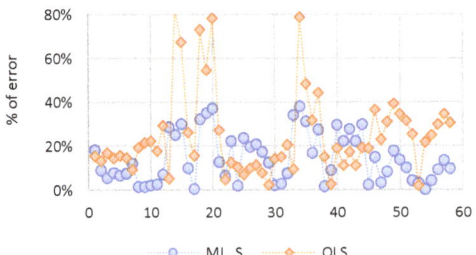

Figure 9. MAPE Stable Model and OLS.

5. Conclusions and Future Work

The start of operations of the WEM includes the generation of a set of market instruments that foster competition, thus creating the short-term market, the clean energy certificates market, the power market, and the financial transmission rights market. A fundamental element for its optimal development is access to reliable information in a timely manner. In this context, information on energy prices is an essential input for market operations; market participants must have robust estimates that allow them to make decisions for the supply and purchase of energy. The market structure has changed rapidly due to the incorporation of new suppliers in contrast to the previous scheme. The opening of the electricity sector seeks to encourage the entry of more competitors to allow competitive prices to be offered and to encourage the clean energies (through the clean energy certificate market). The behavior of prices observed in the first year of operations shows high volatility in prices, partly attributable to the behavior of fuel costs and the costs of losses.

Taking into consideration the above, this paper proposes a method of estimating energy prices based on alpha-stable regressions, considering the presence of impulsivity in the series of prices and electricity consumption, which result in errors with heavy tails. The main effect of estimating a linear model in these conditions lies in the limitation of the inference power of the model, since there is the possibility that the variance of the errors is infinite. The results of the estimations show the presence of impulsivity in the behavior of the price and consumption of electricity ($\alpha_{LMP} = 1.53$ and $\alpha_C = 1.78$), as expected given the nature of the consumption of electricity with peaks of demand according to the hour, day of the week, and month of consumption. The stable model fits better in the presence of extreme values; however, it does not capture the nonlinear behavior of the relationship between price and electricity consumption identified. MAPE shows a better performance of the stable model compared to OLS, 14%, and 25%, respectively.

Given the short time of operation of the Mexican wholesale electricity market, there is no open availability of more variables that could help improve the specification of the model. It is proposed as future work to integrate additional information to the model as well as to increase the frequency of the data at daily and hourly prices. Similarly, it is recommended to explore the integration of stable models with other nonlinear treatment methodologies. The present work is a first step towards the generation of technical evidence that allows the correct development of the Mexican electric market.

Author Contributions: Conceptualization, R.R.-A. and J.A.M.-S.; Data curation, J.A.M.-S. and B.R.-B.; Formal analysis, R.R.-A. and B.R.-B.; Methodology, R.R.-A. and J.A.M.-S.; Writing – original draft, B.R.-B.

Funding: This research received no external funding.

Conflicts of Interest: The authors declare no conflict of interest.

References

1. OECD; International Energy Agency. *Mexico Energy Outlook—World Energy Outlook Special Report*; International Energy Agency: Paris, France, 2016; Available online: https://www.iea.org/publications/freepublications/publication/MexicoEnergyOutlook.pdf (accessed on 30 June 2018).

2. OECD; International Energy Agency. *Energy Policies Beyond Countries*; International Energy Agency: Paris, France, 2017; Available online: https://www.iea.org/publications/freepublications/publication/EnergyPoliciesBeyondIEACountriesMexico2017.pdf (accessed on 1 May 2018).
3. Fundación Colosio. *Reforma Energética: Motor de Crecimiento Económico y Bienestar*; Fundación Colosio: Mexico City, Mexico, 2014; Available online: http://fundacioncolosio.mx/content/media/2015/02/1%20Reforma%20energ%C3%A9tica.pdf (accessed on 15 April 2018).
4. JATA. *Practical Handbook and Introduction to Mexico's Energy Sector*; JATA: San Pedro Garza Garcia, Mexico, 2016; Available online: http://jatabogados.com/publications/articles/JATA-Introduction_to_Mexicos_Energy_Sector.pdf (accessed on 14 Jun 2018).
5. Frain, J.C. *Maximum Likelihood Estimates of Regression Coefficients with α-Stable Residuals and Day of Week Effects in Total Returns on Equity Indices*; Working Paper; Department of Economics, Trinity College Dublin: Dublin, Ireland, 2006; Available online: https://www.tcd.ie/Economics/staff/frainj/Stable_Distribution/weekday.pdf (accessed on 20 October 2018).
6. Bassiou, N.; Kotropoulos, C.; Koliopoulou, E. Symmetric α-stable sparse linear regression for musical audio denoising. In Proceedings of the 8th International Symposium on Image and Signal Processing and Analysis (ISPA), Trieste, Italy, 4–6 September 2013; Available online: http://poseidon.csd.auth.gr/papers/PUBLISHED/CONFERENCE/pdf/2013/Bassiou13.pdf (accessed on 5 November 2018).
7. Koutrouvelis, I.A. Regression-type estimation of the parameters of stable laws. *JASA* **1980**, *75*, 918–928. [CrossRef]
8. Marmolejo-Saucedo, J.A.; Rodríguez-Aguilar, R.; Cedillo-Campos, M.G.; Salazar-Martínez, M.S. Technical efficiency of thermal power units through a stochastic frontier. *Dyna* **2015**, *82*, 63–68. [CrossRef]
9. Secretaría de Energía. *Prospectiva del Sector Eléctrico 2013–2027*; Mexican Ministry of Energy: Mexico City, Mexico, 2013. Available online: http://sener.gob.mx/res/PE_y_DT/pub/2013/Prospectiva_del_Sector_Electrico_2013-2027.pdf (accessed on 18 July 2018).
10. Secretaría de Energía. *Bases del Mercado Eléctrico*; Mexican Ministry of Energy: Mexico City, Mexico, 2018. Available online: https://www.cenace.gob.mx/Docs/MarcoRegulatorio/BasesMercado/Bases%20del%20Mercado%20El%C3%A9ctrico%20Acdo%20Sener%20DOF%202015%2009%2008.pdf (accessed on 22 August 2018).
11. Abirami, A.; Manikanda, T. Locational Marginal Pricing approach for a deregulated electricity market. *Int. Res. J. Eng. Technol.* **2015**, *2*, 348–354.
12. CENACE. *Mathematical Formulation of the Model of Assignment of Units with Security Restrictions and Calculation of Locational Marginal Prices and of Related Services in the Market of a Day in Advance*; Mexican Mnistry of Energy: Mexico City, Mexico, 2016. Available online: https://www.cenace.gob.mx/Docs/MercadoOperacion/Formulaci%C3%B3n%20Matem%C3%A1tica%20Modelo%20AU-MDA%20y%20PML%20v2016%20Enero.pdf (accessed on 5 May 2019).
13. Nolan, J.P. *Modeling Financial Data with Stable Distributions*; Department of Mathematics and Statistics, American University: Washington, DC, USA, 2005; Available online: http://academic2.american.edu/~jjpnolan/ (accessed on 3 April 2018).
14. Andrews, B.; Calder, M.; Davis, R. Maximum Likelihood Estimation for α-Stable Autoregressive Processes. *Ann. Stat.* **2009**, *37*, 1946–1982. [CrossRef]
15. Logan, B.F.; Mallows, C.L.; Rice, S.; Shepp, L.A. Limit distributions of self-normalized sums. *Annals Probab.* **1973**, *1*, 788–809. [CrossRef]
16. DuMouchel, W.H. On the asymptotic normality of the maximum likelihood estimate when sampling from a stable distribution. *Ann. Stattistics* **1973**, *1*, 948–957. [CrossRef]
17. McCulloch, J.H. Linear regression with stable distributions. In *A Practical Guide to Heavy Tails: Statistical Techniques and Applications*; Adler, R.J., Feldman, R.E., Taqqu, M.S., Eds.; Birkhauser: Basel, Switzerland, 1998.
18. Fama, E.F.; Roll, R. Some properties of symmetric stable distributions. *J. Am. Stat. Assoc.* **1968**, *63*, 817–836.
19. Weron, R. Electricity price forecasting: A review of the state of the art with a look into the future. *Int. J. Forecast.* **2014**, *30*, 1030–1081. [CrossRef]
20. Shahidehpour, M.; Yamin, H.; Li, Z. *Market Operations in Electric Power Systems: Forecasting, Scheduling, and Risk Management*; John Wiley & Sons, Inc.: New York, NY, USA, 2002.
21. Weron, R. *Modeling and Forecasting Electricity Loads and Prices: A Statistical Approach*; John Wiley & Sons Ltd.: Chichester, UK, 2006.

22. Zareipour, H. *Price-Based Energy Management in Competitive Electricity Markets*; VDM Verlag Dr. Müller: Saarbrücken, Germany, 2008.
23. Bunn, D.W. *Modeling Prices in Competitive Electricity Markets*; John Wiley & Sons Ltd.: Chichester, UK, 2004.
24. Burger, M.; Graeber, B.; Schindlmayr, G. *Managing Energy Risk: An Integrated View on Power and Other Energy Markets*; John Wiley & Sons, Inc.: New York, NY, USA, 2007.
25. Huisman, R. *An Introduction to Models for the Energy Markets*; Risk Books: London, UK, 2009.
26. Weber, R. *Uncertainty in the Electric Power Industry*; Springer: Berlin, Germany, 2006.
27. Batlle, C.; Barquín, J. A strategic production-costing model for electricity market price analysis. *IEEE Trans. Power Syst.* **2005**, *20*, 67–74. [CrossRef]
28. Koritarov, V.S. Real-world market representation with agents. *IEEE Power Energy Mag.* **2004**, *2*, 39–46. [CrossRef]
29. Ventosa, M.; Baíllo, Á.; Ramos, A.; Rivier, M. Electricity market modeling trends. *Energy Policy* **2005**, *33*, 897–913. [CrossRef]
30. Bunn, D.W. Forecasting loads and prices in competitive power markets. *Proc. IEEE* **2000**, *88*, 163–169. [CrossRef]
31. Garcia-Martos, C.; Rodriguez, J.; Sanchez, M.J. Forecasting electricity prices by extracting dynamic common factors: Application to the Iberian market. *IET Gener. Transm. Distrib.* **2012**, *6*, 11–20. [CrossRef]
32. Amjady, N.; Hemmati, M. Day-ahead price forecasting of electricity markets by a hybrid intelligent system. *Eur. Trans. Electr. Power* **2009**, *19*, 89–102. [CrossRef]
33. Abedinia, O.; Amjady, N.; Shayanfar, H.A. A hybrid artificial neural network and VEPSO based on day-ahead price forecasting of electricity markets. In Proceedings of the International Conference on Artificial Intelligence (ICAI), Las Vegas, CA, USA, 21–24 July 2014.
34. Amjady, N. Day-ahead price forecasting of electricity markets by a new fuzzy neural network. *IEEE Trans. Power Syst.* **2006**, *21*, 887–996. [CrossRef]
35. Aggarwal, S.K.; Saini, L.M.; Kumar, A. Electricity price forecasting in Ontario electricity market using wavelet transform in artificial neural network-based model. *Int. J. Contr. Autom. Syst.* **2008**, *6*, 639–650.
36. Chan, S.C.; Tsui, K.M.; Wu, H.C.; Hou, Y.; Wu, Y.-C.; Wu, F. Load/price forecasting and managing demand response for smart grids. *IEEE Signal Process. Mag.* **2012**, *29*, 68–85. [CrossRef]
37. Chan, K.F.; Gray, P.; van Campen, B. A new approach to characterizing and forecasting electricity price volatility. *Int. J. Forecast.* **2008**, *24*, 728–743. [CrossRef]
38. Garcia-Martos, C.; Conejo, A.J. Price forecasting techniques in power systems. In *Wiley Encyclopedia of Electrical and Electronics Engineering*; John Wiley & Sons, Inc.: Hoboken, NJ, USA, 2013; pp. 1–23.
39. Hyndman, R.; Koehler, A.B. Another look at measures of forecast accuracy. *Int. J. Forecast.* **2006**, *22*, 679–688. [CrossRef]
40. Garcia-Ascanio, C.; Mate, C. Electric power demand forecasting using interval time series: A comparison between VAR and iMLP. *Energy Policy* **2010**, *38*, 715–725. [CrossRef]
41. Jonsson, T.; Pinson, P.; Nielsen, H.A.; Madsen, H.; Nielsen, T.S. Forecasting electricity spot prices accounting for wind power predictions. *IEEE Trans. Sust. Energy* **2013**, *4*, 210–218. [CrossRef]
42. Misiorek, A.; Trück, S.; Weron, R. Point and interval forecasting of spot electricity prices: Linear vs. non-linear time series models. *Stud. Nonlinear Dyn. E.* **2006**, *10*, 1–36. [CrossRef]
43. Nogales, F.J.; Conejo, A.J. Electricity price forecasting through transfer function models. *J. Oper. Res. Soc.* **2006**, *57*, 350–356. [CrossRef]
44. Weron, R.; Misiorek, A. Forecasting spot electricity prices: A comparison of parametric and semiparametric time series models. *Int. J. Forecast.* **2008**, *24*, 744–763. [CrossRef]
45. Nolan, J.P.; Ojeda-Revah, D. Linear and nonlinear regression with stable errors. *J. Econ.* **2013**, *172*, 186–194. [CrossRef]
46. Nolan, J.P. Multivariate elliptically countered stable distributions: Theory and estimation. *Comput. Stat.* **2013**, *23*, 2067–2089. [CrossRef]

 © 2019 by the authors. Licensee MDPI, Basel, Switzerland. This article is an open access article distributed under the terms and conditions of the Creative Commons Attribution (CC BY) license (http://creativecommons.org/licenses/by/4.0/).

Article

Environmental Impacts of Reusable Transport Items: A Case Study of Pallet Pooling in a Retailer Supply Chain

Riccardo Accorsi *, Giulia Baruffaldi, Riccardo Manzini and Chiara Pini

Department of Industrial Engineering, Alma Mater Studiorum–University of Bologna, 40136 Bologna, Italy; giulia.baruffaldi2@unibo.it (G.B.); riccardo.manzini@unibo.it (R.M.); chiara.pini@luxottica.com (C.P.)
* Correspondence: riccardo.accorsi2@unibo.it; Tel.: +39-051-20-93-415

Received: 23 April 2019; Accepted: 30 May 2019; Published: 4 June 2019

Abstract: Manufacturing, storage, and transportation processes are typically facilitated by pallets, containers, and other reusable transport items (RTIs) designed to guarantee many cycles along a lifespan of several years. As a consequence, both supply and reverse transportation of RTIs need to be managed to avoid stockout along the supply chain and the unsustainable production of new tools from virgin materials. This paper focuses on the business of pallet management by analyzing the transport operations of a pallet pooling network serving a large-scale nationwide retailer. The pooler is responsible for supplying, collecting, and refurbishing pallets. The combination of the pooler's management strategies with different retailer network configurations results in different pooling scenarios, which are assessed and compared in this paper through a what-if analysis. The logistical and environmental impacts generated by the pallet distribution activities are quantified per each scenario through a tailored software incorporating Geographic Information System (GIS) and routing functionalities. Findings from this analysis suggest how to reduce vehicle distance traveled (vehicles-km) by 65% and pollutant emissions by 60% by combining network infrastructures and pooling management strategies—identifying an empirical best practice for managers of pallet businesses.

Keywords: pallet management strategy; reusable handling system; closed-loop network; what-if analysis; environmental impact

1. Introduction

Manufacturing, storage, and transportation processes are typically facilitated by pallets, containers, and other reusable transport items (RTIs), which permeate supply chain operations [1]. As a standardized platform for unit loads, pallets are widely diffused in manufacturing facilities, warehouses and distribution centers, and stores and are thereby considered to be company assets. As they aid manufacturing and logistical operations, the number of pallets in circulation grows as countrywide economic and trade indicators (e.g., GDP) increase [2].

The management of pallets entails several processes along their life cycle and throughout logistics networks, such as production, distribution, and refurbishing. Pallets are therefore typically designed to guarantee many cycles over a lifespan of several years. According to the circular economy paradigm, the reuse of packaging and RTIs is justified to reduce waste, raw materials exploitation, associated costs, and environmental impacts [3,4]. Nevertheless, the use of RTIs is far more complex than more traditional disposable package management due to the increase of reverse logistics [5].

For this reason, managers of supply chains more often opt for the implementation of closed-loop networks (CLNs) that integrate forward distribution operations with reverse logistics, including handling, storage, and collection of reusable items [6]. In such networks, the trade-off between the emissions released by transportation and the reduction in the use of virgin materials has to be identified in order to achieve environmental sustainability targets [7–9]. This aim requires a deep understanding of the observed network, highlighting the crucial role of the data collection process in order to enhance the accuracy of the obtained result [10].

Scholars have already pointed out how alternative pallet management strategies across the CLNs can generate different environmental impacts associated with the whole pallet life cycle, as well as different logistical costs [11]. Among the most diffused strategies, this paper focuses on two, both based on pallet re-circulation. The first is the buy–sell program, in which an equal number of pallets received by a node of the supply chain is shipped back to the previous node through a back-haul trip. The entire pallet supply and collection process is hence assigned to the carriers or the truck fleet owners, who are implicitly responsible for balancing the overall pallet population throughout the supply chain. An improved version of this program consists of consolidating pallets and arranging full-loaded back-trips.

A second strategy, namely the pooling program, involves a new actor, the pooler, who is responsible for supplying ready-to-use pallets to vendors, manufacturers, and other logistic nodes; collecting pallets from customers or downstream; refurbishing damaged pallets; and holding inventory within its facilities until new pallet orders are placed. While the pooling program typically guarantees more efficiency than the buy–sell program (e.g., fewer pallets lost, reduction of stockout risk), it compels advanced decision-making on where to locate pooler facilities, how to set facilities' capacities, and how to distribute transport flows (i.e., collection, inventory balancing, and supply) across the network [12]. Such decisions contribute to establishing a new reverse network able to provide pallet management services to multiple forward supply chains. Nevertheless, the way forward and pooling networks interact with each other may affect the logistical and environmental performance of a pallet management service.

These statements motivate this paper, which focuses on the assessment and comparison of the environmental impact and logistical performance caused by alternative pallet management strategies combined with different configurations of CLNs. Specifically, the two aforementioned strategies are evaluated with respect to different configurations of a retailer network operating countrywide in Italy. The combination of strategies and networks generates several logistical scenarios whose impacts are calculated through a developed GIS-driven decision support tool for the punctual calculation of the carbon footprint (CO_2eq) from transportation and the vehicles-km within the network.

The outline of this paper is as follows. Section 2 briefly summarizes the literature on pallet management services to which this paper provides a novel contribution. Section 3 describes the methodology, the area of analysis, and the characteristics of the pallet management scenarios. Sections 4 and 5 illustrate and discuss the results and summarize the findings through a multi-scenario comparison. Finally, Section 6 concludes the paper, proposing guidelines to practitioners as well as suggestions for future research.

2. Literature Review

Although pallets are widely distributed worldwide in many industrial sectors, relatively few scientific papers focus on the economic and environmental performance of pallet management services, and these are provided by a narrow group of scholars. According to Glock [1], who provided an exhaustive survey of the field, the decision support models and methods for managing RTI networks belong to four different categories (i.e., packaging system comparison, RTI return forecast, RTI inventory management, and operations management and optimization). This paper falls within the first category, which compares the performance of alternative RTI systems/scenarios. Examples of these are provided by Ray et al. [13], who compared the cost per pallet trip generated by purchasing or rental, and Carrano et al. [14], who compared the environmental impacts of three pallet management strategies, namely single-use, reusable buy–sell, and pooling strategies, over the entire pallet life cycle. They accounted for the carbon emissions released by the production, usage, refurbishing, and disposal phases and used optimization to find the optimal environmental strategy.

With respect to RTIs, Hassanzadeh Amin et al. [15] explored barriers and benefits of the reverse logistics of plastic pallets in Canada, whereas Bengtsson and Logie [16] and Carrano et al. [17] focused on wood pallets. Specifically, the latter were the first to estimate the carbon footprint of a wood pallet life cycle. Elia and Gnoni [18] used discrete-event simulation to aid in the design of a CLN for pallet management in order to meet cost and time efficiency targets. The environmental impact and the cost of pallets were measured by Bilbao et al. [11], who supported the selection of pallet management systems and of the materials composing pallets. Katephap and Limnararat [19] developed a mathematical model to calculate the cost of alternative reverse logistics arrangements and used the Life Cycle Assessment (LCA) methodology to estimate the resulting impacts on the environment.

Others have focused on optimizing pallet management operations within a CLN for pallet pooling [12]. Ren et al. [20] dealt with uncertainty while optimizing the whole pallet allocation process, including buying (leasing), distribution, repositioning, and recovery. Assuming the pooler's perspective, Zhou et al. [21] optimized the delivery and collection routes from the pooler's facility to the clients of the pallet management service. Tornese et al. [22] quantified the effects of repair facility location and pallet service conditions (i.e., pallet handling and loading conditions) on a pooling system in accordance with economic and environmental performance.

As underlined by Glock [1], although several analytical approaches to studying RTI systems and networks can be found in the literature, few applications of such approaches and methods in real-world scenarios have been exemplified. Among the few who have discussed the adoption of RTI in specific industrial realities, Palsson et al. [23] reported a case study of the company Volvo, assessing the cost and environmental impacts of alternative packaging systems. The reason behind this lack of real-world examples is due to the high complexity of some industrial applications, which results in a great amount of data and parameters to collect and variables to formulate. For example, the management of the transport flows in both analytical models and discrete-event simulation methods is indeed complicated by the number of arcs between the nodes of the reverse network, as well as the total number of shipments observed over a planning horizon (e.g., a year). Therefore, the impacts of transportation flows are often neglected and estimated instead of calculated.

Based on these statements, this paper exploits a tailored GIS-driven decision support tool to calculate the logistical and environmental impacts caused by the pallet usage and transportation phases in real-world integrated retailer and pallet pooling networks. A multi-scenario analysis conducted on a significant real-world case study is used to assess and identify the most sustainable network configuration and pallet management strategy. Findings from this work provide guidelines to the managers of both retailers and poolers.

3. Methods and Materials

The methodology of this study included the following steps. First, we selected a significant countrywide case study consisting of two integrated networks of a large-scale retailer and a pooler, respectively. The observed networks included more than 1500 nodes located all over Italy, where more than half of goods are transported by truck [24]. The case study was approached by drawing arcs (i.e., physical connections) between logistic nodes and by examining how these were affected by the choice of the pallet management strategy. As the analysis combined the perspectives of both the pooler and the retailer, different retailer network configurations were assessed. The combination of a pallet management strategy and a network configuration generated a scenario.

Second, an extensive data collection campaign was conducted by interfacing with the companies' Enterprise Requirement Planning Systems (ERPs) and other data repositories of the logistic nodes of both networks. A significant horizon of analysis (i.e., 1 year) was considered. Within this, the involved network nodes' locations, shipping documents, and loading truck records were collected. This step allowed us to build a SQL database with the aim of supporting the accounting of the transport impacts and the Key Performance Indicators (KPIs) evaluation for the scenarios.

The calculation of logistical and environmental KPIs [25] was carried out through a tailored GIS-driven decision support tool developed for the scope of this study. The logistical tasks performed within the observed network were formalized and coded as 'methods' of the software application. This was able to interface with the SQL database and with a GIS plug-in to quantify the routing distance from each couple of nodes. The tool, written in C# .NET, was then used to virtualize the goods transport and pallet handling tasks and quantify the logistical and environmental performance of each shipment and scenario in agreement with a what-if analysis. The graphical user interface (GUI) tool is exemplified in Figure 1, whilst more details on how it was designed and the input data required are given in the work of Accorsi et al. [26]. The dot plot in the GUI of Figure 1 reports the carried load and the vehicle-km per route necessary to quantify the released greenhouse gases (GHGs) and air pollutants for each shipment.

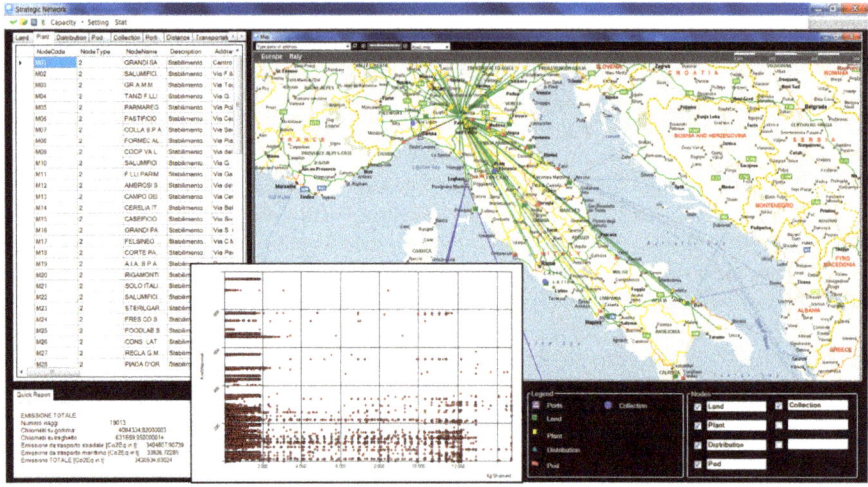

Figure 1. Graphical user interfaces (GUIs) of the transport carbon emission calculator. The main GUI of the software provides the list and the type (e.g., reusable transport item (RTI) depot, distribution hub, refurbishing depot, etc.) of the network nodes (**on the left**) and the map of the logistic flows (**on the right**). Another GUI (**shown above**) represents the dot plot of the weight for load-vehicle-km elaborated per each shipment.

The reminder of this section goes into detail on the generation of the scenarios and provides an overview of the data architecture.

3.1. Pallet Management Strategies and Network Configuration

The observed retailer supply chain includes three different type of actors, as illustrated in Figure 2: the vendors (1), which supply the regional retailer depots (2) that, in turn, directly serve the retailer stores (3). The as-is pallet management strategy adopted to provide the network with the required number of pallets is the so-called extended-use or buy–sell program. This outsourced pallet management strategy makes the carriers responsible for collecting empty pallets and using back-haul trips to replenish the vendors' stocks. According to this policy, the pallets received by a regional depot equal those shipped back to the vendors exploiting the carrier backhaul trips (see the as-is buy–sell scenario in Figure 2). As the perspective undertaken is that of the retailer, we set the boundaries of the analysis to those processes that directly affect the logistical and environmental impacts of the retailers and the vendors. We, therefore, did not include the pallet facilities owned by the carriers.

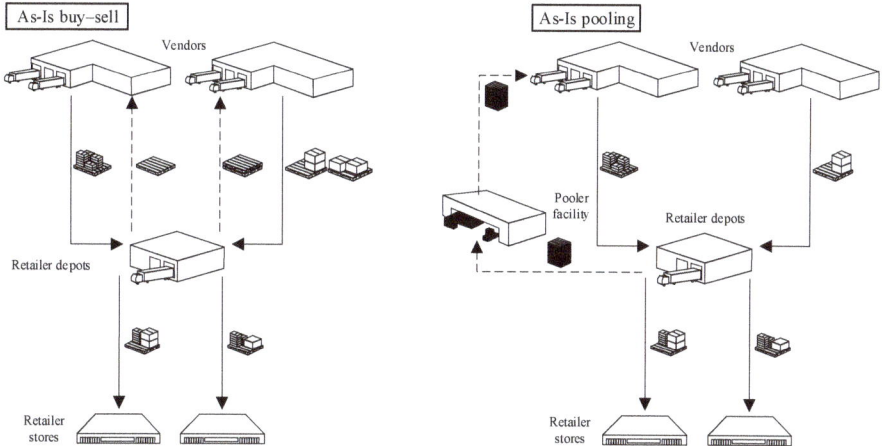

Figure 2. The two-stage retailer supply chain according to the exchange and pooling pallets programs.

In the few last years, an alternative pallet management strategy, namely the leased pallet pooling or pooling program [20], has been adopted by the retailer supply chain (see the as-is pooling scenario in Figure 2). This develops around a pooler, who is responsible for collecting empty pallets from the retailer depots and serving the retailer supply chain. The main benefit generated by pallet pooling is the consolidation of shipments for pallet collection, which enhances truck capacity utilization. It is worth noting that, although Figures 2 and 3 do not show pallets flows to and from remanufacturing facilities [27], these movements were accounted for in the analysis.

Based on these two initial pallet management strategies, beyond the vendors, retailer depots, and retailer stores, several potential actors and logistics nodes can contribute to pallet circulation throughout a retailer supply chain. These are, for instance, centralized retailer distribution centers (i.e., hubs), carriers, and pooler facilities. For this reason, the analyzed supply chain has become interested in the evaluation and benchmarking of the economic and environmental performance of alternative scenarios, which combine different pallet management strategies and network configurations.

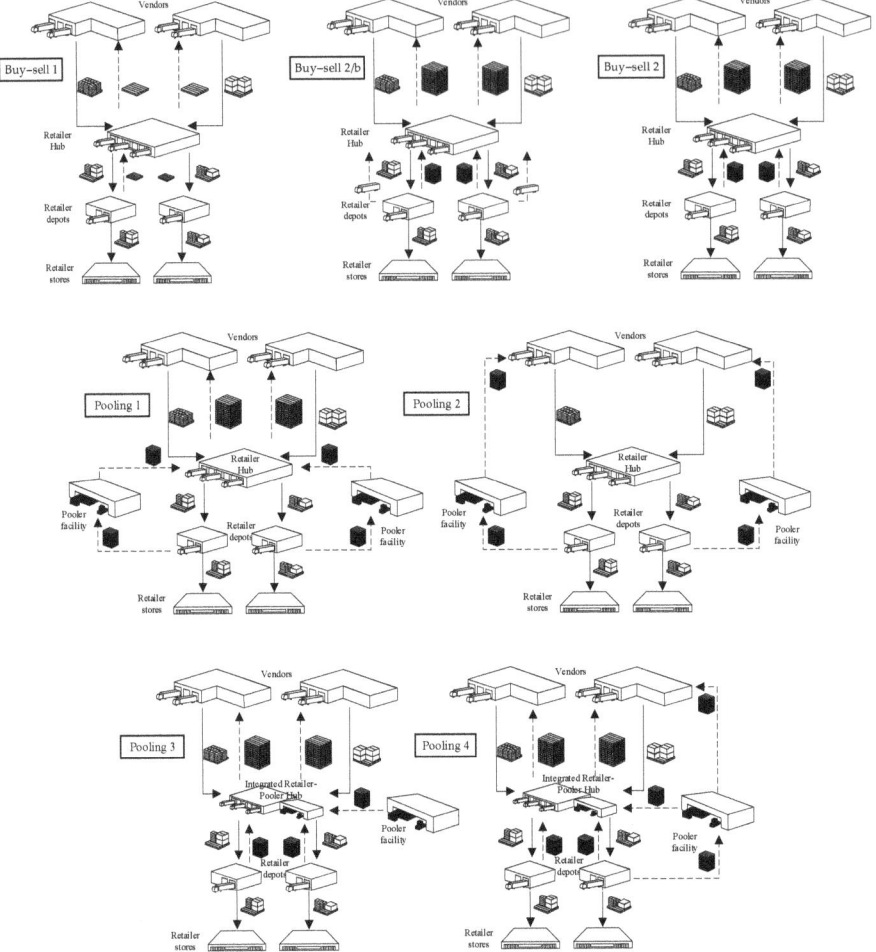

Figure 3. Management strategies and network configurations scenarios.

3.2. Simulated Scenarios

By combining alternative pallet management strategies and different network configurations, seven what-if scenarios were generated and assessed. These resulted from adopting buy–sell or pooling programs and redesigning the network according to two aims: (1) the establishment of a new retailer hub between vendors and depots and (2) integration with the pooler's network. Each scenario accounted for the forward distribution of loaded pallets and for reverse flows of empty pallets in different ways. Characteristics of the what-if scenarios, in comparison with the ones introduced in Figure 2, are summarized in Table 1 and Figure 3.

Table 1. Pallet logistic scenarios (Legend: ●/○ as yes/no).

Scenario	Mng. Strategy	Retailer Hub	Pooler's Net.	Pooler-in-Hub	Description
Buy–sell: as-is	Buy–sell	○	○	○	The business-as-usual scenario.
pooling: as-is	Pooling	○	●	○	The recently implemented scenario without the retailer hub.
What-if scenarios					
buy–sell 1	Buy–sell	●	○	○	The hub receives full pallets from the suppliers and allows for consolidation of the shipments to the depots. Pallet exchange is guaranteed both upstream and downstream.
buy–sell 2	Buy–sell: cons.	●	○	○	The hub also allows for consolidating empty pallets that are shipped back by full-loaded trucks. Back-trips are not accounted for.
buy–sell 2b	Buy–sell: cons.	●	○	○	The hub also allows for consolidating empty pallets that are shipped back by full-loaded trucks. Back-trips are accounted for.
pooling 1	Pooling	●	●	○	Empty pallets are collected from the pooler, who replenishes the stock of the hub. Exchange policy is followed upstream with the suppliers.
pooling 2	Pooling	●	●	○	The pooler closes the empty pallets cycle from the depots to the suppliers.
pooling 3	Pooling	●	●	●	The integrated hub–pooler facility allows for the coupling of the forward goods network with the reverse pallet network.
pooling 4	Pooling	●	●	●	This optimal scenario also enables the whole pooler's network to collect from depots and serve directly the hub or the suppliers in order to minimize traveling.

3.3. Input Data

As the retailer network is distributed countrywide, estimating the average traveling distances to describe the pallet distribution process could be misleading. In order to investigate the impacts of the pallet management service, a tailored database was filled by actual transport records and to–from routing charts. The dataset, illustrated and exemplified in Table 2, was collected from the companies' ERPs and organized through a relation SQL database. This database included information regarding nodes; distances; the to–from chart of the routing distances; trips, including all the shipment records throughout the network within the observed horizon; unit loads, describing the size of the pallets; and transport, including mean performance in terms of loading capacity and pollutant emissions.

Table 2. The database reflecting information on nodes, distance, trips, unit loads, and transport means.

Node				Distance		
Node Code	Node Type	Latitude	Longitude	Node from	Node to	Route [km]
002	Vendor	44.035	12.188	002	006	15
006	Retailer store	45.151	10.836	002	008	80

Trip						
Ship. code	Node from	Node to	Date	Product	Weight [kg]	Transport mean
2534	002	008	8 September 2018	Apples	630	Delivery lorry
2535	010	003	8 October 2018	Kiwi	600	Full trailer comb.

Unit Load				
Code	Length	Width	Height	Weight
{1200 × 1000 × 1000}	1200	1000	1000	28
{1200 × 800 × 1000}	1200	800	1000	20

Transport Means						
Transport Mean	Emission Class	Load Capacity (t)	CO (g/tkm)	NO_x	...	CO_{2eq}
Delivery lorry	EURO 1	3.5	0.58	1.50	...	146.05
Full trailer	EURO 5	40	0.00425	0.1	...	31.17

Emissions parameters were obtained from the Lipasto emission factor database [28] and scaled per loaded ton and vehicle-km. As the proposed analysis focused on the comparison between different scenarios and not on pure accounting for the specific network's impact, the adoption of such an emission factor database was therefore acceptable.

Furthermore, this source is more complete than others (e.g., Keller and de Haan [29]), as it provides records for different transport modes (e.g., roadways, seaways). Some relevant inputs and parameters of the proposed case study are reported in Table 3.

Table 3. Input data.

Network Nodes			Processes/Handling Units					
Vendor/supplier nodes		84	Time horizon (TH)			4 February 2015–31 March 2016		
Retailer regional depots		17	Shipments/TH (As-Is)			29,802		
Retailer hub		1						
Pooler's facilities		16	Unit load/pallets			4 {800 × 1200 × 1000; 600 × 1200 × 800; 800 × 600 × 600; 1000 × 1200 × 1000}		
Pallet refurbishing facilities		4						
Transport Mean	Emission class	Load capacity (t)	CO (g/tkm)	NO_x	PM	CH_4	NH_3	CO_2eq
Full trailer	EURO 5	40	0.00425	0.1	0.0011	0.00005	0.00088	31.17
Semi trailer	EURO 5	25	0.0012	0.016	0.0001	0.000062	0.0015	10.92

The dataset fueled a tailored decision support tool that allowed us to assess and quantify the logistical and environmental performance of the current CLN and to further manipulate shipment records to virtualize and generate several what-if pallet management service and network scenarios. For a detailed description of the tool, its functionalities, and GUIs refer to other papers [26].

4. Results

With the aim to assess the environmental and logistical impacts associated with the pallet service CLN, the analyzed scenarios were illustrated using the GIS-driven tool. Figure 4 shows three network flow maps, which provide an overview of the obtained results and demonstrate how alternative scenarios generate different impacts on pallet distribution and collection flows. All the flow maps resulted from the implementation of the pooling program, although the network configuration varied in accordance with the type of actors involved and the transport tasks.

In the maps, Euclidean arcs connect the network's nodes according to the shipment profile, and the different colors correspond to different transport tasks. In Figure 4a, the map represents the as-is pooling scenario, as illustrated in Figure 2, where the absence of a hub generates the direct interchange between the vendors and the retailer depots, represented by purple lines, whereas light-blue arcs represent shipments from the pooler's facilities to the vendors. The choice of establishing the hub near Parma allows the conveyance of freight from the vendors into a unique node before supplying the retailer depots located nationwide, thus consolidating both deliveries and pallet supplies. Figure 4b,c represents the pooling 2 and pooling 1 scenarios, respectively, in which the orange arcs represent shipments from the pooler's facilities to the retailer hub, the red arcs represent shipments from the hub to the vendors, the yellow arcs represent shipments from the hub to the retailer depots, and the light-blue arcs represent shipments from the pooler's facilities to the vendors. In such scenarios, the typical triangular connections of the CLN are highlighted in the map. In comparison with the pooling 1 scenario, the pooling 2 scenario partly contributed to reducing the total traveling, because the pooler's facility directly supplied pallets to the vendors located nearby.

In Figure 5, the traveling distance (vehicles-km) was calculated for the different transport tasks of the CLN, and carbon emissions were determined for each scenario. Even though the buy–sell 2 scenario is reported in Figure 5, it did not consider the empty back trips of the trucks from the retailer depots. Therefore, it excluded some of the externalities the retailer generated in the environment and was thus incomparable with the other scenarios.

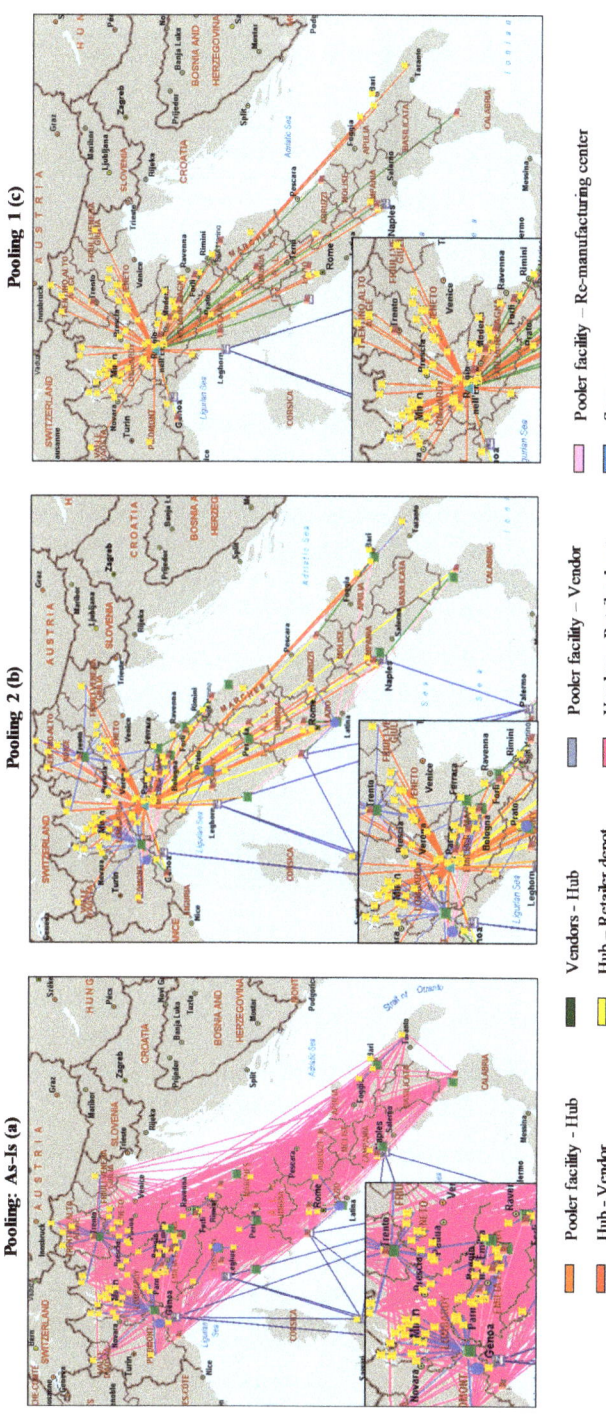

Figure 4. Transport flow maps: (a) pooling: as-is, (b) pooling 2, and (c) pooling 1.

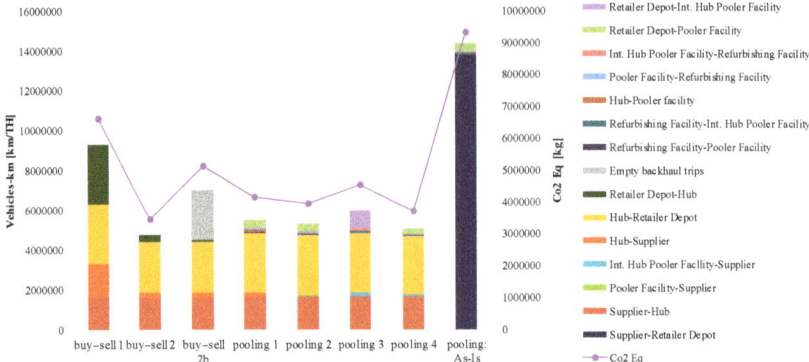

Figure 5. Performance of the scenarios: logistical and environmental indicators.

As shown in Figure 5, it was determined that, except for the as-is pooling scenario characterized by direct interchange between the vendors and the retailer depots, the pooler (i.e., the pooling 1, 2, 3, and 4 scenarios) evidently allowed for the reduction of transport and CLN environmental impacts. The comparison between the pallet management scenarios demonstrates how the CLN benefits from the presence of a centralized hub that allows for the consolidation of forward shipments of goods and, thus, of RTIs as well.

5. Discussion

To further investigate the effects of selecting the proper pallet management strategy and network configuration on the CLN's environmental and logistical performance, factors such as the vehicles-km and the amount of CO_2eq, the total number of shipments, the total transported tertiary package weight (i.e., pallet weight), and the total number of handled pallets were evaluated, as shown in Figure 6.

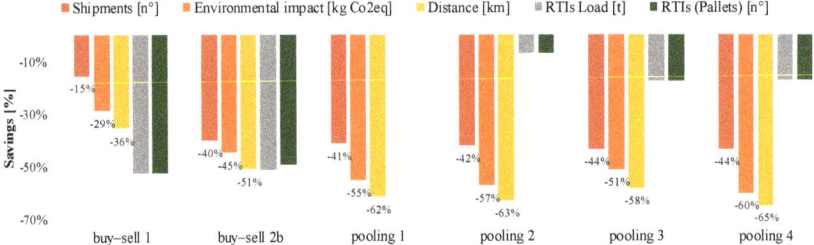

Figure 6. Comparison of the performance indicators in the RTI management scenarios: number of shipments, environmental impact, vehicles-km, total transported pallet weight, and number of pallets.

As shown in Figure 6, histograms were developed to report the savings obtained per each metric with respect to the current scenario (i.e., the as-is pooling scenario). This comparison reveals that there were greater environmental benefits, i.e., the highest drop of CO_2eq emissions, when implementing the pooling management strategies throughout the retailer hub. Moreover, both the buy–sell strategies performed better than the as-is pooling network for all the observed performance indicators. This result indicates that the distribution of supply chain depots, and particularly of centralized hubs, was more important than the management strategy (i.e., buy–sell vs. pooling) in reducing the economic and environmental impacts resulting from the RTI system.

The reduction of the shipped load of RTIs and of the overall number of pallets handled was derived from the consolidation opportunity offered by the retailer hub. According to the pooling scenarios, it is clear that such savings were intensified when the pallets management service was handled by the pooler entirely from the retailer depots upstream to the vendors (i.e., the pooling 2 and 4 scenarios). Furthermore, transportation benefits from overlapping the pooler and retailer networks (i.e., establishing a pooler facility within the hub) occurred in the best-performing scenario (i.e., the pooling 4 scenario).

Another analysis was carried out to compare the performance of the different scenarios based on the profile of the vehicle-km and the total RTI load per shipment. Specifically, Figure 7 shows the same scenarios illustrated in Figure 4 but via a class-based frequency plot analysis using vehicle-km and vehicle-kg as the coordinates of a class. This analysis was based on the plot chart calculated by the software and illustrated in Figure 1.

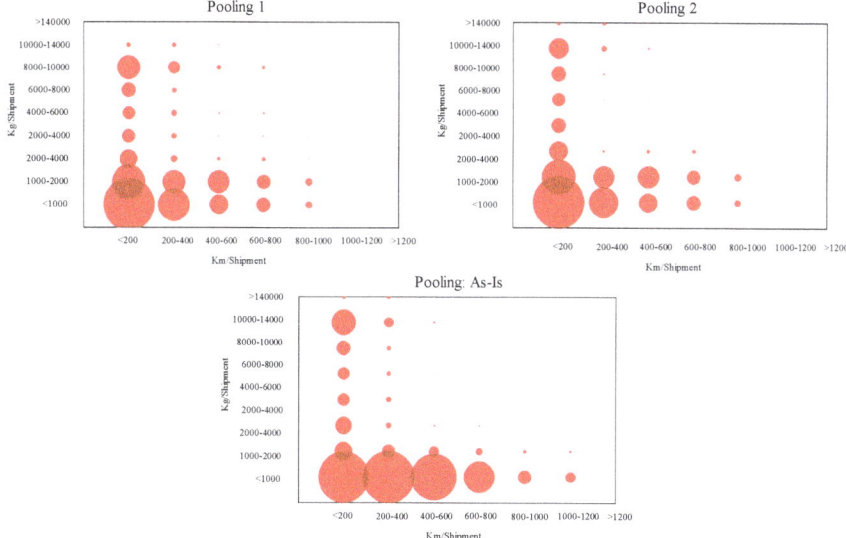

Figure 7. Frequency analysis of the total vehicles-km and carried load per shipment.

The results of this analysis reveal how, whilst the pooling 1 and pooling 2 scenarios showed similar behaviors, the as-is pooling scenario required more long-range shipments characterized by a scarce utilization of the trucks. This impact resulted from the absence of a centralized hub, which forced the vendors to travel longer routes in order to supply the retailers. Conversely, the presence of the hub reduced the length of the routes, allowed for the consolidation of the loads, and enhanced the truck utilization, as highlighted by the larger diameter of the bubbles between the 1000- and 2000-kg shipments in Figure 7. In comparison with the pooling 1 scenario, the pooling 2 scenario presented a higher number of shipments with loads between 10,000 and 14,000 kg. This behavior was due to the location of the pooler's facilities, which directly replenished the vendors with full-loaded shipments of pallets.

6. Concluding Remarks and Findings

This paper conducted a comparative analysis of alternative RTI management service scenarios applied to a case study of pallet pooling in a nationwide retailer supply chain. This paper contributes to a gap in the literature, as we investigated the combination of two pallet management strategies (i.e., buy–sell and pooling strategies) with several network configurations of a complex real-world case, involving up to 120 nodes and around 30,000 deliveries per year, and we quantified the associated logistical and environmental impacts using a tailored GIS-driven tool. Although there were limitations caused by using a single case study, this case of a real countrywide network of a pooler and a large-scale Italian retailer [30] might lead to useful insights for pallet service providers, clients, carriers, and policymakers.

First, the environmental performance of CLNs for RTIs, and specifically for pallets, was extremely affected by the transport phase. In developing a strategy for pallet service distribution, the characteristics and distribution of the observed network infrastructure, the node distribution, and the shipment profile must be taken into account carefully, as they were shown to play a crucial role in this study. Thus, the study of the environmental and logistical performance of these networks require detailed analysis and calculation of quantitative and objective KPIs. The average estimation of traveling during the whole RTI life cycle can be misleading and should instead be replaced by the actual calculation of the vehicle-km and vehicle-kg per each shipment of pallets.

Secondly, the pooler contributed significantly to reducing the carbon footprint of the pallet service CLN but was useless if not coupled with an efficient distribution route for the supply chain nodes that needed to be serviced. Traveling and environmental savings were indeed affected by the level of integration between the pooler and the retailer networks and by the presence of a centralized hub, which allowed for a reduction of CO_2eq emissions of 45% even with the buy–sell management strategy. Furthermore, the greatest traveling and carbon emission reduction relating to the RTI system, up to 65% and 60%, respectively, compared with the as-is scenario, was obtained when the pallet management service was handled by the pooler entirely from the retailer depots upstream to the suppliers.

In future studies, further developments are expected to be achieved with respect to modeling RTI management services and network configurations and using optimization [8,31] to find the optimal trade-off between the environmental performance of the CLN and the costs experienced by the pooler, the retailer, and the carriers, as well as the optimal location of CLN depots for the management of RTI life cycles.

Author Contributions: R.A. and R.M. conceptualized the project with the involved companies guaranteeing supervision. The software was developed by C.P. and R.A., who also both contributed to its validation. C.P. collected data and implemented the analysis while G.B. organized the results. G.B. and R.A. wrote the paper.

Funding: This research received no external funding.

Acknowledgments: The authors would like to sincerely thank the company, CPR System, which was involved in the study. In particular, the authors would like to thank Enrico Frigo, Monica Artosi, and Sabrina Pagnoni by name for their valuable support in this research project. The project also benefited from the support of Andrea Mantelli and Alessandra Lograsso from CONAD Soc. Coop.

Conflicts of Interest: The authors declare no conflict of interest.

References

1. Glock, C.H. Decision support models for managing returnable transport items in supply chains: A systematic literature review. *Int. J. Prod. Econ.* **2017**, *183*, 561–569. [CrossRef]
2. Waqas, M.; Dong, D.; Ahmad, N.; Zhu, Y.; Nadeem, M. Critical Barriers to Implementation of Reverse Logistics in the Manufacturing Industry: A Case Study of a Developing Country. *Sustainability* **2018**, *10*, 4202. [CrossRef]
3. Tonn, B.; Frymier, P.D.; Stiefel, D.; Skinner, L.S.; Suraweera, N.; Tuck, R. Toward an infinitely reusable, recyclable, and renewable industrial ecosystem. *J. Clean. Prod.* **2014**, *66*, 392–406. [CrossRef]
4. Elia, V.; Gnoni, M.G.; Tornese, F. Measuring circular economy strategies through index methods: A critical analysis. *J. Clean. Prod.* **2017**, *142*, 2741–2751. [CrossRef]
5. Koskela, S.; Dahlbo, H.; Judl, J.; Korhonen, M.-R.; Niininen, M. Reusable plastic crate or recyclable cardboard box? A comparison of two delivery systems. *J. Clean. Prod.* **2014**, *69*, 83–90. [CrossRef]
6. Bottani, E.; Casella, G. Minimization of the Environmental Emissions of Closed-Loop Supply Chains: A Case Study of Returnable Transport Assets Management. *Sustainability* **2018**, *10*, 329. [CrossRef]
7. Ross, S.; Evans, D. The environmental effect of reusing and recycling a plastic based packaging system. *J. Clean. Prod.* **2003**, *11*, 561–571. [CrossRef]
8. Accorsi, R.; Manzini, R.; Pini, C.; Penazzi, S. On the design of closed-loop networks for product life cycle management: Economic, environmental and geography considerations. *J. Transp. Geogr.* **2015**, *48*, 121–134. [CrossRef]

9. Soysal, M. Closed-loop Inventory Routing Problem for returnable transport items. *Transp. Res. Part D* **2016**, *48*, 31–45. [CrossRef]
10. Auvinen, H.; Clausen, U.; Davydenko, I.; Diekmann, D.; Ehrler, V.; Lewis, A. Calculating emissions along supply chains—Towards the global methodological harmonization. *Res. Transp. Bus. Manag.* **2014**, *12*, 41–46. [CrossRef]
11. Bilbao, A.M.; Carrano, A.L.; Hewitt, M.; Thorn, B.K. On the environmental impacts of pallet management operations. *Manag. Res. Rev.* **2011**, *34*, 1222–1236. [CrossRef]
12. Govindan, K.; Soleimani, H.; Kannan, D. Reverse logistics and closed-loop supply chain: A comprehensive review to explore the future. *Eur. J. Oper. Res.* **2015**, *240*, 603–626. [CrossRef]
13. Ray, C.D.; Michael, J.H.; Scholnick, B.N. Supply chain system costs of alternative grocery industry pallet systems. *For. Prod. J.* **2006**, *56*, 52–57.
14. Carrano, A.L.; Thorn, B.K.; Woltag, H. Characterizing the carbon footprint of wood pallet logistics. *For. Prod. J.* **2014**, *64*, 232–241. [CrossRef]
15. Hassanzadeh Amin, S.; Wu, H.; Karaphillis, G. A perspective on the reverse logistics of plastic pallets in Canada. *J. Remanuf.* **2018**, *8*, 153–174. [CrossRef]
16. Bengtsson, J.; Logie, J. Life Cycle Assessment of one-way and pooled pallet alternatives. *Procedia CIRP* **2015**, *29*, 414–419. [CrossRef]
17. Carrano, A.L.; Pazour, J.; Roy, D.; Thorn, B.K. Selection of pallet management strategies based on carbon emissions impact. *Int. J. Prod. Econ.* **2015**, *164*, 258–270. [CrossRef]
18. Elia, V.; Gnoni, M.G. Designing an effective closed loop system for pallet management. *Int. J. Prod. Econ.* **2015**, *170*, 730–740. [CrossRef]
19. Katephap, N.; Limnararat, S. The operational, economic and environmental benefits of returnable packaging under various reverse logistics arrangements. *Int. J. Intell. Eng. Syst.* **2017**, *10*, 210–219. [CrossRef]
20. Ren, J.W.; Zhang, X.Y.; Zhang, J.; Ma, L. A multi-scenario model for pallets allocation over a pallet pool. *Syst. Eng. Theory Pract.* **2014**, *34*, 1788–1798.
21. Zhou, K.; He, S.; Song, R. Optimization for service routes of pallet service center based on the pallet pool mode. *Comput. Intell. Neurosci.* **2016**, *2016*, 5691735. [CrossRef] [PubMed]
22. Tornese, F.; Pazour, J.; Thorn, B.K.; Roy, D.; Carrano, A.L. Investigating the environmental and economic impact of loading conditions and repositioning strategies for pallet pooling providers. *J. Clean. Prod.* **2018**, *172*, 155–168. [CrossRef]
23. Palsson, H.; Finnsgard, C.; Wänström, C. Selection of packaging systems in supply chains from a sustainability perspective: The case of Volvo. *Packag. Technol. Sci.* **2013**, *26*, 289–310. [CrossRef]
24. Evangelista, P. Environmental sustainability practices in the transport and logistics service industry: An exploratory case study investigation. *Res. Transp. Bus. Manag.* **2014**, *12*, 63–72. [CrossRef]
25. Mishra, D.; Gunasekaran, A.; Papadopoulos, T.; Hazen, B. Green supply chain performance measures: A review and bibliometric analysis. *Sustain. Prod. Consum.* **2017**, *10*, 85–99. [CrossRef]
26. Accorsi, R.; Cholette, S.; Manzini, R.; Tufano, A. A hierarchical data architecture for sustainable food supply chain management and planning. *J. Clean. Prod.* **2018**, *203*, 1039–1054. [CrossRef]
27. Tornese, F.; Carrano, A.L.; Thorn, B.K.; Pazour, J.A.; Roy, D. Carbon footprint analysis of pallet remanufacturing. *J. Clean. Prod.* **2016**, *126*, 630–642. [CrossRef]
28. VTT Technical Research Centre of Finland Ltd. LIPASTO Unit Emissions Database. Available online: http://lipasto.vtt.fi/yksikkopaastot/ (accessed on 8 October 2018).
29. Keller, M.; de Haan, P. *Handbuch Emissionsfaktoren des Straßenverkehrs 2.1*; Dokumentation: Graz, Austria; Essen, Germany; Bern, Switzerland; Heidelberg, Germany, 2004.
30. Deloitte. *Global Powers of Retailing 2018: Transformative Change, Reinvigorated Commerce*; Deloitte: Boston, MA, USA, 2018.
31. Bottani, E.; Montanari, R.; Rinaldi, M.; Vignali, G. Modeling and multi-objective optimization of closed loop supply chains: A case study. *Comput. Ind. Eng.* **2015**, *87*, 328–342. [CrossRef]

© 2019 by the authors. Licensee MDPI, Basel, Switzerland. This article is an open access article distributed under the terms and conditions of the Creative Commons Attribution (CC BY) license (http://creativecommons.org/licenses/by/4.0/).

Article

Comparison of Four Environmental Assessment Tools in Swedish Manufacturing: A Case Study

Sasha Shahbazi *, Martin Kurdve, Mats Zackrisson, Christina Jönsson and Anna Runa Kristinsdottir

RISE IVF, 43 122 Mölndal, Sweden; Martin.kurdve@ri.se (M.K.); mats.zackrisson@ri.se (M.Z.); christina.jonsson@ri.se (C.J.); anna.runa.kristinsdottir@ri.se (A.R.K.)
* Correspondence: sasha.shahbazi@ri.se; Tel.: +46-70-780-61-25

Received: 18 March 2019; Accepted: 4 April 2019; Published: 11 April 2019

Abstract: To achieve sustainable development goals, it is essential to include the industrial system. There are sufficient numbers of tools and methods for measuring, assessing and improving the quality, productivity and efficiency of production, but the number of tools and methods for environmental initiatives on the shop floor is rather low. Incorporating environmental considerations into production and performance management systems still generally involves a top-down approach aggregated for an entire manufacturing plant. Green lean studies have been attempting to fill this gap to some extent, but the lack of detailed methodologies and practical tools for environmental manufacturing improvement on the shop floor is still evident. This paper reports on the application of four environmental assessment tools commonly used among Swedish manufacturing companies—Green Performance Map (GPM), Environmental Value Stream Mapping (EVSM), Waste Flow Mapping (WFM), and Life Cycle Assessment (LCA)—to help practitioners and scholars to understand the different features of each tool, so in turn the right tool(s) can be selected according to particular questions and the industrial settings. Because there are some overlap and differences between the tools and a given tool may be more appropriate to a situation depending on the question posed, a combination of tools is suggested to embrace different types of data collection and analysis to include different environmental impacts for better prioritization and decision-making.

Keywords: sustainable manufacturing; environmental assessment tool; green lean

1. Introduction

Examining different sustainability definitions associated with industrial systems, such as sustainable manufacturing [1], sustainable production [2] or corporate sustainability [3], the importance of the manufacturing in sustainable development is perceived. However, environmental considerations in manufacturing and performance management systems mostly involve an aggregated top-down approach to the entire plant, that often emanating from a separate environmental department. This approach contrasts with the core areas of old-school production systems, such as productivity, quality, delivery precision and cost efficiency, which are considered via both bottom-up and top-down strategies. Additionally, environmental operations improvement has received insufficient attention i.e., there are many tools and methods to measure, assess and improve quality, productivity and efficiency in manufacturing, while relatively few tools and methods exist that specifically target environmental initiatives on the shop floor [4,5]. Green lean studies have attempted to fill this gap with some tools for environmental initiatives on an operational level [6,7], but a lack of detailed methodologies and selection of practical tools when improving the environmental operations and sustainability performance of manufacturing remains.

This paper reports on the application of four environmental assessment tools: Green Performance Map (GPM), Environmental Value Stream Map (EVSM), Waste Flow Mapping (WFM) and Life Cycle

Assessment (LCA), which are commonly used among Swedish manufacturing companies. A short description of each tool is given in Section 2.2. These tools were applied at the same manufacturing process at the same company to solve a particular environmental (and operational) issue related to material efficiency and metal scrap generation of the vehicle frame production process. The authors have earlier applied these tools on several companies of various sizes with different industrial challenges, but none have compared the performance of these tools on the same case. Therefore, the objective of this paper is to help practitioners and scholars to understand the different features of each tool, so in turn (a) right tool(s) can be selected according to particular questions and the industrial settings i.e., variables such as the industrial activity, data accessibility, and staff engagement level and competence. In this paper, these tools are compared to investigate their application and to enable a more precise understanding of the issues involved; however, the intention is not to select a superior tool. Moreover, this study neither deals with material selection nor product design, rather with selecting the best environmental assessment tool for the operational problem posed. Although the empirical focus was on material efficiency and metal scrap reduction, other environmental aspects such as energy use and hazardousness were considered when evaluating, comparing and discussing the tools. Therefore, metal scrap generation in this paper could be seen as an example (demonstrator) for deploying these tools. To fulfill the research objective, two main research questions were posed: (1) what features enables environmental improvements in manufacturing and how those fit into the selected tools? (2) how can these tools be applied in practice?

2. Theoretical Framework

2.1. Moving from Sustainability Concepts to Practical Tools

Although manufacturing companies have been adopting environmental considerations in their production systems since the 1960s [8], environmental improvement thinking and decision-making still most commonly use a top-down approach and environmental data are aggregated for the entire plant on a yearly basis, primarily for the purpose of environmental reporting to authorities and external stakeholders [9], or on a product basis for eco-design purposes, or as an evaluation for marketing. Environmental improvement actions are usually implemented as projects by the environment department and not as an integrated part of continuous improvement activities in work units. However, Cherrafi et al. [10], Dües et al. [11] report tendencies toward a change e.g., in new ISO standards and operative green lean developments.

Environmental assessment and management can be related to different levels of the decision hierarchy [12]. Tools and methods can be linked to a conceptual level with broad and long-term goals, to the supply chain and local industry level, to an operation or a process, or to the lowest level of operational improvements on the shop floor. Previous studies have shown that there are few practical tools appropriate for the operational level that can regularly assess, manage and improve the environmental sustainability of an operation via a bottom-up approach [4,13]. Therefore, developing tools should focus on the operational level and shop floor, where resources are consumed and actual manufacturing is performed, causing a variety of environmental impacts. Bridging the gap between operational and environmental management requires a set of tools to support collaboration between different functions (internal and external), systematic work procedures and problem solving to promote easy learning, and time efficiency [14,15]. These criteria are essential for mutual understanding, intra-organizational communication, improving performance and becoming a learning organization [16]. However, most environmental assessment tools and methods are complex and require expert knowledge of environment management, making it difficult to integrate and apply them into daily continuous improvement work (Kaizen). Studies on green lean manufacturing have (to some degree) bridged this gap by integrating environmental sustainability goals into lean-based production systems [6,17]. Lean principles, such as continuous improvement (Kaizen), visualization, go-to-gemba (i.e., going to the shop floor where problems occur), simplicity in use and learning, and increased engagement of different functions, have been successfully applied and have improved operations and environmental performance in manufacturing (see e.g., Zokaei et al. [7], Dües et al. [11], Wu et al. [18] and Diane et al. [19]). Therefore, a production system based on an integrated green lean philosophy is likely to have high potential for environmental improvements, as the culture of continuous improvement, engagement and waste elimination are already inherited from lean production [9,20]. This goes hand-in-hand with selected tools which are mainly based on lean principles with focus on shop floor.

Overall, the differences in utility of tools and the overlaps between the available tools in previous literature motivated us to compare the most commonly used environmental assessment tools among large international manufacturing companies located in Sweden (in line with the goal of the study). The selected tools have been previously used by the authors in different studies in several manufacturing companies in Sweden (see e.g., Kurdve et al. [21]) for WFM, Zackrisson et al. [22] for LCA, Shahbazi and Wiktorsson [23] for GPM and Kurdve et al. [24] for EVSM). The selected tools are mainly based on lean principles, have bottom-up approach and are mainly focused on shop floor. Therefore, there are some attributes similar to other tools presented in previous literature, see Table 1. For instance, GPM and WFM are similar to tools developed by Zokaei et al. [7]. Among the selected tools, LCA has a more holistic view of environmental aspects and their impact, is not entirely based on lean principles and is slightly more complicated than the other tools (although screening LCA can be used for simpler use), but it was deliberately selected to broaden the analysis.

Table 1. Similar tools identified in the literature.

Tools Applied	Common Level of Use *	Tools with Some Level of Similarity
LCA	Product level	Chemical Risk Assessment, Eco-strategy Wheel, Energy Mapping
WFM	Line/site level	Material Flow Analysis, Logistic Handling, Material Handling Analysis, Material Energy Waste Map
GPM	Cell level	Material Flow Cost Accounting, System Boundary Map and Green Impact Matrix
EVSM	Line level	Green Big Picture

* All tools might be applied at all levels; however, the table indicates the common level of use.

Previous research on green lean and environmental assessment tools, suggests that a tool benefits from the following essential features to enable environmental improvements in manufacturing:

- Being hands-on and operational [25], supporting collaboration and understanding between different internal and external actors [26,27];
- Being easy to learn and implement, visualization [28], time efficiency, continuous improvement and engagement [29];
- Including root cause analysis [30], be harmonized with ISO 14001 [31,32] and support the go-to-gemba concept [33];
- Being goal-oriented, supporting measurements [34], and being focused on a limited area of influence while supporting systematic work procedures (standardized work) [35].

2.2. Summary of Included Tools

Here, the selected tools are briefly described. A deeper description of each tool is given in the Supplementary Material.

2.2.1. Green Performance Map

The GPM [36] is a hands-on tool based on lean principles and green manufacturing concepts aiming to support team level improvements. The tool follows the input and output model to identify, analyze, assess and visualize a variety of environmental aspects of an operation. The flow of input material is divided into productive materials (which are the primary product material), process materials, energy and water consumption; the flow of output is divided into emissions and noise, consumed water, products and residuals. GPM has been reported to be effective for Swedish manufacturing companies, e.g., by Kurdve and Wiktorsson [37]; Shahbazi and Wiktorsson [23], while is similar to tools used internationally by e.g., Pampanelli et al. [38], Sawhney et al. [39] and Zokaei et al. [7].

2.2.2. Environmental Value Stream Mapping

EVSM [40] not only maps the operation aspects, such as process, information flows, inventory levels and time associated with production (such as lead time and takt time), but also visualizes environmental aspects, such as material and waste flows and energy and water consumption. It also considers correlated environmental impacts according to the company's environmental management system and annual targets. Examples of studies using EVSM include Torres and Gati [41], Müller et al. [42], Posselt et al. [43], Gunduz and Fahmi Naser [44] and Dadashzadeh and Wharton [45], among others.

2.2.3. Waste Flow Mapping

WFM [21] resembles material flow analysis but applied on a micro-level (site, line or cell). This method follows current state analysis principles and includes three phases: (1) mapping of waste generation points and fractions, on-site data collection through observations, interviews, deploying eco-mapping [46] and sorting analysis [47] and logistics and waste-handling data collection; (2) material efficiency analysis for each segment, including waste hierarchy, material segmentation, and root cause analysis to determine the causes of material losses; (3) analysis of efficiency in each waste management sub-process and assessment of waste material handling analysis [48], showing losses in terms of equipment and workers utilization.

2.2.4. Life Cycle Assessment

LCA can be considered the broadest environmental assessment tool with the smallest connection to lean principles among the selected tools. LCA assesses significant environmental aspects and their impact from extraction to production, use and end-of-life phase via a holistic approach with the life cycle perspective. LCA can be deployed in ways ranging from a quick LCA using generic data from databases to a full LCA with a high level of detail. In this study, screening LCA was performed which is relevant to the shop level (in line with other tools) and allows fair comparison of the tools. Input and output data across the system boundaries were collected, validated and normalized. Examples of studies deploying LCA in manufacturing include Cheung et al. [49]; Zhang and Haapala [50]; Thammaraksa et al. [51], among others.

3. Materials and Methods

This research is based on a single case study at a large manufacturing company in Sweden, which provided an in-depth understanding [52]. The selection of the case company was based on the company's environmental management system, environmental goals, reputation and interest in achieving sustainability improvements in operation. Additionally, one of the current environmental goals of this company is to increase resource efficiency. For empirical data collection and analysis, a process of manufacturing vehicle frames (consisting of a pair of beams) was selected. The process uses metal as productive material and generates metal scrap and limited amounts of combustible waste, including plastic, bio-waste, paper and wood. The operation process includes metal sheet bending, punching, plasma cutting, blasting, phosphating, painting, heat treatment, and shipping to customers. Finished vehicle frames are sent to an in-plant assembly line or shipped to external plants worldwide. The four shift teams with nine operators and a team leader each, three technicians, a production planner, a production manager, the plant environmental manager, a purchaser and a quality controller were consulted in this case.

Data were collected on the material value chain and information flow, as well as scrap volumes, costs and revenues, statistics from an external waste management entrepreneur, final treatment options, and transportation modes; other necessary data were obtained from environmental reports. Empirical data collection involved multiple sources of evidence including participant observations, on-site walkthroughs, archival review, in-depth interviews (between 30–90 min) and discussions with experts, ensuring data triangulation. Two authors spent two weeks in the studied manufacturing process, participated in meetings and morning briefs, reviewed internal environmental and operational reports and discussed and interviewed different functions including internal environmental manager at the factory level, external environmental manager at the enterprise level, operators, team manager, production manager, waste management entrepreneurs, production planner and production technicians. The document reviews helped to realize a basic insight into companies, their overall strategy and environmental targets and current improvement projects. To be able to deploy the tools, a case study focus based on the current environmental and operational problem at the company was selected i.e., material efficiency and metal scrap generation. The case study focus was supposed to provide answers to question including which sub-process has the largest environmental impact? How much scrap is generated in this process and the respective sub-processes and why? And what improvements can be made to move towards circularity, decreasing scrap generation and environmental impact?

In addition to the empirical study, a structured literature review of environmental assessment tools was carried out. The literature search was conducted in both the scientific databases and grey literature. The literature selection method used a keyword search regarding relevant tools, followed by abstract review and full-text reading. The search incorporated the keywords "lean and green" "green lean" and "environmental assessment", along with combinations with terms "tool", "approach" and "methods". The search was focused on papers published between 2005 and 2018 addressing a situation similar to that in the Swedish manufacturing industry; however, papers outside of this scope were also included in the study. This search was compounded with a qualitative upstream and downstream search of the references in the selected articles.

Data collection and analysis were conducted iteratively to ensure the necessary adjustments. The collected data were continuously summarized in case study protocols and project members were consulted. The empirical findings were also compared with existing literature to enhance understanding of the similarities and differences relative to other studies. To ensure validity and reliability, measures suggested by Yin [52] were adopted, including collection of data from a variety of sources, such as observation, document reviews, interviews and discussions using different tools. The validity was further strengthened by peer examination of collected data and results in different time stages by authors, industrial practitioners, project members and research fellows. The structure followed a logical design with a defined problem statement composed by the company and researchers. Figure 1 depicts the research process.

The main implication of this article and the single case study is qualitative and include analytic generalizations [52]. Thus, the replication of achieved results at similar manufacturing companies is expected. Furthermore, the empirical data in this article relate to a large global manufacturing company in the automotive industry located in Sweden; hence, while the results may not be generalizable to completely different industries or to similar manufacturing companies outside of Sweden, it can be assumed that the achieved results represent relevant empirical evidence. In addition, environmental costs, standards and regulations as well as organizational factors such as environmental consciousness and manufacturing culture, vary greatly over time and across geographic areas. Consequently, results of similar studies could differ over time and in different locations. Additionally, in line with analytic generalization in qualitative research, the single case study represents a case with an opportunity to observe, test and analyze a phenomenon (here environmental assessment tools) that few have studied before [53]. A single case study with an informative approach captures the circumstances and different criteria (see Table 2; Table 4) of an everyday situation (here, manufacturing) [52].

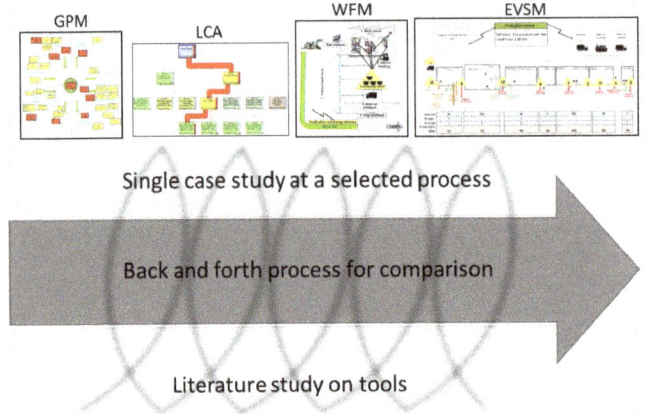

Figure 1. Research process of this study.

4. Empirical Deployment of Tools

This section presents the results and comparison of tools in terms of use experience and features, which are summarized in Table 2. A complete description of tool deployment is given in the Supplementary Material.

4.1. Green Performance Map

GPM was used according to the structure presented by Bellgran et al. [36] to identify and assess various environmental aspects, although with some limitations in quantification. The overall operation process was divided into seven sub-processes to be studied individually in detail. Hence an overall process level GPM as well as seven detailed sub-process GPMs were mapped. Figure 2 illustrates the GPMs at the two levels. This categorization was based on the types of processes, human resourcing and the production layout. The sub-process GPMs and an overall GPM data were then compared, and differences, inputs and outputs were aligned to capture missing information. Collecting quantitative data on costs and environmental aspects was challenging, particularly at a detailed level for sub-processes. For instance, determining the energy consumption at each sub-process was impractical, although the overall number was available. This number for material and scrap-related data was relatively easier to obtain on a sub-process level. It was also challenging to prioritize improvement actions considering the environmental impacts, costs, and resource requirements. From a conventional operations improvement perspective, although GPM improvement reduces cost, it did not improve lean aspects such as production flow, inventory or information flow.

Table 2. Cross-comparison of tools in terms of use experience and features.

Tools	GPM	EVSM	WFM	LCA
Result type	• Overview of material and energy flows at process and sub-process level • Qualitative results in the form of environmental aspects for each sub-process	• Focus on the amount, location and type of scrap • Overview of the operation • Information flow regarding production • Supplier and customer information	• Total amount of waste • Cost of waste bins, handling and transport • Sorting degree of different material fractions • Categorization, quantification and localization of scrap	• Quantitative results at the site level in the form of calculated environmental impacts, e.g., climate impacts • Overview of material and energy flows at the site level • Transportation and end-of-life information
Operation level (site, process, cell)	Overall process and sub-process/cell	Entire process from coil to finished frame	Entire process from coil to finished frame	Entire process from coil to finished frame (cradle-to-gate model from site/process and database data)
Environmental aspects included	All flows (mainly those seen on the shop floor)	Specific selected material flows, (in this case metal scrap)	All types of materials and waste, but in this case with a focus on metals	All types of resources, usually with focus on significant environmental impacts
Time required for data collection and analysis (this study)	2–4 h for an expert 30 man-hours of operators' time	2–4 h for an expert 20 man-hours of operators/technician's time	2 days for an expert 35 man-hours of technician's time	5 days for an expert, excluding most data collection 10 man-hours of technician time, in addition to the use of data from other tools
End-of-life scenario	Partially included	Not included	Included	Included
Software demand (price)	No software	No software needed but e.g., Visio recommended for drawings	No software; Microsoft Excel needed for calculations	LCA software (SimaPro/Gabi/Open LCA) and databases
Visualization type	• provide quick understanding of processes and correlated environmental aspects • One-page input and output for material and energy flow	• Process flow and one environmental parameter	• Ecomap shows waste generation points • Waste-sorting analysis via pie chart • Waste-handling logistics via spaghetti diagram	• Eco-profile • System boundary figure showing process flow • Environmental impacts at midpoint or endpoint level in absolute or relative terms • Software-dependent graphs
Guidance documents	Handbook available	Reports by the US-EPA	Handbook available	ISO 14044, ILCD Handbook
Ease of learning (knowledge requirements and days)	Easy to learn and implement. Needs • Workshop leader • Lean experience • One-day introduction	Easy to use. Needs • Workshop leader • One-week training	Slightly difficult due to variety of tools. Needs • Environmental manager or similar function • One to two days	Difficult. Needs • Expert • Several days
Supporting Go-to-gemba	Takes place at shop floor via walkthroughs	Requires shop floor visit	Requires shop floor visit	Data normally not found at shop floor, but a factory visit is recommended to understand and complement data
Employee engagement	Increased engagement in improvement actions on the shop floor	Increased engagement in improvement actions on the shop floor	Increased engagement in improvement actions on the shop floor	

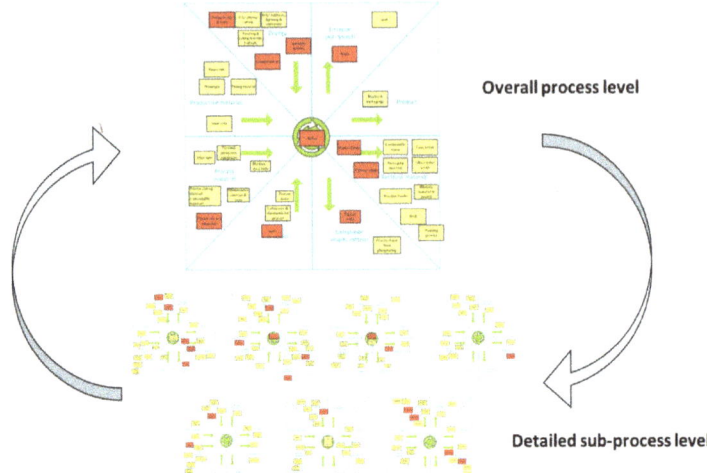

Figure 2. Performance Map Performed in this study at two levels. This figure does not intend to present the identified environmental impact and prioritization, but to present a schematic overview of GPM and how two different levels correlate.

4.2. Environmental Value Stream Mapping

EVSM was conducted following the structure presented by the EPA [40] with a focus on scrap generation. The EVSM also used the same sub-process categorization as the GPM. Based on further investigations, causes of scrap were divided into three categories: (1) design, (2) set up and processing and (3) quality. A notable drawback of using EVSM relates to the number of environmental aspects considered; it was challenging to include an additional environmental aspect (e.g., energy consumption) due to the difficulty of collecting additional data, along with the complexity of visualization. Therefore, Figure 3 depicts only metal scrap generation.

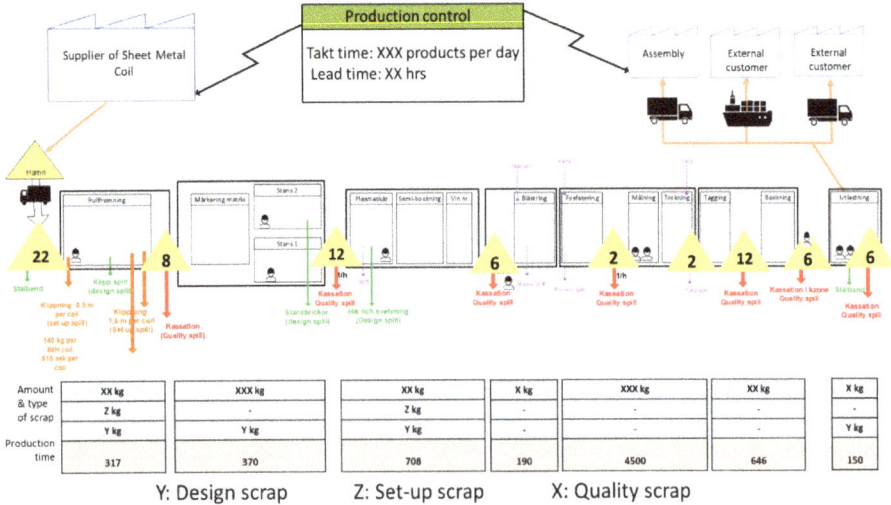

Figure 3. Value Stream Map performed in this study for the productive material. This figure does not intend to present why or how much scrap generated, but to present a schematic overview of EVSM for metal scrap generation.

Design scrap (Y) included designed waste material that is inevitable to avoid, such as holes from a punching machine to fulfill design requirements. Set-up scrap (Z) included scrap due to the machine's set-up or manufacturing processes (technology). Some examples are scrap from a plasma-cutting machine or a from steel-forming machine during the first round of changing the product's specifications to fulfill the process capability. Quality scrap (X) included all scrap due to quality failures, insufficient inspections and human errors. The EVSM also provided a flow analysis with lead time and buffer sizes.

4.3. Waste Flow Mapping

WFM was performed according to Kurdve et al. [21] to study different waste and material flows, with focus on the metal segment. In addition to data from internal waste management, waste management entrepreneurs, quality, maintenance and purchasing systems, data from 2015 and 2016 were also included to analyze an overview of the operation, scrap generation rate, bins and their contents. The process was divided into the same sub-processes as EVSM and GPM. An eco-map was created to understand the machinery, equipment and production flow as well as to localize the different types of waste bins (Figure 4). In addition, bins and containers contents were inspected and analyzed in terms of amount and sorting degree, while quality scrap generation points received extra analysis. Root cause analysis was then performed for the major scrap sources. Moreover, the transportation infrastructure was investigated and waste handling losses were found.

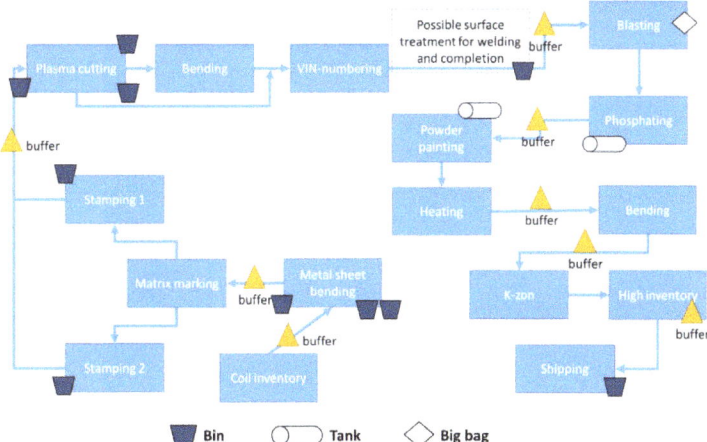

Figure 4. A schematic eco-map overview of the process. This figure only intends to present a schematic overview eco-map.

4.4. Life Cycle Assessment

A screening LCA was conducted for the entire vehicle frame manufacturing process with focus on impacts from the use of materials, energy, water, waste and hazardous materials. The screening LCA involved only the core data on the manufacturing process. Material composition and assumptions about the included phases of the life cycle are specified, but generic datasets from databases were used for all background data. While the LCA consisted of the required four stages: (1) goal and scope, (2) data inventory, (3) impact assessment, and 4) interpretation, the execution was modified from ISO 14044 [54] to benefit from joint data collection. Hence, the first two stages, i.e., goal and scope and data inventory, were performed using previously collected data from GPM, EVSM and WFM. Then additional data were collected to achieve a holistic scope of the environmental aspects. In the interpretation stage, results from other tools were also included to have a system perspective.

Figure 5 depicts the scope of the performed LCA for one vehicle frame, where the functional unit was "one vehicle frame". The system boundary includes the input material production (cradle) as well as the vehicle frame production and waste handling at the factory (gate). In coherence with the general rules of the EPD system [55], a cut off is done for recycled waste, i.e., the recycled waste belongs to the next product system. This practice allows to add product life cycles without double-counting emissions and resources. Internal and external transportation shown by arrows was included based on the availability of data. The product owner was involved in the interpretation stage but to a lesser degree than usual, instead operational practitioners reviewed the results and supported the interpretation.

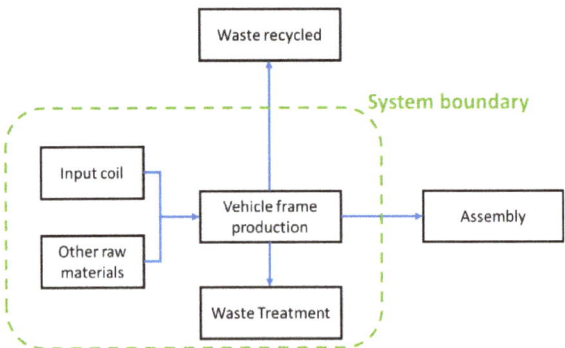

Figure 5. System boundary of performed LCA on the process in this study.

4.5. Comparison on Tool Deployment

Comparing the experience of using the tools, Table 2 shows overlaps and differences in several features. LCA differs the most from the other tools, since (1) it is traditionally focused on products (instead of processes), (2) it usually includes a broader scope of environmental aspects and therefore requires more time, (3) it requires more expertise and knowledge, (4) it is less associated with operational management and lean principles such as go-to-gemba, and (5) it requires specialized software and supporting databases. The overlapping features mainly include: (1) using quantitative and qualitative data (2) visualizing the environmental issues of concern on the same level, (3) including the entire process, although GPM can be used even at the cell or sub-process level, and (4) being based on a handbook or a guide to application.

GPM and LCA can focus on all environmental aspects and costs associated with them, whereas WFM concentrates only on waste and material flows. EVSM can consider various environmental aspects; however, our experience showed that it was cumbersome and confusing to consider more than one at the time. In terms of material and waste flows, EVSM did not include an end-of-life scenario i.e., how waste is treated afterwards such as recycling, reuse, and remanufacturing.

Comparing the tools in term of visualization, EVSM and WFM integrate the manufacturing process with environmental aspects for a better understanding, whereas GPM includes an input-output visualized model and LCA shows the environmental impacts at the site level. Some level of expertise was needed for all the tools, although this was much higher for LCA. Furthermore, fewer resources, including time, man-hours and expertise were required to carry out GPM, EVSM and WFM compared to LCA, i.e., the former tools required a workshop leader and several hours of practice to understand data collection and analytical methods. GPM was more dependent on operators' participation and shop floor walkthroughs than the other tools, even though walkthroughs (go-to-gemba) are essential for EVSM and WFM, but unnecessary for LCA. GPM, EVSM and WFM in addition to go-to-gemba, engage management and employees on daily continuous improvement (Kaizen) of the problem while LCA does not. Using GPM, EVSM and WFM for environmental sustainability improvements in manufacturing is more in harmony with lean principles than LCA, but those do not result in a holistic product/process view.

In addition, there is a potential risk that if the GPM, EVSM and WFM tools are not supported with an impact assessment when selecting or prioritizing environmental aspects, then the improvement actions will not be focused on the most important matters. It is vital to match the complexity of a given problem with the precision and completeness of a method or tool.

5. Case Results and Discussion

This section reports and discussed empirical results achieved applying the tools in practice with regards to our case study focus: Material efficiency and metal scrap generation. Therefore, this section presents results we achieved using each tool to answer case study questions including which sub-process has the largest environmental impact? How much scrap is generated in this process and the respective sub-processes and why? And what improvements can be made to move towards circularity, decreasing scrap generation and environmental impact?

5.1. Green Performance Map

Analyzing the production process using GPM, helped to identifying different types of environmental aspects and their relative importance based on the company's operational and environmental goals, in both the overall process and detailed sub-process levels. Focusing on only one level would have caused neglecting some environmental aspects. The results suggested five areas to start with improvement actions (labelled with red tags):

- Energy consumption for compressed air, hydraulic systems, painting and heat treatment;
- Hazardous materials, including chemicals and lubricants;
- Processed water from heat treatment, including phosphating and blasting;
- Scrap generation and waste of productive material;
- Noise from punching machines.

In addition to these five areas, GPM identified other environmental aspects with less critical impact (labelled with yellow tags), such as waste of painting powder, incorrect sorting, packaging issues, health and safety risk issues, and waste of consumable equipment (e.g., safety gloves and glasses). At the overall operation process level—from coil to shipping to customer—detailed quantitative data on scrap generation could be obtained, but quantitative data related to noise and material consumption remained unobtainable. At the sub-process level, it was even more difficult to obtain quantitative data, for example on material consumption and waste generation, energy consumption and noise made.

According to the GPMs performed on a sub-process level, metal sheet bending and plasma cutting produce the most scrap. This was based on a qualitative speculation, because it was challenging to use GPM for collecting quantitative data on the number of scrapped frames and their environmental impacts. Therefore, the qualitative speculation on the most scrap generation sub-processes turned out to be not entirely correct. It became clear later (via using EVSM) that GPM did not correctly prioritize punching machines as the most scrap generating sub-process (however plasma cutting and metal sheet bending are the third and fourth most scrap generating sub-processes) and did not locate the root cause of scrap generation; for example, scraps generated due to quality failure in surface treatment were identified in the shipping stage, meaning that a pair of vehicle frames that should have been scraped after the surface treatment, had gone through all the other processes wasting energy and materials as well as production time. Nevertheless, GPM engaged personnel in improvement actions.

5.2. Environmental Value Stream Mapping

Visualizing different material flows on EVSM and then adding energy made the EVSM complicated; therefore, only metal scrap flows were taken into consideration in this tool. However, energy statistics were collected using energy mapping tools. From an environmental perspective with focus on one aspect—scrap generation—, EVSM helped to understand the type, volume and reasons for scrap generation in each sub-process, i.e., localization (where), quantification (how much), categorization to quality, set up or design scrap (why). This provided a good starting point for improvement measures. The results further determined that quality scrap should be prioritized as it had the highest scrap generation proportion and associated cost; almost all sub-processes generate quality scrap. The most quality scraps were generated by the punching machines (27%), followed by surface treatment (22%), plasma cutting (19%), and metal sheet forming (17%). Surface treatment included blasting, phosphating, painting and paint curing oven. Reasons for scrap generation within these operations were identified afterwards, e.g., half of the frames scrapped after surface treatment had paint lumps. Although the focus was on material, the investigation showed that surface treatment consumes one-third of total plant energy consumption. In addition to environmental data, EVSM also showed traditional production-related data such as takt time, lead time, first time through (95% with 3.5% product reworked and 1.5% scrap) and overall equipment effectiveness (OEE). The process has a high OEE (~80%), however it is expected for such a semi-automated, low-complexity and high-volume process.

5.3. Waste Flow Mapping

To a large extent, the results achieved by WFM overlapped with those from GPM and EVSM, particularly concerning scrap volumes and waste handling at each sub-process. However, WFM also analyzed inefficiencies in waste management and transportation. It indicated a high degree of waste-sorting with an average of 88%, with improvement potential to 95%. Moreover, metal scraps are sent (with lower revenue) to open-loop recycling, resulting in a 100% recycling rate. Figure 6 illustrates the waste-sorting analyses. The pie chart shows the waste segments produced in the studied process with their respective percentages. Data for scrap generation and production rate during the period between 2015 and 2016 were used to estimate the volumes and analyze the root cause. The calculated scrap generation rate is shown in Table 3.

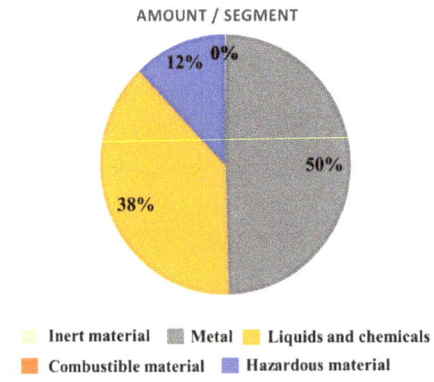

Figure 6. Waste-sorting analyses performed via WFM.

A vertical analysis of the handling and costs of waste management was performed based on interviews. The bins used were typically "dumping hoppers" for metals and transparent plastic bins for plastic waste, wood and combustibles. There were essentially no mixed metals bins. Other scrap metal collection points contained obsolete material, scrapped equipment and construction material. The metal scraps are transported to the environmental zone by two internal forklifts (one large and one regular), and then from there external transport of scraps is performed on demand every other day by the waste management entrepreneur. Shipping containers off-site typically use half of the maximum load of the truck. It is estimated that the forklifts are used two to three hours per week for waste management purposes. With 144 h of operating time per week, the relative cost is estimated to be 2%

of the annual investment, plus 3 h/week of operator time. The average load of trucks is low, especially for one type of cutting (3.9 tons/truck load). Hazardous material is also internally transported by a forklift to the environmental zone, and phosphating baths are collected in tanks and emptied by pipes to tank truck transport. Some fraction required expensive handling vehicles on site i.e., a suction tank and crane trucks.

Table 3. Metal scrap generation analysis.

	2015		2016	
	% of Total Scrap	% of Total Material Consumption in Vehicle Frame Production	% of Total Scrap	% of Total Material Consumption in Vehicle Frame Production
Design scrap	59%	4.5%	64%	5.4%
Set-up scrap	1%	0.1%	1%	0.1%
Quality scrap	40%	3.1%	35%	3%
Total		7.73%		8.5%

5.4. Life Cycle Assessment

LCA used previously collected data via the other tools and complemented them with additional data mainly from environmental data systems. Simapro software and the Ecoinvent database were used to analyze and quantify different environmental aspects including climate impact, eutrophication, acidification and smog. Figure 7 illustrates the climate impacts of the studied process from the cradle (material production) to the gate at the factory. The thickness of the arrows corresponds to the climate impact measured in carbon dioxide equivalents from each part. In the lower left corner of the process boxes, emissions in kilograms of CO_2 equivalents are given. As shown in Figure 7, the production phase of steels coil (raw material) has the greatest climate impact. However, concentrating on the manufacturing phase only (vehicle frame production), quality scrap generation has the greatest environmental impact. For instance, both design scrap and quality scrap have greater climate impacts than production of the coating powder. The set-up scrap has slightly lower climate impacts than coating powder production. Energy use during production has a slightly lower climate impact than inbound steel transportation. Conclusions above are consistent for other impact categories of ground-level ozone, acidification and eutrophication.

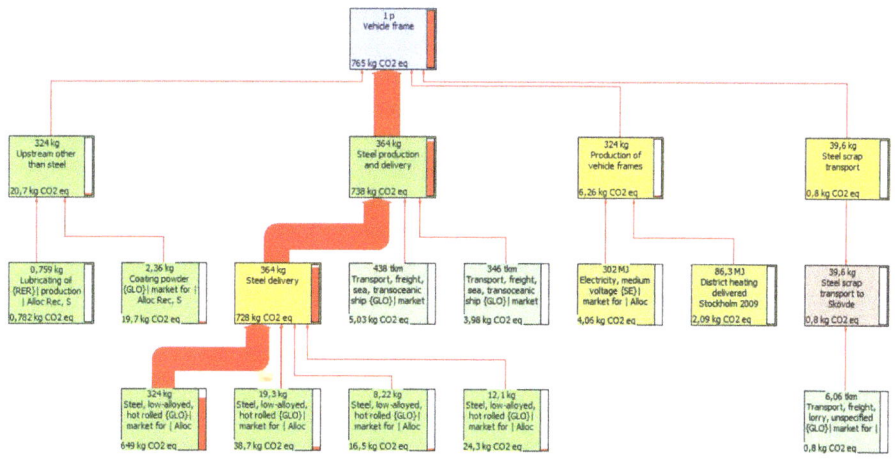

Figure 7. Climate impacts for a frame from cradle to gate.

5.5. Cross Comparison

Comparing the empirical results achieved by applying the tools in practice with regard to our case study focus—material efficiency and metal scrap generation— indicates differences between the tools in both system boundaries (how much of the production system was considered) and which impact category/-ies was/were considered. For example, with regards to transportation: LCA includes transportation and the associated impacts from all operations within the scope; the WFM includes waste management transportation but not inbound material; and GPM and EVSM can include transportation if it is included as an operation in the process or as an environmental aspect. With regard to the impact category, WFM focuses on waste and material flows; LCA and GPM handle many different impact categories; EVSM considers various environmental aspects, but our experience showed that it was cumbersome and confusing to do so while collecting and analyzing data. In addition, EVSM did not include an end-of-life scenario, i.e., how waste is treated afterwards, such as recycling, reuse, and remanufacturing, which directly correlates to environmental impact. Therefore, our experience concludes that with GPM and LCA, different environmental impact categories can be studied, whereas WFM focuses on one impact category (waste), and with EVSM, it is better to focus on one impact category at a time.

As mentioned earlier, this paper does not intend to select one tool as superior to the others. The results suggest that there is no such thing as "one right tool". Instead, we advocate using a combination of tools, each of which has different strengths and weaknesses, include different types of data collection and analysis, and different information on environmental impacts. Combining tools can lead to better prioritization, decision-making and increased engagement. Based on the predetermined improvement goal, one tool per se might not deliver the desired outcome. Thus, combinations of tools that support each other should be considered e.g., the overall GPM for the entire vehicle frame operation could have been quantified and together with a screening LCA could be used to verify and prioritize the most significant environmental impacts and engage personnel in improvement actions on the shop floor. Because EVSM and WFM were the only methods capable of identifying the root cause of quality scrap generation, one of these methods was necessary for our particular case study. Figure 8 illustrates the integration of environmental assessment tools discussed in this paper, based on four stages of LCA.

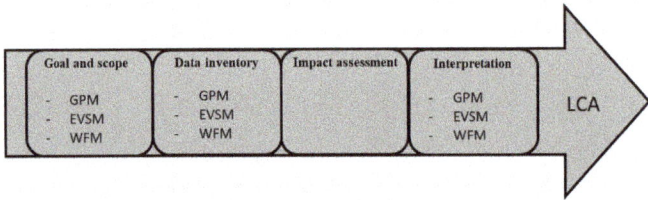

Figure 8. Integration of environmental assessment tools.

Stepping outside the system boundary, a limited LCA investigation focusing on alternatives to the current steel scrap recycling was conducted. Most of the generated scrap is currently transported 165 km for cast iron foundry. However, transporting the scrap to a steel supplier (452 km) could result in better use of existing alloy elements. The climate cost for the longer transport to the steel supplier is 2.3 kg CO_2 (see Figure 9), whereas it is 0.8 kg CO_2 for transport to foundry. This can be compared to the climate cost of the ferro-nickel alloy in 39.6 kg of product scrap which is 9.65 kg CO_2. This is significantly greater than the climate cost of extra transport to the steel supplier. Nevertheless, the generic dataset used to model the steel does not exactly match the actual steel quality, so this potential improvement action must be confirmed with actual steel qualities.

Table 4 provides answers to the question posed by the company in this study.

Table 4. Main answers of the industrial questions posed in the case study.

Focus of the Case Study	GPM	EVSM	WFM	LCA
Which sub-process has the most environmental impact?	GPM pinpointed several environmental aspects e.g., scrap generation in metal sheet bending, plasma cutting and k-zone; high energy consumption at hydraulic systems and a pneumatic truck in metal sheet bending; high energy consumption at heat treatment; hazardous ash from blasting and plasma cutting; water waste and chemicals generated from phosphating; and other sustainability issues such as safety risks in surface treatment and high noises from punching machines. However, it was challenging to quantify the environmental aspects and correlating impact.	EVSM concluded that the punching machines produces the greatest number of quality scraps and was thus the most important environmental aspect. However, due to complexity, EVSM did not consider environmental aspects such as noise and water and energy consumption.	According to the WFM, the most scraps are generated from punching machines, plasma cutting, surface treatment, and metal sheet forming. However, quality scraps from the punching machines had the greatest environmental impact. In addition to scrap generation, environmental aspects such as waste segregation, transportation and hazardousness were considered. For instance, 38% of process fluids waste from surface treatment are sent to destruction, which indicates that this process has large environmental impacts and costs.	According to the LCA, from a product life cycle perspective, it was found that steel coil production has by far the greatest environmental impacts (climate impact, eutrophication, acidification and smog). Therefore, scrap generation during production is of large importance. Furthermore, quality and design scraps have greater environmental impacts than energy use in the plant.
How much scrap is generated in the vehicle frame production process and in the respective sub-processes and why?	Metal sheet bending, plasma cutting and k-zone produced the most scrap. However, it was challenging to collect quantitative data and determine environmental impacts, and therefore GPM did not correctly prioritize punching machines. GPM also failed to localize the root cause scrap generation points.	The most scraps are generated from punching machine plasma cutting, surface treatment, and metal sheet forming. EVSM localized and quantified scrap generation throughout the process and identified the reason for the scrap generation at sub-process level.	WFM identified that quality scrap from the punching machine has the greatest environmental aspect. The proportion of scrap generation compared to production was also calculated.	Data on all major waste flows were quantified with LCA; first for the whole process output and then based on their relative impact, improvement potentials are sought. It is not unusual to identify hitherto unknown flows during LCA since mass balances are often used to check inputs with outputs.
What improvements can be made to move towards circularity, decreasing scrap generation and environmental impact?	GPM identified various environmental aspects and pinpointed their origin in the process but not specifically suggest any improvement.	Quality scrap should be improved with a better inspection and logging system.	Waste segregation potentials should be improved.	The comprehensive LCA could find and prioritize the greatest environmental aspects based on their relative environmental impact. The improvement potential for improved metal recycling could also be quantified but a detailed inventory of the process and sub-processes is needed to point out specific process steps to be modified for better material efficiency.

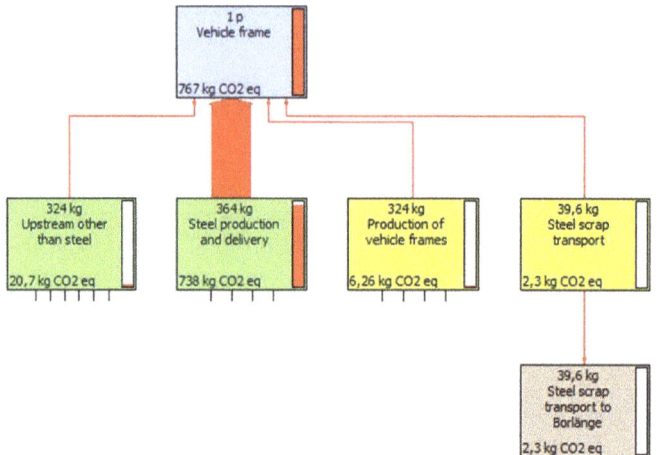

Figure 9. Climate impacts for a frame from cradle to gate for scrap transport to a steel supplier.

6. Conclusions

Previous studies have shown that there is an insufficient use of practical tools and methods on the manufacturing shop floor for regular environmental performance assessment and improvement. Recent studies have shown that the green lean concept has begun to address this gap. Therefore, in this paper, four common environmental assessment tools used in Swedish manufacturing industry —GPM, EVSM, WFM and LCA—were used to assess the environmental aspects of a vehicle frame production process but with a focus on material efficiency and metal scrap generation. These tools were compared in terms of use experience and features as well as results achieved to fulfill the objectives of this study: To help practitioners and scholars to understand the different features of each tool, so in turn, the right tool(s) can be selected according to particular questions and the industrial settings. Although this paper focused on material efficiency and metal scrap generation, other environmental aspects such as energy use, material consumption and chemicals (hazardousness) were also considered when applied the tools.

A literature review revealed that a tool benefits from the following essential features to enable environmental improvements in manufacturing:

- Being hands-on and operational, supporting collaboration and understanding between different internal and external actors;
- Being easy to learn and implement, visualization, time efficiency, continuous improvement and engagement;
- Including root cause analysis, being harmonized with ISO 14001 and supporting the go-to-gemba concept;
- Being goal-oriented, supporting measurements, and being focused on a limited area of influence while supporting systematic work procedures (standardized work)

The results of deploying the selected tools in terms of use experience and feature comparison on the same manufacturing process are presented in Table 2. The tools are compared in different features such as result type, operation level (site, process, cell), resource type, time required for data collection and analysis, end-of-life scenario, software demand (price), visualization type, guidance documents, ease of learning (knowledge requirements and days), and go-to-gemba.

Table 4 presents the application of these tools in a real-time industrial case study with focus on material efficiency and metal scrap generation. According to the results, the tools have overlaps and differences. Each tool has strengths and weaknesses that depend on variables such as the question posed, expected results, the level of evaluation (site, process or cell), and the accessibility of data. GPM can effectively generate an input-output model, providing a visual overview of environmental problems; EVSM shows selected environmental aspects alongside production-related data and information flows, which enhances understanding; WFM provides a detailed analysis of material and waste flows along the waste management supply chain; LCA can help understand the degree of environmental impact associated with different environmental aspects and therefore is essential for correct prioritization and to avoid sub-optimization. It is also concluded that data collection for the various tools can be performed at the same time, allowing for parallel application of more than one tool with minimal extra time and resource efforts. Combining tools in this way can provide a superior answer to a specific question and better guide and prioritization of shop floor solutions.

The results indicate that there is no such thing as "one right tool" superior to others. Rather, a combination of tools is suggested for different types of data and analysis, and to assess various environmental impacts, which results in better prioritization and decision-making and increasing the effectiveness and efficiency of the operation, environmental performance and the value stream in terms of lean and green. Future research might deploy the tools at more manufacturers with different variables relating to, for example, company size, industry type, product type, and auxiliary and residual material types. Future research can also include SMEs which generally have fewer resources to monitor environmental issues. Additionally, the case study focus was carried out at a company using metal as a primary product material. Future research could replicate this research in other industries that use different primary product materials, such as plastics.

Supplementary Materials: A deeper description of each tool is available online at http://www.mdpi.com/2071-1050/11/7/2173/s1.

Author Contributions: The authors contributed in this article as following, conceptualization and methodology, C.J., M.K., M.Z., S.S., investigation, M.K., M.Z., S.S. formal analysis A.R.K., C.J., M.K., M.Z., S.S., validation S.S., M.K., M.Z., discussion and concluding A.R.K., C.J., M.K., M.Z., S.S.

Funding: This research was funded by Mistra Closing the loop II and XPRES.

Acknowledgments: This article is connected to the research project CiMMRec (Circular Models for mixed and Multi material Recycling in Manufacturing extended Loops). The authors acknowledge the funding received for the project from Mistra Closing the loop II and XPRES.

Conflicts of Interest: The authors declare no conflict of interest.

References

1. Garetti, M.; Taisch, M. Sustainable manufacturing: Trends and research challenges. *Prod. Plan. Control* **2011**, *23*, 83–104.
2. Veleva, V.; Bailey, J.; Jurczyk, N. Using sustainable production indicators to measure progress in ISO 14001, EHS System and EPA achievement track. *Corp. Environ. Strat.* **2001**, *8*, 326–338.
3. IISD, International Institute for Sustainable Development. *Business Strategies for Sustainable Development*; The International Institute for Sustainable Development in conjunction with Deloitte & Touche and the World Business Council for Sustainable Development: Winnipeg, MB, Canada, 1992.
4. Smith, L.; Ball, P. Steps towards sustainable manufacturing through modelling material, energy and waste flows. *Int. J. Prod. Econ.* **2012**, *140*, 227–238.
5. Bey, N.; Hauschild, M.Z.; Mcaloone, T.C. Drivers and barriers for implementation of environmental strategies in manufacturing companies. *CIRP Ann. Manuf. Technol.* **2013**, *62*, 43–46.
6. Kurdve, M.; Zackrisson, M.; Wiktorsson, M.; Harlin, U. Lean and green integration into production system models – experiences from Swedish industry. *J. Clean. Prod.* **2014**, *85*, 180–190.
7. Zokaei, K.; Lovins, H.; Wood, A.; HInes, P. *Creating a Lean and Green Business System: Techniques for Improving Profits and Sustainability*; CRC Press, Taylor and Francis Group: London, UK, 2013.
8. Noren, G.; Strömdahl, I. *Näringslivets miljöarbete och Sveriges miljömål*; SvensktNäringsliv: Stockholm, Sweden, 2007.

9. Shahbazi, S. Material Efficiency Management in Manufacturing. Licentiate Thesis, Mälardalen University, Västerås, Sweden, 2015.
10. Cherrafi, A.; Elfezazi, S.; Govindan, K.; Garza-Reyes, J.A.; Benhida, K.; Mokhlis, A. A framework for the integration of Green and Lean Six Sigma for superior sustainability performance. *Int. J. Prod. Res.* **2017**, *55*, 4481–4515.
11. Dües, C.M.; Tan, K.H.; Lim, M. Green as the new Lean: How to use Lean practices as a catalyst to greening your supply chain. *J. Clean. Prod.* **2013**, *40*, 93–100.
12. Shahbazi, S.; Wiktorsson, M.; Kurdve, M.; Jönsson, C.; Bjelkemyr, M. Material efficiency in manufacturing: Swedish evidence on potential, barriers and strategies. *J. Clean. Prod.* **2016**, *127*, 438–450.
13. Lindahl, M. Engineering Designers' Requirements on Design for Environment Methods and Tools. Ph.D. Thesis, KTH, Stockholm, Sweden, 2005.
14. Bergendahl, M.N. Stödmetoder och samverkan i produktutveckling (Supporting Methods and Collaboration in Product Development). Ph.D. Thesis, KTH Royal Institute of Technology, Stockholm, Sweden, 1992.
15. Jönbrink, A.K.; Kristinsdottir, A.R.; Roos, S.; Sundgren, M.; Johansson, E.; Nyström, B.; Nayström, P. Why use ecodesign in the industry 2013? A Survey regarding barriers and opportunities related to ecodesign. In Proceedings of the EcoDesign 2013 International Symposium, Tokyo, Japan, 4–6 December 2013.
16. Ellinger, A.E.; Daugherty, P.J.; Keller, S.B. The relationship between marketing/logistics interdepartmental integration and performance in US manufacturing firms: An empirical study. *J. Bus. Logist.* **2000**, *21*, 1–22.
17. Iranmanesh, M.; Zailani, S.; Hyun, S.S.; Ali, M.H.; Kim, K. Impact of Lean Manufacturing Practices on Firms' Sustainable Performance: Lean Culture as a Moderator. *Sustainability* **2019**, *11*, 1112.
18. Wu, L.; Subramanian, N.; Abdulrahman, M.D.; Liu, C.; Lai, K.-H.; Pawar, K.S. The Impact of Integrated Practices of Lean, Green, and Social Management Systems on Firm Sustainability Performance—Evidence from Chinese Fashion Auto-Parts Suppliers. *Sustainability* **2015**, *7*, 3838–3858.
19. Mollenkopf, D.; Stolze, H.; Wendy, H.T.; Ueltschy, M. Green, lean, and global supply chains. *Int. J. Phys. Dist. Logist. Manag.* **2010**, *40*, 14–41.
20. Garza-Reyes, J.A. Lean and green—A systematic review of the state of the art literature. *J. Clean. Prod.* **2015**, *102*, 18–29.
21. Kurdve, M.; Shahbazi, S.; Wendin, M.; Bengtsson, C.; Wiktorsson, M. Waste flow mapping to improve sustainability of waste management: A case study approach. *J. Clean. Prod.* **2015**, *98*, 304–315.
22. Zackrisson, M.; Avellán, L.; Orlenius, J. Life cycle assessment of lithium-ion batteries for plug-in hybrid electric vehicles—Critical issues. *J. Clean. Prod.* **2014**, *18*, 1519–1529.
23. Shahbazi, S.; Wiktorsson, M. Using the Green Performance Map: Towards material efficiency measurement. In Proceedings of the 23th EurOMA Conference, Trondheim, Norway, 17–22 June 2016.
24. Kurdve, M.; Wiktorsson, M. *Green Performance Map: Visualizing Environmental KPI's*; European Operations Management Association (EurOMA): Dublin, Ireland, 2013.
25. Baumgartner, R.J. Managing Corporate Sustainability and CSR: A Conceptual Framework Combining Values, Strategies and Instruments Contributing to Sustainable Development. *Corporate Soc. Responsib. Environ. Manag.* **2014**, *21*, 258–271.
26. Simpson, D.F.; Power, D.J. Use the supply relationship to develop lean and green suppliers. *Supply Chain Manag. Int. J.* **2005**, *10*, 60–68.
27. Vachon, S.; Klassen, R.D. Environmental management and manufacturing performance: The role of collaboration in the supply chain. *Int. J. Prod. Econ.* **2008**, *111*, 299–315.
28. Faulkner, W.; Badurdeen, F. Sustainable Value Stream Mapping (Sus-VSM): Methodology to visualize and assess manufacturing sustainability performance. *J. Clean. Prod.* **2014**, *85*, 8–18.
29. Benn, S.; TEO, S.T.T.; Martin, A. Employee participation and engagement in working for the environment. *Pers. Rev.* **2015**, *44*, 492–510.
30. Banawi, A.; Bilec, M.M. A framework to improve construction processes: Integrating Lean, Green and Six Sigma. *Int. J. Constr. Manag.* **2014**, *14*, 45–55.
31. Souza, J.P.E.; Alves, J.M. Lean-integrated management system: A model for sustainability improvement. *J. Clean. Prod.* **2018**, *172*, 2667–2682.
32. Puvanasvaran, P.; Tian, R.K.S.; Suresh, V.; Muhamad, M. Lean principles adoption in environmental management system (EMS)-ISO 14001. *J. Ind. Eng. Manag.* **2012**, *5*, 406–430.

33. Seth, D.; Seth, N.; Dhariwal, P. Application of value stream mapping (VSM) for lean and cycle time reduction in complex production environments: A case study. *Prod. Plan. Control* **2017**, *28*, 398–419.
34. Ruben, B.; Viondh, S.; Asokan, P. State of art perspectives of lean and sustainable manufacturing. *Int. J. Lean Six Sigma* **2018**, *10*, 234–256.
35. Mĺkva, M.; Prajová, V.; Yakimovich, B.; Korshunov, A.; Tyurin, I. Standardization—One of the Tools of Continuous Improvement. *Procedia Eng.* **2016**, *149*, 329–332.
36. Bellgran, M.; Höckerdal, K.; Kurdve, M.; Wiktorsson, M. *Green Performance Map—Handbook*; Mälardalen University: Eskilstuna, Sweden, 2012.
37. Kurdve, M.; Hanarp, P.; Chen, X.; Qiu, X.; Zhang, Y.; Stahre, J.; Laring, J. *Use of Environmental Value Stream Mapping and Environmental Loss Analysis in Lean Manufacturing Work at Volvo*; Swedish Production Symposium, SPS11; Swedish Production Symposium: Lund, Sweden, 2011.
38. Pampanelli, A.B.; Found, P.; Bernardes, A.M. A Lean & Green Model for a production cell. *J. Clean. Prod.* **2014**, *85*, 19–30.
39. Sawhney, R.; Teparakul, P.; Bagchi, A.; Li, X. En-Lean: A framework to align lean and green manufacturing in the metal cutting supply chain. *Int. J. Enterprise Netw. Manag.* **2007**, *1*, 238–260.
40. EPA. Lean and Clean Value Stream Mapping. 2015. Available online: www.epa.gov/e3 (accessed on 15 December 2018).
41. Torres, A.S.; Gati, A.M. Environmental Value Stream Mapping (EVSM) as sustainability management tool. In Proceedings of the Portland International Conference on Management of Engineering Technology, PICMET 2009, Portland, OR, USA, 2–6 August 2009; pp. 1689–1698.
42. Müller, E.; Schillig, R.; Stock, T.; Schmeiler, M. Improvement of Injection Moulding Processes by Using Dual Energy Signatures. *Procedia CIRP* **2014**, *17*, 704–709.
43. Posselt, G.; Fischer, J.; Heinemann, T.; Thiede, S.; Alvandi, S.; Weinert, N.; Kara, S.; Herrmann, C. Extending Energy Value Stream Models by the TBS Dimension—Applied on a Multi Product Process Chain in the Railway Industry. *Procedia CIRP* **2014**, *15*, 80–85.
44. Gunduz, M.; Fahmi Naser, A. Cost Based Value Stream Mapping as a Sustainable Construction Tool for Underground Pipeline Construction Projects. *Sustainability* **2017**, *9*, 2184.
45. Dadashzadeh, M.D.; Wharton, T.J. A value stream approach for greening the IT department. *Int. J. Manag. Inf. Syst.* **2012**, *16*, 12. [CrossRef]
46. Engel, H.W. *Ecomapping—A Visual, Simple and Practical Tool to Analyse and Manage the Environmental Performance of Small Companies and Craft Industries*; EcoMapping Network: Brussels, Belgium, 2002.
47. Hogland, W.; Stenis, J. Assessment and system analysis of industrial waste management. *Waste Manag.* **2000**, *20*, 537–543.
48. Muther, R.; Haganas, K. *Systematic Handling Analysis*; Management and Industrial Research Publications: Kansas City, MO, USA, 2002.
49. Cheung, W.M.; Leong, J.T.; Vichare, P. Incorporating lean thinking and life cycle assessment to reduce environmental impacts of plastic injection moulded products. *J. Clean. Prod.* **2017**, *167*, 759–775.
50. Zhang, H.; Haapala, K.R. Integrating sustainable manufacturing assessment into decision making for a production work cell. *J. Clean. Prod.* **2015**, *105*, 52–63.
51. Thammaraksa, C.; Wattanawan, A.; Prapaspongsa, T. Corporate environmental assessment of a large jewelry company: From a life cycle assessment to green industry. *J. Clean. Prod.* **2017**, *164*, 485–494.
52. Yin, R.K. *Case Study Research: Design and Methods*, 5th ed.; SAGE Publications, Inc.: London, UK, 2010.
53. Saunders, M.; Lewis, P.; Thornhill, A. *Research Methods for Business Students*; Financial Times Prentice Hall: Upper Saddle River, NJ, USA, 2009.
54. *ISO 14044. Environmental Management—Life Cycle Assessment—Requirements and Guidelines*; International Organization for Standardization: Geneva, Switzerland, 2006.
55. EPD®, General programme instructions for the international epd® system. Version 3.0. 2017-12-11. Stockholm, Sweden, 2017.

© 2019 by the authors. Licensee MDPI, Basel, Switzerland. This article is an open access article distributed under the terms and conditions of the Creative Commons Attribution (CC BY) license (http://creativecommons.org/licenses/by/4.0/).

Article

Parameter Setting for a Genetic Algorithm Layout Planner as a Toll of Sustainable Manufacturing

Martin Krajčovič [1], Viktor Hančinský [2], Ľuboslav Dulina [1], Patrik Grznár [1,*], Martin Gašo [1] and Juraj Vaculík [3]

[1] Faculty of Mechanical Engineering, University of Žilina, Univerzitná 8215/1, 010 26 Žilina, Slovakia; martin.krajcovic@fstroj.uniza.sk (M.K.); luboslav.dulina@fstroj.uniza.sk (Ľ.D.); martin.gaso@fstroj.uniza.sk (M.G.)
[2] GE Aviation s.r.o., Beranových 65, 199 02 Prague 9, Letňany, Czech Republic; kpi@fstroj.uniza.sk
[3] Faculty of Operation and Economics of Transport and Communications, University of Žilina, Univerzitná 8215/1, 010 26 Žilina, Slovakia; juraj.vaculik@fpedas.uniza.sk
* Correspondence: patrik.grznar@fstroj.uniza.sk; Tel.: +421-41-513-2733

Received: 7 March 2019; Accepted: 3 April 2019; Published: 8 April 2019

Abstract: The long-term sustainability of the enterprise requires constant attention to the continuous improvement of business processes and systems so that the enterprise is still competitive in a dynamic and turbulent market environment. Improvement of processes must lead to the ability of the enterprise to increase production performance, the quality of provided services on a constantly increasing level of productivity and decreasing level of cost. One of the most important potentials for sustainability competitiveness of an enterprise is the continuous restructuring of production and logistics systems to continuously optimize material flows in the enterprise in terms of the changing requirements of customers and the behavior of enterprise system surroundings. Increasing pressure has been applied to projecting manufacturing and logistics systems due to labor intensity, time consumption, and costs for the whole technological projecting process. Moreover, it is also due to quality growth, complexity, and information ability of outputs generated from this process. One option is the use of evolution algorithms for space solution optimization for manufacturing and logistics systems. This method has higher quality results compared to classical heuristic methods. The advantage is the ability to leave specific local extremes. Classical heuristics are unable to do so. Genetic algorithms belong to this group. This article presents a unique genetic algorithm layout planner (GALP) that uses a genetic algorithm to optimize the spatial arrangement. In the first part of this article, there is a description of a framework of the current state of layout planning and genetic algorithms used in manufacturing and logistics system design, methods for layout design, and basic characteristics of genetic algorithms. The second part of the article introduces its own GALP algorithm. It is a structure which is integrated into the design process of manufacturing systems. The core of the article are parameters setting and experimental verification of the proposed algorithm. The final part of the article is a discussion about the results of the GALP application.

Keywords: sustainability; genetic algorithm; layout planning; model

1. Introduction

Today, the sustainable enterprise needs to use approaches and concepts that allow rapid adaptability of business processes and systems for dynamic environs change. An important concept that creates conditions for quick design reconfiguration of processes and systems is the digital factory.

The digital factory is a concept including a network of digital models, methods, and tools, such as simulation and 3D-visualisation, which are integrated through comprehensive data and flexible modules management. Products, processes, and resources are modeled based on actual data, in a virtual factory. Based on the actual data and models the planned products and production processes can be improved by use of virtual models until the processes are fully developed, extensively tested, and mostly error-free for their use in a real factory [1].

Manufacturing system's design and layout planning and are basic activities in the digital factory. Their main task is to judge relations of each production system element regarding time and spatial requirements as well as work, technological, handling, control, and other activities inevitable for the rational production process and suitable spatial and time structure of production process.

According to [2] the manufacturing system's design in a digital factory needs three groups of input data. The first group of data is data about the products which will be produced in the production system (assortment, production volume, product structure, physical characteristics of products, demand timeframe, etc.). The second is data about processes used in the production of products (technologies used in the production, steps of processes, operating times, etc.) The last group is data about available/needed resources for production (production machines and facilities, transportation, handling and storage devices, handling units, etc.).

When planning and building manufacturing systems, it is possible to use several approaches whose applicability is dependent on a particular case. Basic approaches according to Lee [3] are as follows:

- Knowledge-based approach: In this approach, the production system is made through gained knowledge, instinct, and common sense. Such systems benefits from the rich knowledge of all current and past employees, but it also has its downside. There is a tendency to use out dated information and overlook new technology and organizational structures. In addition, this approach could be highly unorganized as various workers could have opposing experiences. When planning complex systems, it is advisable to collect knowledge from different perspectives on issues and use them after close examination.
- Cloning: This approach duplicates the existing production system, or which are a part of it. If the existing performance production system is satisfactory and conditions for the planning system are equal, it will be possible to build this production system in a relatively short time, which is the main advantage. However, for the majority of cloning there is limited contribution because the place, process, and people or legislation within each production system could be very different.
- Bottom-Up: The bottom-up and top-down approach starts with details and then it moves step by step up to the level of a whole production plant. The approach is suitable under known conditions and how they should be integrated into a larger group without any change. These conditions are mostly fulfilled in smaller companies in a stable environment. However, the disadvantage of this approach is a long solution period, and until we get to a final layout and plant construction, all details must be integrated. In more complex projects overload of details might occur which leads to worse project clarity.
- Systematic Approach (SLP—Systematic Layout Planning): This approach uses procedures, conventions, and phases which helps the planner to know exactly what to do in each project step. This approach introduces system and structure in planning and also contribution, such as time and work reduction. Primarily, the layout of each block in space is solved.
- Strategic Approach (Top-Down approach): This approach puts emphasis on aims and sets technology and organization in a way they would support each other. This approach starts with a company strategy, plant location selection, and moves forward towards the detailed layout of each element. This approach is direct and with clear aim enabling each project assistant to proceed with same–mutual direction.
- Dominant approach: This approach focuses on a certain form of presentation or company advertisement via a planned plant. It uses an interior or exterior to portray technological innovation, artistic feeling or financial company support.

The layout planning (as a sub-activity of manufacturing system design) is a complex activity involving the optimization of the positions of machines, transportation systems, and workstations [4].

Until recently, classical heuristic approaches (e.g., CORELAP, ALDEP, PLANET, CRAFT, BLOCPLAN) were preferred in optimizing spatial arrangement. Nowadays, layout optimization has been made more efficient by using information technology tools and advanced optimization methods. These methods called metaheuristics have higher quality results compared to classical heuristic methods. Their advantage is the ability to leave specific local extremes and to find a better solution than classical heuristics. One group of metaheuristics are genetic algorithms [5].

Genetic algorithms (GA) are a useful tool for the solution for different tasks in practice. First and foremost, genetic algorithms are used in experiments and simple problems. After verifying the usability of genetic algorithms and the increase of computer technological performance, genetic algorithms started to be used in more complicated tasks. One of the areas of GA practical applications is the design of manufacturing and logistics systems.

One of the first application areas of genetic algorithms in production systems design was production and assembly line balancing and design. In Rekiek [6], the problem of design and optimization of the assembly line was analyzed in detail. However, newly presented multiple objective grouping algorithm (MO-GGA, Multiple Objective Grouping GA) is based on grouping genetic algorithms (GGA) and on the method multiple objective decision PROMETHEE II (Preference Ranking Organization METHod for Enrichment Evaluation). The main difference between grouping and classical genetic algorithms is in gene structure and approach of operators (crossover, mutation, and inversion) of these genes. In GGA, there is a group of objects encrypted in a gene. Apart from an ordinary algorithm, there is the object itself which is encrypted. Thus, there is an optimization of OptiLine software presented in a publication which uses a genetic algorithm.

Hnat [7] underlined a question of assembly line balance through the help of genetic algorithms. In addition, he emphasized a decoder in suitable and proposed technology. Furthermore, use of this application illustrated that a chromosome will not contain an encoded solution, but the information gathered could be proposed as a solution. Moreover, by selecting a suitable representation, it is possible to create a new space instead of the original one.

Kothari [8] designed a genetic algorithm GENALGO for machine arrangement of a given length to line. This algorithm periodically performs a local search for individuals and their suitability increase. Specific function is given as a product sum of all distances between machines and their intensity.

Genetic algorithms for optimizing manufacturing facility layout work [5] summarizes adapted scientific works dealing with various problem solutions of layout optimization. This type of algorithm is used as a solution.

- Suresh with his colleagues used the genetic approach to problem solving of a structural device arrangement with the aim to keep costs for interaction between individual departments to a minimum. Device arrangement problem focuses on finding the best cell arrangement and not solely relying on machines and devices.
- Gupta with his colleagues used a genetic algorithm to distribute products to families and design arrangement between cells. The developed algorithm is focused on cell system arrangement or area arrangement of a production hall rather than the arrangement inside the cells. Neither personal arrangement of machines in cells nor relations between them were considered.
- Two-step hierarchical decomposition approach has been developed, too. First, it is the decision of each object arrangement. For this so-called task, the greedy genetic algorithm is used. Second, it is finding the best disposition arrangement. A genetic algorithm has been used for space solution search. Authors state that for less complex problems, the algorithm offers optimal results and for more serious problems, it overcomes existing methods in speed and quality of the solution.

Kia [9] introduced the model of genetic algorithm use for multi-story objects. In this model, manufacturing cells for pre-defined slots are allocated, but transport between each level has to be considered—for example, with the help of elevators fixed in the layout. This particular solution does not take into account real machine dimensions in the proposal, nor does it consider relationships between each workplace. However, it is possible to define cell capacity—meaning how many machines a cell can contain. The algorithm, in its fitness function, evaluates transport performance in a cell, between cells and floors. Thus, the algorithm can work in more time periods when it evaluates the price for additional purchase/non-utilization of purchased machines.

Apart from the creation of disposition itself, a genetic algorithm was created for manufacturing cell creation, too. Wu [10] described an algorithm using a two-layered hierarchical scheme, to encode information about machines, products, and also information for dispositional arrangement creation.

Heglas in his work [11] describes the use of discrete simulation together with the use of the evolution method for optimization of the manufacturing system. The output of the work is a designed and verified simulation and an optimization system concept together with a genetic algorithm which is presented by an application form Gasfos2. The application is programmed in the VisualBasic 6.0 language. Thus, it uses a core created by a Galib library (freely accessible library, used for academic purposes, programmed in C++ language which supports working with genetic algorithms) and works with simulation software Witness for optimization manufacturing systems.

The described solutions are primarily focused on the individual cell or department arrangement. They do not take into account real restrictions, showing the inner object arrangement and its entry–exit places. During the search there was no complex system for appropriate integration of such solutions to design process found.

The reason for choice genetic algorithms for the planning of the production layout disposition is dependent on characteristic preferences. It is an especially attribute of genetic algorithms that they leave specific local extremes and find a better solution than classical heuristics.

The author's workplace has long focused on one area of research as well as on the use of genetic algorithms in various areas of industrial engineering. The mentioned workplace has experience with the application of genetic algorithms in the, for example, balancing production lines, parametric simulations, etc. Thus, production layout optimization is another area of research on the application of genetic algorithms.

2. Materials and Methods

2.1. Methods for Layout Design

The process of layout design requires data from construction and technological preparation of production. Data for the manufacturing and logistics systems design can be divided into two basic groups [12]:

- Numerical data—is mainly used to describe conditions in which the system will operate. They are the basic input for output analyses of the manufacturing and logistics system in compliance with a digital factory concept and the numerical data are structured in three key areas [2]: information about products, which will be made and transported in the manufacturing system (product types, piece lists, construction parameters, planned production volumes, etc.); information about processes of their production (operations, manufacturing and assembling processes, used technologies, time norms, etc.); information about resources for product manufacture (manufacturing machines and equipment, tools, workers, transport, and handling machines, handling units, storage premises, etc.).
- Graphical data—represents a visual display of individual elements of the manufacturing and logistics system which are used mainly in layout design, modeling and simulation of the resultant system.

When we know the need for individual resources of the designed system, material flows and other connections among individual elements, we can begin to design an ideal spatial arrangement of the manufacturing or logistics system.

When proposing an ideal arrangement it is advantageous to use optimization methods and algorithms, which can be classified into [13] four groups:

- Graphical methods—are suitable for the solution of simple problems because when looking for an optimal solution, a graphical presentation of spatial arrangement is used. The following methods belong to this group: The Sankey chart, spaghetti diagram, and relationship diagram, etc.
- Analytical methods—are represented by optimization methods of operational analysis. They are characterized by a mathematical model that describes an objective function and boundary conditions of the problem solution. Their disadvantages are a high demand for calculation, complicated and often impossible mathematical description of real conditions in the system, and low interactivity of a designer with a proposed solution. This group consists of methods of linear and non-linear programming, transport problem, and methods of dynamic programming, etc.
- Heuristic methods—are based on simple algorithms for solution and investigation into the fulfillment of criteria (conditions) given by a particular algorithm. They feature relative simplicity, low demand for computing and high interactivity with a designer (the designer can engage with the solution in any phase). However, there is no guarantee that they will find the global optimum and they are usually unable to determine how close the found solution is to the optimum. Heuristic methods for the proposal of spatial arrangement are divided into construction, change, and combined procedures. Construction procedures are based on gradual insertion of system elements to the layout (it begins with the elements with the highest intensity of transport or with the strongest couplings). The following methods belong to this group: CORELAP, ALDEP, PLANET, MAT, MIP, INLAYT, FLAT, etc. Change procedures go out from the original placement and try to improve through the object interchange. Some examples of the methods are as follows: CRAFT, MCRAFT, MULTIPLE, H63, FRAD, COFAD, etc. Combined procedures use a combination of two approaches mentioned above (it is usually a construction procedure proposing the initial placement and a change procedure for its improvement). Examples of methods: BLOCPLAN, LOGIC, etc.
- Metaheuristic methods—These methods produce results of a much higher quality than classical heuristics. Their advantage is the ability to leave—under certain conditions—found local extrema, which classical heuristics cannot do. The following methods belong to this group: genetic algorithms, simulated annealing, tabu search, Ant Colony optimization, etc.

2.2. Genetic Algorithms

The genetic algorithm (GA) belongs to one of the basic stochastic optimization algorithms with distinctive evolutionary features. Nowadays, it is the most used evolutionary optimization algorithm with a wide range of theoretical and practical applications [14,15].

The general procedure of genetic algorithm Figure 1:

1. Initialization—a creation of the initial (zero) population, that usually consists of randomly generated individuals.
2. Start of a cycle—thanks to a certain selection method, a few individuals with a high fitness function are selected from a zero population
3. New individuals—they are generated from selected individuals via the use of basic methods (crossover, mutation, and reproduction), and a new generation is created.
4. Competence calculation of new individuals (fitness function calculation)
5. End of a cycle—a decision-making unit:

    ○ As long as the finishing criterion is not completed, move on to the point no. 2
    ○ If the finishing criterion is finished, the algorithm is completed.

6. End of an algorithm—the individual with the highest competence represents the main algorithm output and the best possible solution found.

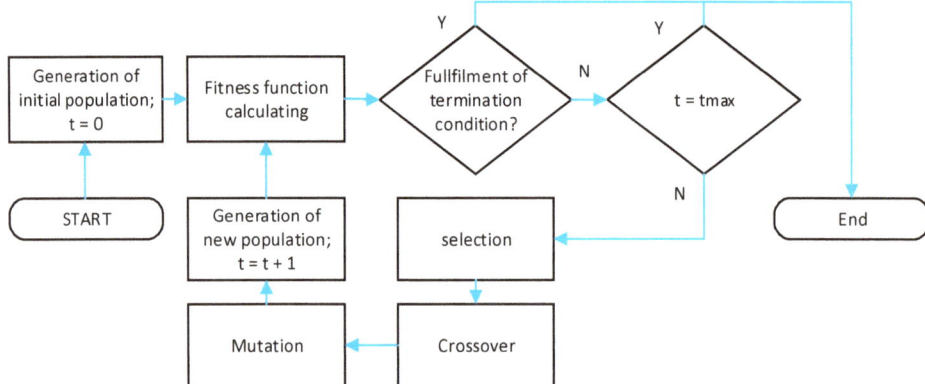

Figure 1. Genetic algorithm.

Selecting the appropriate presentation of a problem is the most important part of an application of a genetic algorithm. In a the case of a genetic algorithm, the space of real problem is transformed into the space of strings. These could be, for example, bit-strings, which were one of the first used representations. Real-valued vector representation is the most commonly used for practical issues. In the case of discrete spaces, integer vector could be selected.

Constant population size is regulated in two ways. It is a so-called generation model that replaces the whole population by offsprings (via mutation or recombination). The second option is to keep one part of the previous generation. This is done by elite selection or in other words by individuals with the highest fitness function. By selecting this way, it is guaranteed that the competence of the best individual will continue to improve [16].

Basic parameters of genetic algorithms are [17]:

1. Selection: The selection process of parents for the creation of a next offspring generation. Finding the correct selection pressure is one of the key aspects when looking for an effective solution. High selection pressure leads to a quicker convergence. However, there is a possibility of the algorithm getting stuck in the local extreme. On the other hand, low pressure prolongs the solution time. Multiple criteria are used in selection:

- Fitness—proportionate selection.
- Stochastic universal sampling.
- Rank selection.
- Elitism.
- Steady-state selection.
- Tournament selection.

2. Offspring creation: Two basic operators are used when creating GA offsprings (Figure 2):

- Crossover: A genetic operator which mutually changes chromosome parts.
- Mutation: A genetic operator, used to keep genetic population diversity; mutation will change one or more in chromosomes, which will prevent early convergence of a solution and will provide possibilities of a random search in a closed area of converged population.

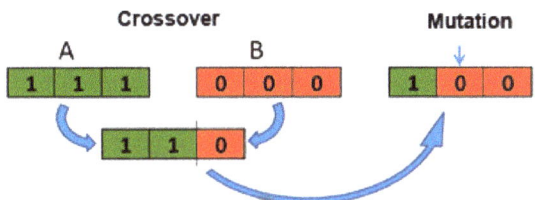

Figure 2. Operators: crossover and mutation.

3. Replacement strategies: These decide which offpsrings will have to be replaced in a new generation. One of them uses so-called non-overlapping populations. This means each generation is made of new individuals who are offerings of individuals from a previous generation. However, when the strategy of overlapping populations is used, it is necessary to decide which individuals will be excluded and replaced by new ones.

4. Solution evaluation: Function formulation which contains information about the individual's skillfulness. This formulation is one of the most important points when applying a genetic algorithm to solve problems. Evaluation of solution quality is usually based on fitness function which always returns the real value for each possible solution. The higher or lower (depending on a problem formulation) the value the better the potential solution.

3. Results

3.1. Genetic Algorithm Integration into the Design Process of Manufacturing Systems

The design approach of manufacturing disposition with the use of genetic algorithms, proposed by authors from this article, requires the realization of the following basic phases as Figure 3 shows:

1. Preparation phase for the disposition arrangement proposal—preparation of numerical data for analysis and layout optimization, graphical data for 2D and 3D model creation of the manufacturing system.
2. Application phase of a genetic algorithm—algorithm core—optimized block layout is its output.
3. Processing phase of designed disposition arrangement in CAD system—the transformation of a proposed block layout into a detailed 3D model of the manufacturing system.
4. Phase of the proposed solution's static verification—verification of a proposed solution based on calculation and analysis of material flows.
5. Phase of proposed solution's dynamic verification—verification of a proposed solution with the use of software simulation [18].

The next chapter of this article contains a detailed description of Phase 2, based on the basic structure of the used genetic algorithm, experimental selection of basic GA parameters and verification of algorithm functionality and comparison of achieved results with a classical heuristics application.

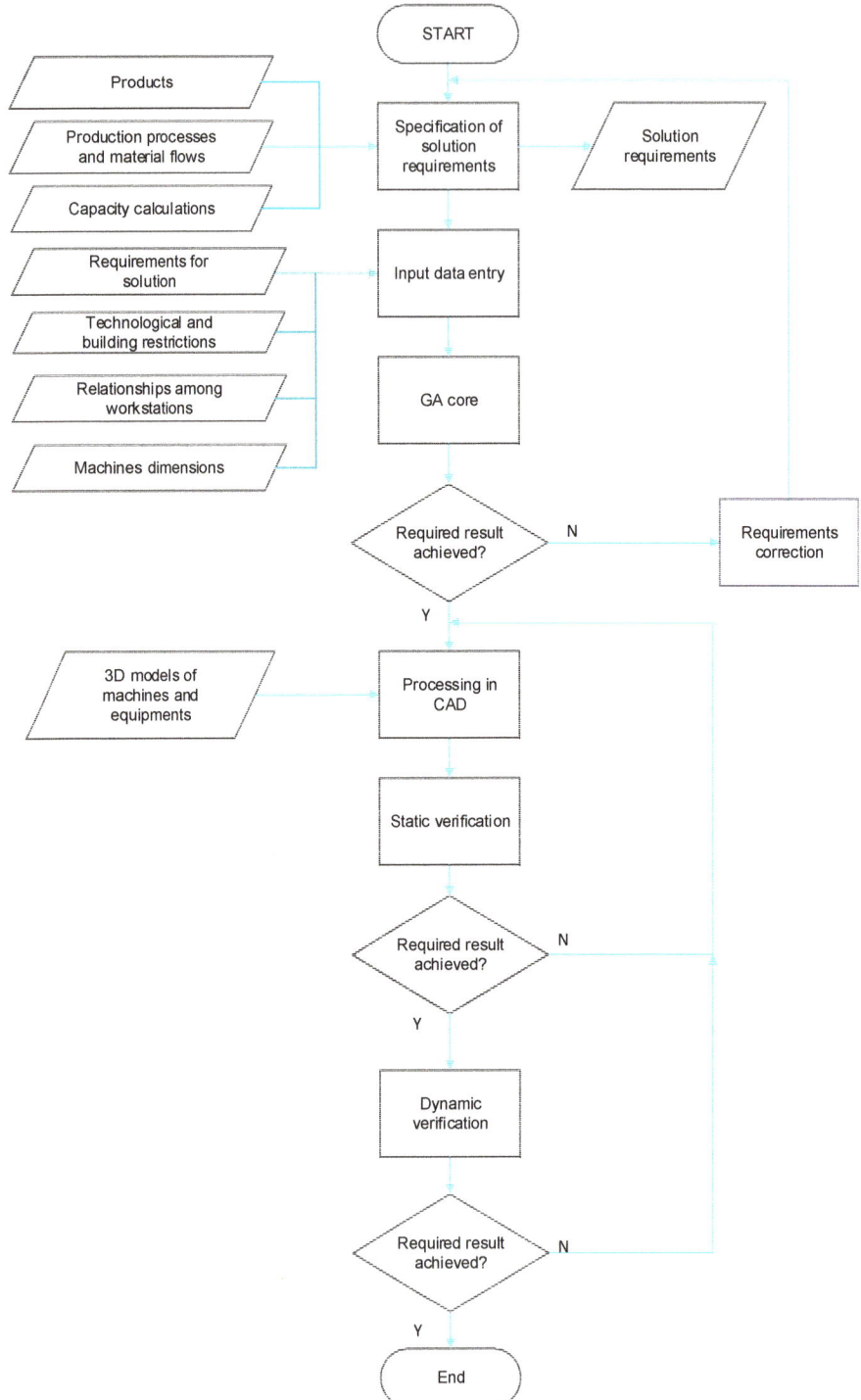

Figure 3. Production layout design using a genetic algorithm.

3.2. Layout Optimization Using a Genetic Algorithm

The proposed genetic algorithm for layout optimization consists of the following steps as Figure 4 shows: The requirement specification and input value assignment for the GA; core of the GA—optimization of space arrangement; GA procedure completion (finishing requirements).

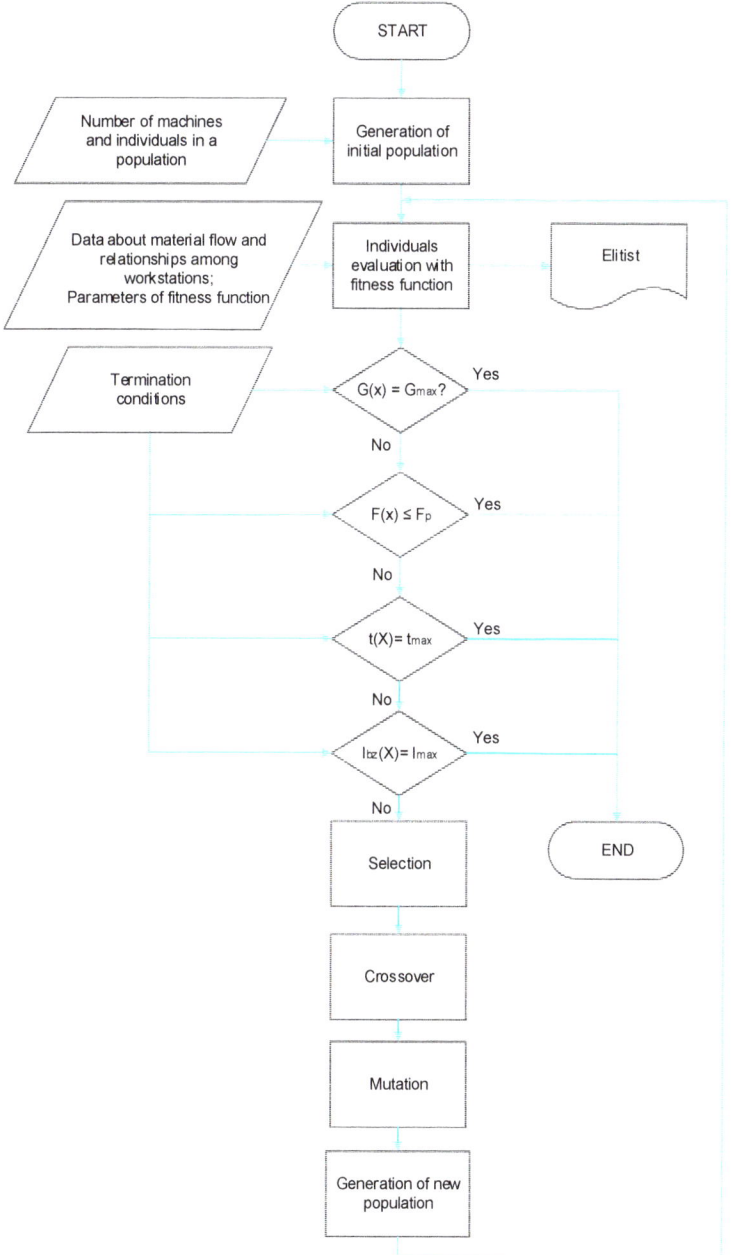

Figure 4. Genetic algorithm for layout optimization.

The methodology of layout optimization using the genetic algorithm described in Figure 3 is implemented into a software solution. Software architecture includes six basic software modules that are interconnected. These interconnections were designed within the design of our own planning procedure with the use of a genetic algorithm and provide component activities in the optimization process of the production system and its spatial arrangement Figure 5 [18]. The first software module is a user interface made in a Microsoft (MS) EXCEL environment that allows input parameter setting and summarizes results of optimization. Our own genetic algorithm is programmed in MATLAB. The MATLAB application takes input data from the MS EXCEL table, makes a genetic algorithm procedure and transfers results (block layout, solution progress, and a final value of fitness function) back into the MS EXCEL application. Simultaneously the algorithm generates block layout for an optimal solution into the AutoCAD environment. If the PC workstation has installed the software module from Siemens Tecnomatix (FactoryCAD, FactoryFlow, Plant Simulation) the next phase of manufacturing system design can be realized. It includes the creation of a 3D model of a manufacturing system (FactoryCAD), static analysis of material flow in manufacturing layout (FactoryFlow), and dynamic verification of proposed solution using computer simulation (Plant Simulation).

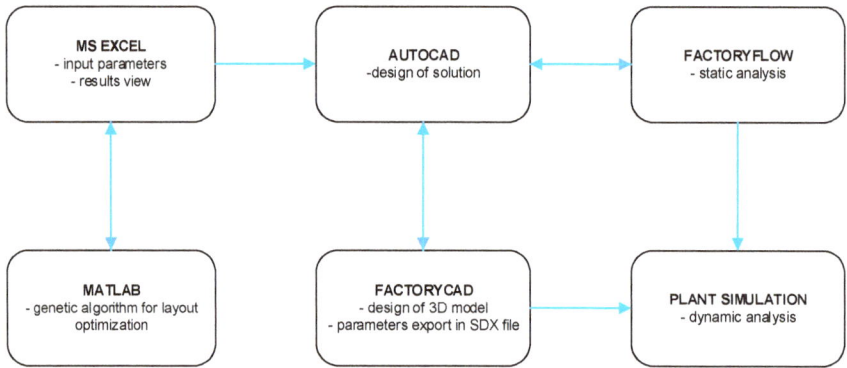

Figure 5. Interconnection in a layout design procedure.

3.2.1. Solution Requirement Specification and Input Value Assigning

In the first part of the solution, it is necessary to define basic requirements for the proposed manufacturing disposition. These requirements come from a previous phase of the process and analysis of input data.

For optimization purposes and GA use, it is necessary to set the following parameters [19]:

- Number of placed workplaces, machines, and devices;
- Mutual relations and intensity among workplaces;
- A,E,I,O,U,X coefficients for relation evaluation;
- Ration of fitness function intensity and mutual relations;
- Specification of entry–exit places of a manufacturing system;
- Specification of machines and devices;
- Specification of hall dimensions and potential construction restrictions (walls, columns, corridors).

It is also necessary to set parameters of a genetic algorithm as [20]:

- Maximum number of generations (iterations);
- Number of individuals (solutions) in a generation;
- Selection types, crossover, and mutation of their probability;
- Required value of fitness function (optional information);

- The maximum solution time (optional information);
- The maximum number of generations without solution improvement.

3.2.2. Genetic Algorithm for Layout Optimization and Its Basic Parameter Setting

After specification of all input data, our own optimization of space arrangement follows with the help of a genetic algorithm. The basic parts of the GA core as shown in Figure 5 will be explained in the following text.

1. Generation of an initial population

The first step is to create a population that represents a group of solutions which will be further developed. In this solution, an individual is created by genes in the quantity that is equivalent to the value of placed machines. These can have a value of 1 up to n, where n is equal to the number of deployed machines. A sequence of individual genes corresponds with a sequence of where machines will be placed in the proposal. Next, there is one gene in each individual reserved for a pattern definition by which workplaces will be included in the proposal as it shows Figure 6.

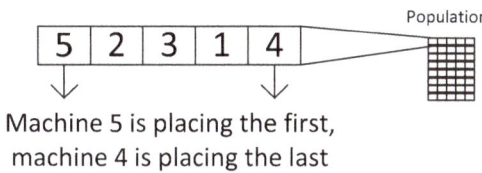

Machine 5 is placing the first,
machine 4 is placing the last

Figure 6. Interpretation of machine order and identification.

The total matrix dimension corresponding to a population in one generation [20] is, therefore, the number of individuals in a generation * (number of placed machines + 1).

2. Individual evaluation of fitness function

After having created a population, it is necessary to evaluate a fitness function. The resulting fitness function was designed as a sum of two components with verified weight. Verification was done according to the intensity of material flow and distance (f_{ID}) and according to relations and distance (f_V).

Equation (1)—Evaluation according to the intensity and distance [19]:

$$f_{ID} = \sum_{i,j=1}^{i,j=n} D_{ij} * I_{ij} \tag{1}$$

where n = number of placed machines; D = right-angle distance between workplaces ($D_{ij} = |x_i - x_j| + |y_i - y_j|$); I = intensity between workplaces i and j.

See note: Right angle distance was chosen for the distance evaluation of a closer real state rather than straightforward distance.

Equations (2) and (3)—Evaluation according to relations and distance [21]:

$$f_V = \sum_{i,j=1}^{i,j=n} V_{ij} * D_{ij} \quad \text{for } V_{ij} \geq 0 \tag{2}$$

$$f_V = \sum_{i,j=1}^{i,j=n} \frac{V_{ij}^2}{D_{ij}} \quad \text{for } V_{ij} < 0 \tag{3}$$

where n = number of placed machines; D = right angle distance between workplaces i–j ($D_{ij} = |x_i - x_j| + |y_i - y_j|$); V = evaluation coefficient of a relation between workplaces i and j (A,E,I,O,U,X).

Equation (4)—Final fitness function value is set as:

$$\min : f = \alpha * f_{ID} + (1 - \alpha) * f_V \tag{4}$$

where α = ratio coefficient of partial fitness functions ($\alpha \in <0;1>$)

Various restrictions are checked in layout construction and the algorithm itself:

- Verification if each object is not mutually overlapping;
- Verification of placed objects in a defined space (dimensions of a production hall);
- Verification of the production hall height restriction;
- Verification of a production hall's height restriction;
- Verification of restrictions regarding transport street arrangement in the production hall;
- Verification of restrictions regarding the definition of the selected object's fixed position in a production hall.

After evaluating all individuals by a fitness function, the best solution is identified and saved in a given generation—an elite individual with his or her reached value and the average value of the fitness population. This data could be displayed during algorithm operation after each generation, to track solution progress. After completion, it is also possible to display a progress graph of average and elite fitness values.

3. Decision-making blocks

In this step, it is necessary to compare specific conditions for algorithm termination in four decision-making blocks. The first condition is to reach the maximum number of generation (iteration) G_{max}. The second condition is to reach or exceed the highest permissible fitness value f_p. The third condition is to reach maximum solution time t_{max}. The last condition is to exceed set iteration number (I_{max}) without improving a reached solution.

The last condition was integrated into the proposal to prevent extensive calculation time if the required or unachievable fitness function value f_p is not set and fitness function value is not improving. Therefore, there is an assumption, that the extreme has been found in a group of solutions.

When meeting any out of four stated conditions, the genetic algorithm is completed.

4. Selection

In case none of the finishing criteria was fulfilled, the algorithm continues by selection, in other words, by selecting individuals who will crossbreed and eventually mutate between each other. For such a solution, the roulette rule was selected. Probability selection was proportional to an individual's achieved suitability. This form was chosen based on a better possibility to search a complex set of solutions when later combining parents and their evaluation as well as their calculation speed [3].

To prevent early convergence, suitability of individuals was integrated into the algorithm via the help of sigma scaling. The average expected number of generated offsprings with sigma scaling is $p_{(i,g)}$ from an individual i in generation g given as Equation (5):

$$p(i,g) = \begin{cases} 1 + \frac{f_{(i,g)} - \bar{f}_{(g)}}{k_s * \sigma_{(g)}} & for\ \sigma_{(g)} \neq 1 \\ 1 & for\ \sigma_{(g)} = 0 \end{cases} \tag{5}$$

where $f_{(i,g)}$ = fitness i of an individual in generation g; $\bar{f}_{(g)}$ = average population fitness in generation g; k_s = coefficient for sigma scaling; $\sigma_{(g)}$ = determinant population deviation in generation g.

For a sigma scaling coefficient $k_s = 1$, an individual rated by the suitability of its standard deviation being closer to a required extreme as the average population suitability will on average produce two offsprings for a new population. The higher the value k_s, the lower the selective pressure.

After remapping, it is possible to select a choice of parents (Figure 7) either by the classical roulette mechanism (generation of random numbers) or by stochastic universal sampling (they are generated uniformly spread indicators that will choose parents in one iteration).

After selecting, pairing follows, where Parent 1 and Parent 2 will be randomly selected from chosen individuals. These should make Offspring 1 and 2.

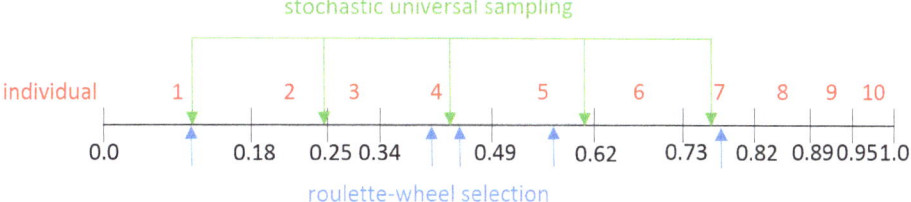

Figure 7. Between roulette-wheel selection and stochastic universal sampling.

5. Crossover

To prevent a duplicate of identical machines in crossover or omission of the same machines from the genetic chain, a mechanism of partially matched crossover was designed. This type of crossover has within its procedure implemented measures. This guarantee that each coded solution will have its machine only once [21].

A procedure of partially matched crossover (Figure 8) is as follows:

1. Generation of two random points delimiting genes, parents will mutually exchange.
2. Pairing of gene values that have been exchanged.
3. Adding parent values to genes where conflicts do not occur.
4. Use of paired values for conflict genes.

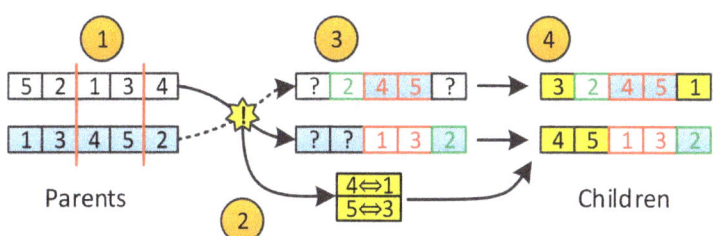

Figure 8. Matched crossing.

It is also necessary to set the crossover's probability and to determine an optimal value range of probability. A series of experiments were carried out. There were two case studies, and the only changed parameter was the probability of crossover. [22] Mutation was switched off and the finishing criterion was set to reach 200 generations. The fitness function and generation value were reached through closely observed parameters. In Experiment 1, 20 machines were placed into a layout, and in Experiment 2, 28 machines were placed. Results of experiments can be seen in Tables 1 and 2.

The optimal crossover probability was set in the range from 75% to 95%, based on a series of experiments. The probability of 100% was not taken into consideration because it was possible for the small part of the previous population to survive.

Table 1. Various crossover probabilities and their results—Experiment 1.

	Number of Crossover	50	55	60	65	70	75	80	85	90	95	100
Fitness	1. run	1,786,250	1,801,250	1,882,500	1,801,250	1,960,000	1,901,250	1,767,212	1,892,500	1,862,500	1,850,000	1,735,000
	2. run	1,815,033	1,927,500	1,946,250	1,843,750	1,982,500	1,806,250	1,771,250	1,825,000	1,851,250	1,755,000	1,817,212
	3. run	1,842,500	1,935,000	1,847,500	1,841,250	1,811,250	1,704,712	1,777,212	1,763,750	1,850,000	1,825,000	1,935,000
	4. run	1,865,000	1,935,000	1,956,250	1,875,000	1,928,462	1,891,250	1,835,962	1,765,962	1,936,250	1,925,962	1,837,500
	5. run	1,877,500	1,840,962	1,853,750	1,860,962	1,742,500	1,793,250	1,825,962	1,916,250	1,654,712	1,825,000	1,866,250
	6. run	1,741,250	1,858,750	1,773,750	1,891,250	1,900,000	1,794,712	1,782,503	1,733,750	1,780,000	1,898,750	1,824,712
	Average	1,821,256	1,883,077	1,876,667	1,852,244	1,887,452	1,815,237	1,793,350	1,816,202	1,822,452	1,846,619	1,835,946
Generation	1. run	145	138	109	105	69	70	102	128	71	77	55
	2. run	136	121	91	97	56	90	93	79	67	100	132
	3. run	128	108	104	96	90	91	69	83	85	75	88
	4. run	117	104	105	98	115	53	93	84	136	91	111
	5. run	112	133	74	132	106	92	79	79	90	102	73
	6. run	94	113	131	121	90	127	125	134	88	58	91
	Average	122.0	119.5	102.3	108.2	87.7	87.2	93.5	97.8	89.5	83.8	91.7

Table 2. Various crossover probabilities and their results—Experiment 2.

	Number of Crossover	50	55	60	65	70	75	80	85	90	95	100
Fitness	1. run	2,120,279	1,973,988	2,074,141	2,065,830	2,014,884	2,087,429	2,197,170	2,241,621	2,034,974	2,027,345	2,036,681
	2. run	2,067,359	1,965,956	2,218,102	2,196,600	2,179,517	2,035,914	2,135,551	1,972,258	2,124,068	2,085,635	2,209,369
	3. run	2,140,407	2,268,313	2,180,144	2,090,574	2,180,138	2,215,854	2,169,616	1,836,980	2,091,044	2,117,620	2,034,550
	4. run	2,258,198	2,090,424	2,277,779	2,194,130	2,250,233	2,126,470	2,180,122	2,105,810	2,025,580	2,099,738	2,157,909
	5. run	2,208,670	2,227,285	2,133,529	2,024,841	2,158,330	2,164,587	2,197,480	2,062,865	2,133,644	2,085,187	2,143,426
	6. run	2,219,823	2,059,236	2,150,834	2,188,592	2,202,106	2,172,660	2,135,509	1,975,300	2,022,912	2,165,858	2,216,239
	Average	2,169,123	2,097,534	2,172,422	2,126,761	2,164,201	2,133,819	2,169,241	2,032,472	2,072,037	2,096,897	2,133,029
Generation	1. run	142	105	84	103	136	122	117	116	82	110	100
	2. run	88	113	115	137	151	90	100	121	92	82	56
	3. run	169	110	120	131	102	81	75	129	74	103	129
	4. run	95	190	68	139	53	105	60	77	144	82	89
	5. run	121	99	161	134	137	98	130	87	134	109	61
	6. run	134	86	160	88	103	140	92	95	87	59	59
	Average	124.8	117.2	118.0	122.0	113.7	106.0	95.7	104.2	102.2	90.8	82.3

6. Mutation

After the crossover, mutation follows. The heuristic insert mutation was adopted to better ensure population diversity and individual qualities after mutation as well as satisfy the process constraints. [18] However, in this type of solution encoding, traditional mutation or in other words value change of a random gene, is out of the question. This would automatically require remedial measures to eliminate duplication or not classified machines. That is why mutation via the help of inversion or exchange was selected in Figure 9. Due to inversion of a rather big intervention into solution, the probability was divided for exchange or inversion in 80:20 ratio.

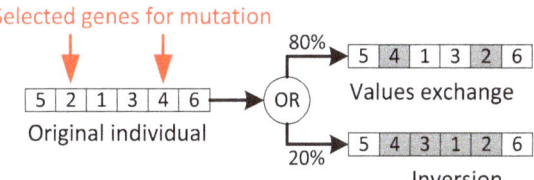

Figure 9. Principle in the proposed algorithm.

Furthermore, it was necessary to set the probability in mutation and the majority of resources state range, depending on the type of problem, from 0.1% to 5%. However, in applications for object arrangement, some resources state probability values of up to 20% [23]. Due to this, the optimal probability ratio of mutation for this specific problem was selected similarly as in crossover. The set of experiments for the same initial conditions (Experiment 1—20 machines, Experiment 2—28 machines) as was the case for selecting optimal crossover probability values were carried out. However, the difference was the mutation probability, which was substituted. In the experiments, crossover probability was set to a constant value of 0.85. The results of the experiments are stated in Tables 3 and 4.

The results of experiments state that with low probability mutation, functioning converges later as it is mainly dependent on randomly generated initial population and crossbreeding in all iterations. Only a small number of individuals is modified by mutation operators. With increasing mutation probability, race converges on average earlier with a higher quality solution, although it is accompanied by a higher generation dispersion of a found solution. This is caused by mutation randomness. The optimal value probability range of mutation was set between 0.05 and 0.15. As we want to avoid the algorithm going into a random search, we do not recommend higher probabilities of initial settings.

One of the conditions of algorithm functioning termination is crossing the fixed number of iterations (I_{max}) without improving the reached solution. In case the extreme has been achieved, it is advised to verify if it is only a local extreme. To verify this, the variable mutation was implemented in the algorithm. This variable mutation increases the probability of its application with an increasing number of interactions, without any improvement. In the basic setting—when functioning finishes after 100 interactions without any improvement, after 70 iterations (I_{bz} = 70), there is a mutation probability increased to 1.5 multiple of the original value. After 80 iterations it is 1.875 multiple of the original value, and eventually, after 90 iterations, it is totally 2.34 multiple of the original value. If there is a different setting for the number of iterations without any improvement, Imax is the variable probability of mutation proportionally recalculated.

7. Making of a new generation

Following the genetic operator activity, parents are replaced by offsprings. In case elitism is used in suitability evaluation and the best possible solution has been saved, this individual replaces one of the offsprings with the worst suitability.

After this step, the algorithm goes back to evaluating new individuals through the help of the fitness function. Furthermore, the algorithm keeps repeating in cycles until one of the finishing conditions is fulfilled in Figure 6.

Table 3. Results of various mutation probabilities—Experiment 1.

	Number of Crossover	0.01	0.02	0.04	0.05	0.08	0.1	0.12	0.15	0.18	0.2	0.22	0.25
Fitness	1. run	1,826,250	1,975,000	1,801,250	1,753,750	1,835,000	1,778,750	1,771,250	1,791,250	1,800,000	1,846,250	1,842,500	1,753,750
	2. run	1,922,500	1,795,000	1,845,000	1,787,500	1,833,750	1,796,254	1,813,750	1,915,000	1,813,750	1,785,000	1,768,750	1,761,250
	3. run	1,841,250	1,892,500	1,872,500	1,770,000	1,783,750	1,817,500	1,836,250	1,752,500	1,771,250	1,820,000	1,760,000	1,752,500
	4. run	1,875,000	1,901,250	1,790,000	1,773,750	1,921,250	1,820,000	1,782,500	1,763,750	1,812,500	1,727,500	1,817,500	1,796,250
	5. run	1,977,500	1,793,750	1,926,250	1,893,750	1,896,250	1,750,000	1,772,500	1,717,500	1,820,000	1,756,250	1,762,500	1,801,250
	6. run	1,893,750	1,897,500	1,852,500	1,855,000	1,743,750	1,905,000	1,797,500	1,847,500	1,772,500	1,682,500	1,717,500	1,687,500
	Average	1,889,375	1,875,833	1,847,917	1,805,625	1,835,625	1,814,251	1,795,625	1,797,917	1,798,333	1,769,583	1,778,125	1,758,750
Generation	1. run	166	166	0:00	92	141	155	81	63	102	76	83	116
	2. run	133	159	69	120	62	111	109	62	119	82	79	76
	3. run	117	94	166	89	68	136	175	99	61	65	83	83
	4. run	73	81	111	122	119	102	112	78	74	86	66	116
	5. run	170	111	138	81	92	130	54	80	65	47	101	57
	6. run	157	110	158	101	117	85	70	149	60	62	75	49
	Average	136.0	120.2	125.5	100.8	99.8	119.8	100.2	88.5	80.2	69.7	81.2	82.8

Table 4. Results of various mutation probabilities—Experiment 2.

	Number of Crossover	0.01	0.02	0.04	0.05	0.08	0.1	0.12	0.15	0.18	0.2	0.22	0.25
Fitness	1. run	2,205,047	2,094,709	2,150,884	2,031,828	2,024,367	2,097,308	2,040,704	2,023,408	1,975,890	2,106,751	2,167,212	2,007,581
	2. run	2,293,938	2,224,044	2,068,203	2,032,667	2,119,026	2,138,655	2,057,521	2,174,757	2,153,584	2,081,075	2,004,717	1,923,141
	3. run	2,174,356	2,223,835	2,236,219	2,108,171	2,107,460	2,183,970	1,984,400	2,184,988	2,230,222	2,078,632	2,073,966	2,030,595
	4. run	2,209,866	2,101,225	2,059,109	2,070,532	2,038,591	2,235,402	2,240,006	2,177,516	2,089,100	2,151,902	2,261,843	2,034,200
	5. run	2,036,578	2,133,150	2,096,508	1,983,330	2,165,419	2,101,584	2,059,324	2,091,831	2,133,639	2,147,631	1,945,574	1,978,386
	6. run	2,002,145	2,073,131	2,286,277	2,131,030	2,098,622	2,165,754	2,128,676	2,244,776	2,095,610	2,150,529	2,173,074	2,001,748
	Average	2,153,655	2,141,682	2,149,533	2,059,593	2,092,248	2,153,779	2,085,105	2,149,546	2,113,008	2,119,420	2,104,398	1,995,942
Generation	1. run	162	92	122	106	144	116	113	81	182	105	81	86
	2. run	106	190	94	105	96	161	97	90	99	75	133	77
	3. run	97	85	158	73	100	74	167	77	99	43	118	126
	4. run	96	83	103	115	81	94	74	61	135	103	95	125
	5. run	177	89	181	91	106	98	107	119	147	86	77	71
	6. run	97	67	89	176	112	124	94	93	142	127	84	132
	Average	122.5	101.0	124.5	111.0	106.5	111.2	108.7	86.8	134.0	89.8	98.0	102.8

8. Genetic algorithm finishing

In decision-making blocks, each genetic algorithm cycle checks whether one of the finishing conditions has not been fulfilled: achieving the maximum number of generations (iterations); achieving or exceeding the highest permissible fitness value; achieving the maximum solution time; exceeding the set number of iterations without improvement.

If some of the finishing conditions were fulfilled, the activity of a genetic algorithm will finish. After completion of its activities, there are generated outputs in the user interface (Figure 10): block layout; achieved fitness value and information in which iteration it was achieved; graph showing the progress of average and elite fitness population values.

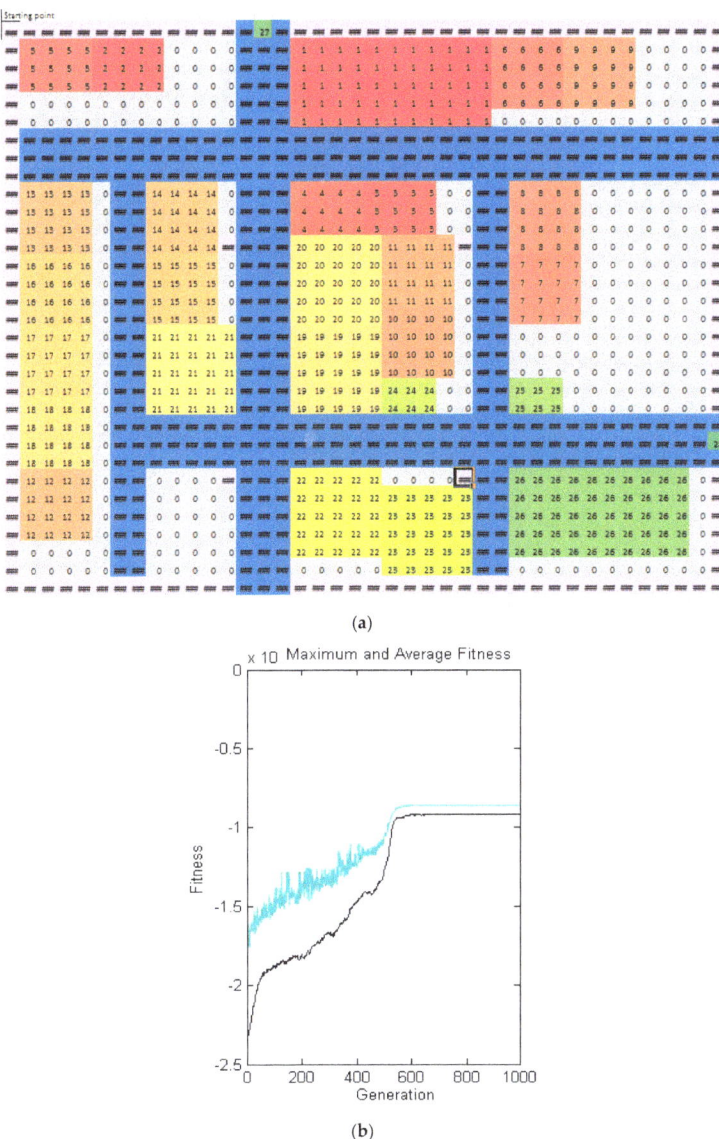

Figure 10. Results of genetic algorithm (**a**) block layout; (**b**) fitness value chart.

In the final phase, the user will decide whether the solution proposed by the genetic algorithm fulfilled all its requirements. If not, it is necessary to closely specify requirements and repeat the generation of an optimal layout. If requirements were fulfilled, the methodology continues by the result of processing in a CAD system.

3.2.3. Experimental Verification of Genetic Algorithm and Result Compared with the Use of Classical Heuristic

To check the functionality of the proposed GALP algorithm (Genetic Algorithm Layout Planner) a series of experiments were carried out. These results were then compared with optimization results with the help of heuristic according to Murat (sequence–pair approach). Heuristic according to Murat has been selected because it is believed the heuristic approach is implemented in Factory PLAN/OPT module, which is a part of Factory Design and Optimization package from Siemens Tecnomatix. The final PLAN/OPT and genetic GALP algorithm proposals were subsequently compared in FactoryFLOW software.

A common characteristic for both algorithms is the block layout output. Both algorithms require a finishing requirement and total time of algorithm functioning. For more complex result comparison, experiments were carried out for 1.5, 10, and 20 min.

Our own experiments were carried out for two types of inputs. Our own experiments were carried out for two types of inputs. In Case 1 a simple manufacturing system that represents a small manufacturing system was generated. This system contains three product families (each of the family has minimum of eight process steps) and 24 workplaces. The number of workplaces was determined based on the planned annual production volume of product families. In Case 2 a complex manufacturing system with nine product families and 60 workplaces. This system represents a large manufacturing system in practice. Of course, there are also much larger manufacturing systems in practice, howe;ver, these systems can be segmented into smaller, mutually independent production groups, based on the classification of the manufactured products.

Evaluation of the application results of the classic heuristics and the GALP algorithm was made on the calculation basis of three basic material flow parameters, which are automatically calculated by Tecnomatix FactoryFLOW software. The first parameter is distance. This parameter determines the total amount of distance traveled in the production layout when considering the right-hand distances and the total number of journeys between different twin of workstations. The second parameter is the cost. This parameter determines the total transport costs on the basis of total distance traveled, the type of transport equipment, and the fixed and variable rates of transport costs. The third parameter, time, specifies total shipment time based on the known distance traveled, the type of transport equipment, defined transport cycle structure (i.e., load–drive–unload) and the time parameters of the transport cycle (i.e., load time, unload time, transportation speed).

Case 1 results are shown in Tables 5–7. The graphical expression of comparison results shown (Figure 11). These experimental results indicate that GALP achieved better results in all cases than the PLAN/OPT algorithm, which, due to unknown reasons, did not keep workplace dimensions in some cases (Figure 12). Case 1 results show that the difference between the classic heuristics (PLAN/OPT) and the GALP algorithm also depends on the calculation time and, hence, on the total number of iterations used by the algorithm. For the shortest calculation time of 1 min, the differences between results at 1.21% (distance and cost savings) and 0.63% (time savings) are in favor of the proposed GALP algorithm. At the longest calculation time (20 min), savings increased to 1.82% (distance and cost savings) and 2.32% (time savings). GALP has also proposed solutions preferring singular direction of material flow with minimum crossing or backward material flow. Due to comparing both algorithms, no restrictions have been imposed on workplace arrangement. However, the GALP algorithm enables basic restriction definition in the layout (production hall dimensions, the height of spaces, material component arrangement, fixed installations or transport corridors in the layout).

Table 5. Experiment results for GALP.

Parameter	Time Calculation	Distance (m)	Costs (EUR)	Time (min)
Achieved results	1 min	571,360.03	25,434.76	69,708.00
	5 min	510,552.24	25,023.22	67,682.66
	10 min	441,472.39	24,825.88	67,536.68
	20 min	430,341.00	24,755.86	67,214.88

Table 6. Experiment results for PLAN/OPT.

Parameter	Time Calculation	Distance (m)	Costs (EUR)	Time (min)
Achieved results	1 min	669,925.47	25,746.26	70,146.72
	5 min	633,664.26	25,668.70	70,281.26
	10 min	548,718.83	25,284.36	68,976.65
	20 min	529,770.15	25,214.38	68,811.54

Table 7. Experiment result comparison GA-PLAN/OPT.

Parameter	Time Calculation	Distance (m)	Costs (EUR)	Time (min)	Distance (%)	Costs (%)	Time (%)
Comparison GA−PLAN/OPT	1 min	−98,565.44	−311.50	−438.72	−14.71	−1.21	−0.63
	5 min	−123,112.02	−645.48	−2598.60	−19.43	−2.51	−3.70
	10 min	−107,246.44	−458.48	−1439.97	−19.54	−1.81	−2.09
	20 min	−99,429.15	−458.52	−1596.66	−18.77	−1.82	−2.32

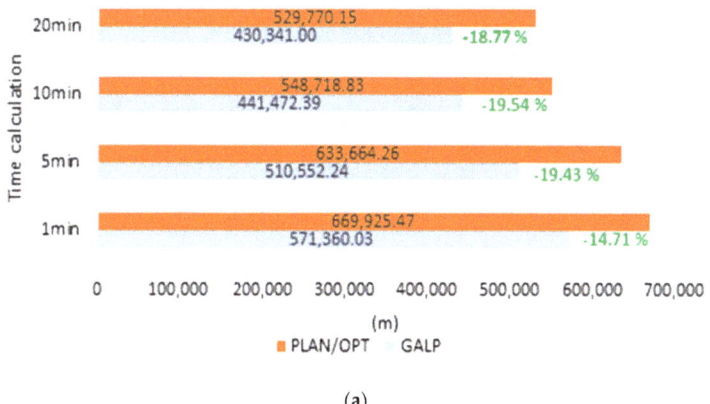

(a)

Figure 11. Cont.

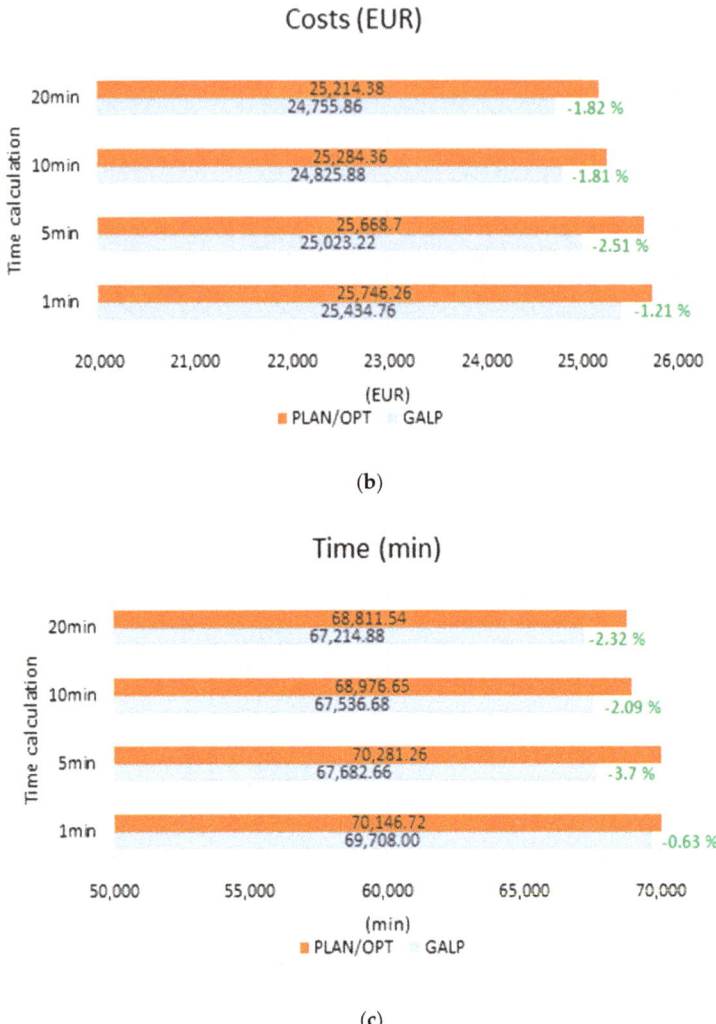

Figure 11. Expression of experiment results PLAN/OPT and GALP: (**a**) distance; (**b**) costs; (**c**) time.

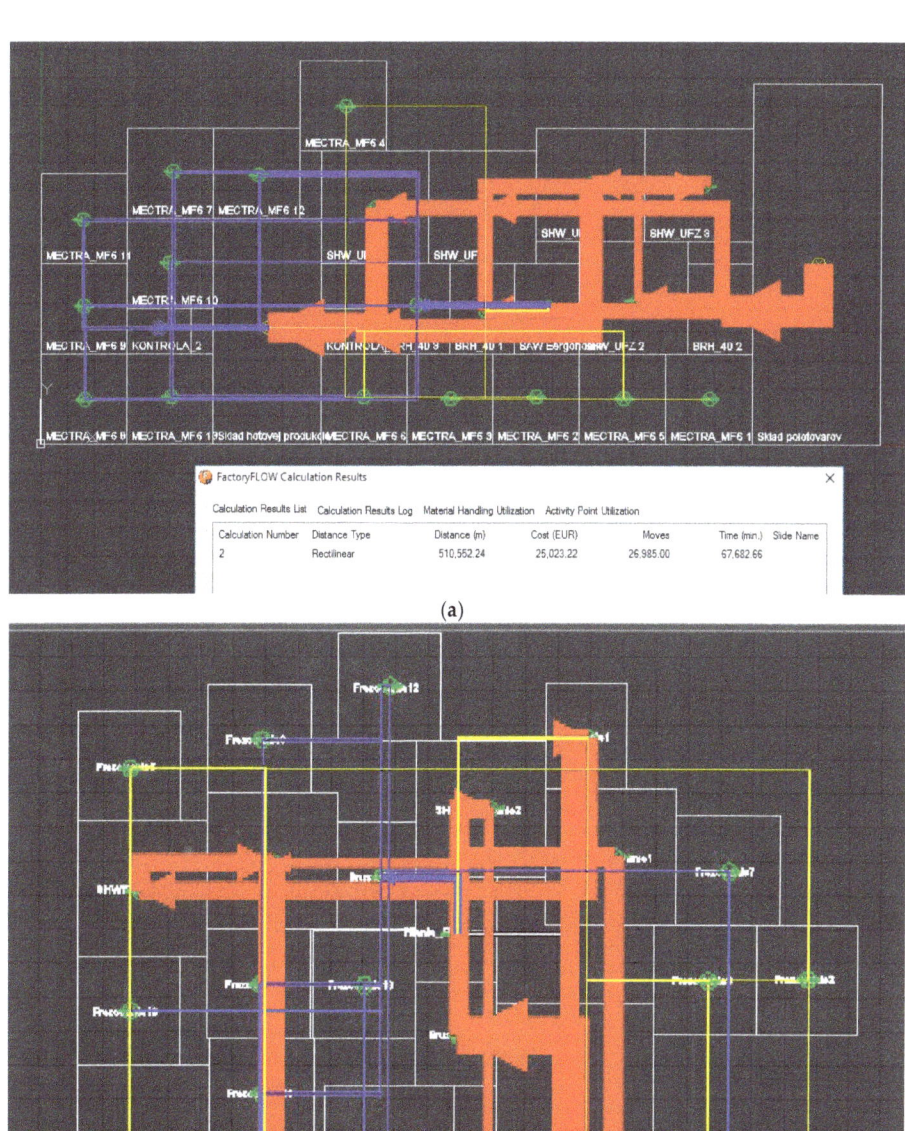

Figure 12. *Cont.*

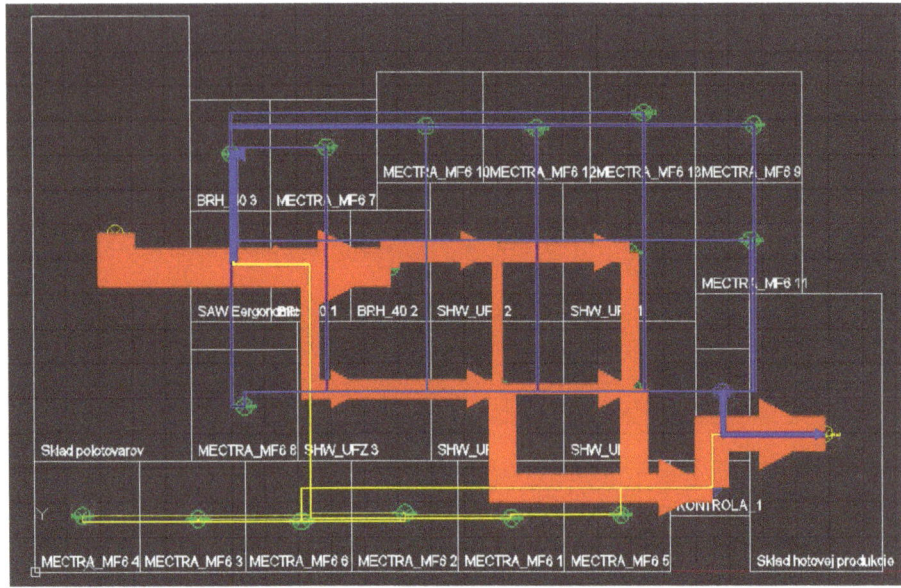

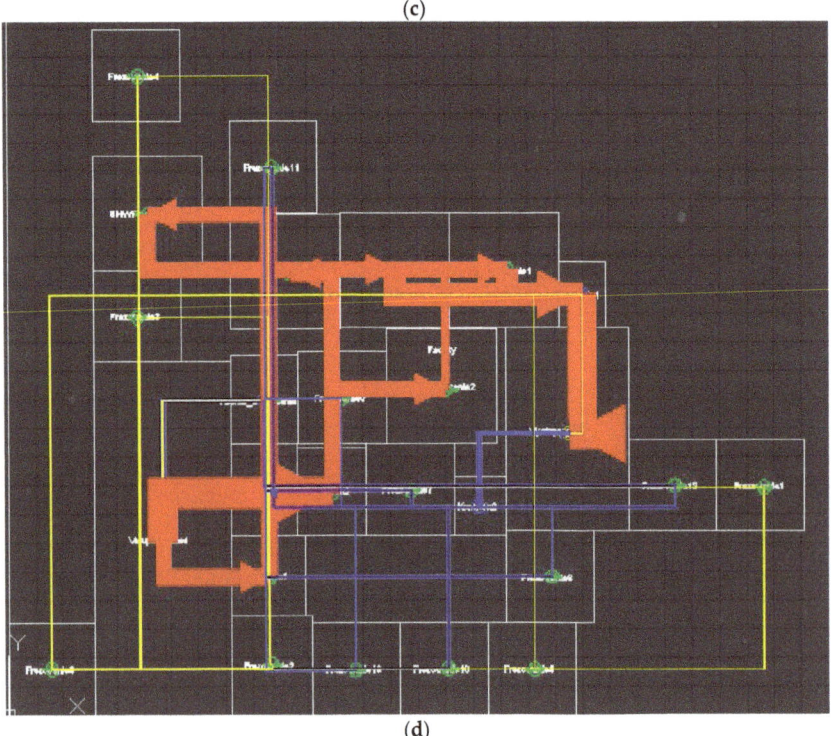

Figure 12. Comparison for GALP and PLAN/OPT heuristics: (**a**) GALP (time limit for optimization = 5 m); (**b**) PLAN/OPT (time limit for optimization = 5 min); (**c**) GALP (time limit for optimization = 20 min); (**d**) PLAN/OPT (time limit for optimization = 20 min).

The graphical expression of comparison results (Figure 11) shows that the proposed GALP algorithm achieved better results than classical heuristics PLAN/OPT. The results support the decision of the choice of a genetic algorithm as a tool for disposition layout planning and, thus, its applicability in industry.

Experiment results were also consequently verified taking into account the complex solution of a manufacturing system (Case 2) and solution time was set to 5.5 h (solution time 1000 task generations in GALP). Advantages of the genetic algorithm became evident in a more extensive problem. Final material flow is directed with a minimum crossing. However, in case of the PLAN/OPT algorithm, there is a crossing, where material flow keeps coming back, and there is not any technology island creation in the manufacturing system. The genetic algorithm proposed layout with a greatly lower value of transportation performance (38.18%) than the heuristic algorithm in a PLAN/OPT module. Experiment result comparison is stated in Table 8. Final block layouts are shown in Figure 13 below.

Table 8. Experiment result comparison for Case 2.

Parameter	GALP	PLAN/OPT	Difference	Difference (%)
Distance covered (m)	2,877,483.27	4,654,622.41	−1,777,139.14	38.18%
Costs (EUR)	83,939.38	91,344.13	−7404.75	8.11%
Time (min)	230,818.14	253,032.38	−22,214.24	8.78%

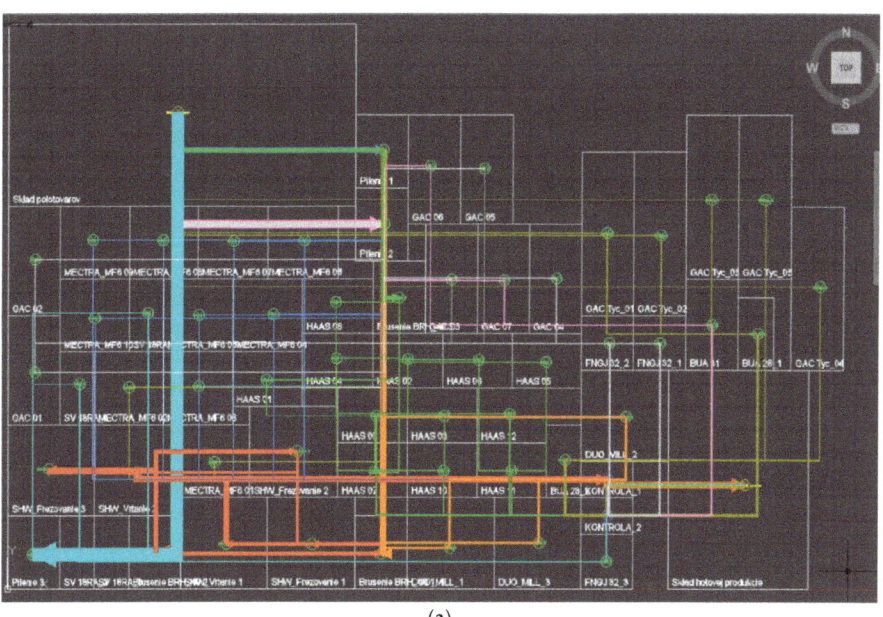

(a)

Figure 13. Cont.

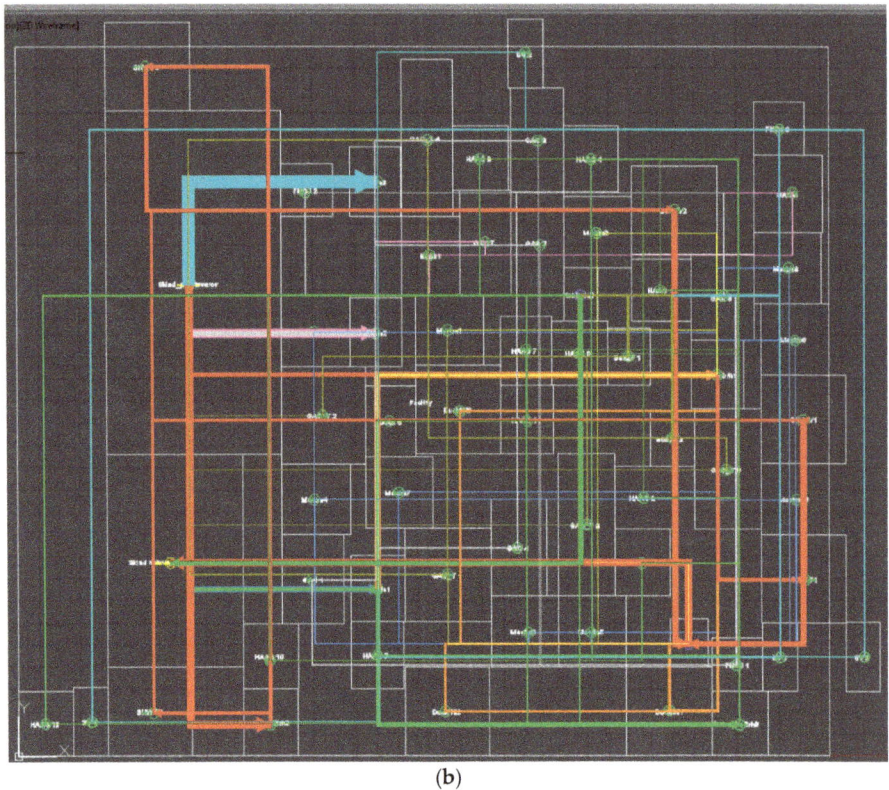

(b)

Figure 13. Block layout comparison for Case 2: (**a**) Layout generated by the algorithm GALP; (**b**) Layout generated by the PLAN/OPT algorithm.

4. Discussion

The main findings of this study are:

- **Application of genetic algorithms for layout design provides better results than traditional heuristics:** As part of the experimental verification, results of the GALP algorithm application were compared with the results of the standard heuristics used for solving layout optimization problems. Heuristics were selected according to Murat (Sequence–Pair approach) because this method is implemented in the software module PLAN/OPT, which is part of the Tecnomatix FactoryFLOW software. Using the above method has enabled the automation of the process of layout variants generation with using traditional heuristics, and software background PLAN/OPT has provided comparable output types as the proposed GALP algorithm. As results of experimental verification of the proposed algorithm show, the GALP application leads to a better layout of the production system. It can be seen on the length of the material flows, on total transport performance, costs and even from point of view of material flow transparency in the resultant layout. In all the GALP experiments led to solutions which resulted in a simple and guided material flow and a lower value of total transport costs. The savings in transport performance and transport costs increase directly in proportion to the growing complexity of the proposed layout.
- **The designed fitness function must be sufficiently comprehensive to ensure that the resulting layout respects the essential requirements for the correct layout:** The fitness function within the

proposed GALP algorithm was designed as a complex function taking into account, on the one hand, the material flow intensity; on the other side relationships among objects (workstations). Such a complex design of fitness functions has allowed the inclusion of other factors in the solution, in addition to the material flow, which affect the design of an optimal spatial structure, such as multi-machine service, teamwork, job rotation, sharing of common documentation, requirements for accuracy and quality of production, requirements for safety and hygiene of work, etc.

- **The proposed GALP algorithm provides a number of advantages in addressing the practical problems associated with the design of the production layout:** Created applications for creating a production disposition that is based on the GALP algorithm simplifies the process of optimizing layout particularly in the variant design phase. A simple user interface for entering input data and visualizing outputs allows designers to realize a set of experiments with different variants of spatial arrangement. The application provides tools for evaluating and comparing individual variants while allowing the static design to be subsequently linked to dynamic solution verification. The practical benefits of the GALP application can be summarized as follows: reducing the time needed to production layout design; reducing the cost for production layout design; verifying a large set of solutions; taking into account restrictive conditions; improving the quality of the proposed layout.

5. Conclusions

The main aim of this article was to describe the use of the genetic algorithm in manufacturing layout optimization for continual and sustainable development of the company. The article described not only the basic algorithm structure but also the experimental setting of optimal algorithm parameter and verification. It also compares algorithm outputs of a traditional heuristic application. Experiment results showed the proposed GA provided a saving of transport performance in the case of less complex problems, which was 15 to 20% compared to classical heuristic results. When the problem complexity increases, savings from the GA use continued to increase (Case 2—saving more than 38%). In addition, the disposal arrangement generated by GA leads to a solution with easier and directed material flow. A proposed genetic algorithm is a part of a complex project methodology of manufacturing dispositions, and its basic steps are described in Chapter 3. Furthermore, this proposed GA enables the user to consider practical restrictions when arranging space in layout optimization, that is a shape and production hall dimensions (length, width, and height) building block placement (e.g., columns), fixed installations, transport corridors, input and output spaces of the manufacturing system, etc. Therefore, this means that a layout has been designed by a genetic algorithm requiring the minimum number of corrections that do not represent significant deviations from optimal parameters of material flows. A layout which is designed by the genetic algorithm is ideal not only for stable development and decrease of cost but also for implementation of those solutions which are sustainable in a long time. That means that when we use an ideal layout, it is much easier to maintain optimization changes making them sustainability in a long time.

Author Contributions: All authors contributed to writing the paper. Documented the literature review, analyzed the data and wrote the paper. All authors were involved in the finalization of the submitted manuscript. All authors read and approved the final manuscript.

Funding: This work was supported by the Slovak Research and Development Agency under contract No. APVV-16-0488.

Conflicts of Interest: The authors declare no conflict of interest.

References

1. Kuehn, W. Digital Factory—Simulation Enhancing the Product and Production Engineering Process. In Proceedings of the 2006 Winter Simulation Conference, Monterey, CA, USA, 3–6 December 2006; pp. 27–39.

2. Krajčovič, M. Modern approaches of manufacturing and logistics systems design. In *Digitálny Podnik [Electronic Source]*; CEIT SK: Žilina, Slovakia, 2011; p. 12. ISBN 978-80-970440-1-5.
3. Lee, Q.; Nelson, W.; Amundsen, A.; Tuttle, H. *Facilities & Workplace Design*, 1st ed.; Engineering & Manufacturing press: Norcross, GA, USA, 1997; ISBN 0-89806-166-0.
4. Centobelli, P.; Cerchione, R.; Murino, T. Layout and Material Flow Optimization in Digital Factory. *Int. J. Simul. Model.* **2016**, *15*, 223–235. [CrossRef]
5. Saleh, N.F.B.; Hussain, A.R.B. Genetic Algorithms for Optimizing Manufacturing Facility Layout. Available online: http://comp.utm.my/pars/files/2013/04/Genetic-Algorithms-for-Optimizing-Manufacturing-Facility-Layout.pdf (accessed on 12 October 2013).
6. Rekiek, B.; Delchambre, A. *Assembly Line Design*; Springer: London, UK, 2006; ISBN 978-1-84628-112-9.
7. Hnát, J. Vyvažovanie Montážnych Procesov s Využitím Nových Prístupov. Dissertation Thesis, UNIZA, Žilina, Slovakia, 2009.
8. Kothari, R.; Ghosh, D. An efficient genetic algorithm for single row facility layout. *Optim. Lett.* **2014**, *1*, 679–690. [CrossRef]
9. Kia, R.; Khaksar-Haghani, F.; Javadian, N.; Tavakkoli-Moghaddam, R. Solving a multi-floor layout design model of a dynamic cellular manufacturing system by an efficient genetic algorithm. *J. Manuf. Syst.* **2014**, *33*, 218–232. [CrossRef]
10. Wu, X.; Chu, C.H.; Wang, Y.; Yan, W. A genetic algorithm for cellular manufacturing design and layout. *Eur. J. Oper. Res.* **2007**, *181*, 156–167. [CrossRef]
11. Heglas, M. Simulácia Výrobného Systému s Využitím Evolučných Metód. Master's Thesis, UNZA, Žilina, Slovakia, 2011.
12. Fusko, M.; Rakyta, M.; Manlig, F. Reducing of intralogistics costs of spare parts and material of implementation digitization in maintenance. *Procedia Eng.* **2017**, *192*, 213–218. [CrossRef]
13. Bubeník, P.; Horák, F. Proactive approach to manufacturing planning. *Qual. Innov. Prosper.* **2014**, *18*, 23–32. [CrossRef]
14. Mičieta, B.; Biňasová, V. Implementation of energy efficient manufacturing to achieve sustainable production. In *The Contemporary Problems of Management—Value-Based Marketing, Social Responsibility and Other Factors in Process of Development—Micro, Meso and Macro Aspect: Monography*; Wydawnictwo Akademii Techniczno-Humanistycznej: Bielsko-Biała, Poland, 2014; pp. 289–310. ISBN 978-83-63713-77-5.
15. Back, T. *Evolutionary Algorithms in Theory and Practice: Evolution Strategies, Evolutionary Programming, Genetic Algorithms*; Oxford University Press: Oxford, UK, 1995; p. 328. ISBN 978-0195099713.
16. Hynek, J. *Genetické Algoritmy a Genetické Programování*; Grada publishing a.s.: Praha, Czech Republic, 2008; ISBN 9788024726953.
17. Krajčovič, M.; Hančinský, V. Production layout planning using genetic algorithms. *Commun. Sci. Lett. Univ. Žilina* **2015**, *17*, 72–77.
18. Hančinský, V. Využitie Evolučných Algoritmov Pri Návrhu Výrobných Dispozícií. Dissertation Thesis, UNIZA, Žilina, Slovakia, 2016.
19. Rajasekharan, M.; Peters, B.A.; Yang, T. A genetic algorithm for facility layout design in flexible manufacturing systems. *Int. J. Prod. Res.* **1998**, *33*, 95–110. [CrossRef]
20. Krajčovič, M.; Bulej, V.; Sapietova, A.; Kuric, I. Intelligent Manufacturing Systems in concept of digital factory. *Commun. Sci. Lett. Univ. Žilina* **2013**, *15*, 77–87.
21. Yang, T.; Zhang, D.; Chen, B.; Li, S. Research on plant layout and production line running simulation in digital factory environment. *Pac. Asia Workshop Comput. Intell. Ind. Appl.* **2008**, *2*, 588–593.
22. Furmann, R.; Štefánik, A. Progressive Solutions Supporting Manufacturing and Logistics Systems Design Developed by CEIT SK s.r.o. *Produkt. Inov.* **2011**, *12*, 3–5.
23. Honiden, T. Tree Structure Modeling and Genetic Algorithm-based Approach to Unequal-area Facility Layout Problem. *IEMS* **2004**, *3*, 123–128.

© 2019 by the authors. Licensee MDPI, Basel, Switzerland. This article is an open access article distributed under the terms and conditions of the Creative Commons Attribution (CC BY) license (http://creativecommons.org/licenses/by/4.0/).

Article

The Role of Green Attributes in Production Processes as Well as Their Impact on Operational, Commercial, and Economic Benefits

José Roberto Mendoza-Fong [1], Jorge Luis García-Alcaraz [2], José Roberto Díaz-Reza [1], Emilio Jiménez-Macías [3] and Julio Blanco-Fernández [4,*]

1. Department of Electrical Engineering and Computing, Universidad Autónoma de Ciudad Juárez, Ciudad Juarez 32310, Mexico; al164438@alumnos.uacj.mx (J.R.M.-F.); al164440@alumnos.uacj.mx (J.R.D.-R.)
2. Department Industrial Engineering and Manufacturing, Universidad Autónoma de Ciudad Juárez, Ciudad Juarez 32310, Mexico; jorge.garcia@uacj.com
3. Department Electrical Engineering, Universidad de La Rioja, 26004 Logroño, Spain; emilio.jimenez@unirioja.es
4. Department Mechanical Engineering, Universidad de La Rioja, 26004 Logroño, Spain
* Correspondence: julio.blanco@unirioja.es; Tel.: +34-941-299-524

Received: 14 November 2018; Accepted: 21 February 2019; Published: 1 March 2019

Abstract: This paper reports a second-order structural equation model composed of four variables: the green attributes before and after an industrial production process, the operating benefits, the commercial benefits, and the economic benefits. The variables are related by means of five hypotheses and are validated statistically with information obtained from 559 responses to a questionnaire applied to the Mexican maquila industry. The model is evaluated using the technique of partial least squares and the results obtained indicate that the green attributes before and after the production process have a direct and positive effect on the obtained benefits, mostly on the operational ones. It is concluded that companies that are focused on increasing their greenness level must monitor and evaluate the existence of green attributes in their production process to guarantee benefits and make fast decisions if required due to deviations.

Keywords: attributes; green manufacturing; benefits; green supply chain; sensitivity analysis

1. Introduction

Nowadays, taking care of the environment is a factor that may influence some industrial activities in a significant way, such as procurement, manufacturing, or distribution processes, as well as the green supply chain (GSC). In addition, companies are looking to incorporate more environmental issues such as industrial pollution prevention and control, sustainability, and investment in initiatives that are environmentally friendly [1]. The previous issues must be part of a long-term competitive strategy [2], since nowadays it is necessary to consider a GSC to minimize or eliminate (if possible) the negative effects that the traditional supply chain (SC) has on the environment.

Moreover, GSC allow used products to return to the production process, creating a cycle that will take advantage of all the available materials, minimizing natural resources used while reducing the environmental impact, which helps the green process in a SC to be progressively improved. In addition, the quality of the product and production processes is enhanced, reducing costs and expanding market fees through customers who are looking for clean manufacturers and products.

Likewise, green manufacturing (GM) can be considered one of the principal driving forces behind sustainable industrialized development. Consequently, researchers and industries have considered GM processes as a vital challenge [3] to develop new market opportunities as well as increase the benefits that may be acquired while focusing on environmental aspects in industries [4]. Furthermore, GM is a system that integrates the design of products and processes that influence the planning and control of manufacturing, identifying, quantifying, evaluating, and managing the flow of environmental waste in order to minimize the impact on the environment [5]. Similarly, a properly designed GM system may reduce operating costs through the efficient usage of raw materials, energy, and work force, which adds value to a product.

GM implementation aims to produce economically viable products with minimal environmental effect, but including a social and economic impact [6]. Thus, GM can be defined as the ability to intelligently use natural resources for manufacturing, which is performed through the creation of products and proposals to achieve economic, environmental, and social objectives; consequently, the environment can be preserved, while continuing to improve the quality of human life. In other words, these processes are possible due to the implementation of new technologies, regulatory measures, and appropriate social behavior [7].

However, implementing a GM process is not an easy task, since some strict regulations and policies must be complied with, especially in developed countries [8]. The manufacturing industry is the main segment for energy consumption and pollution, being responsible for 84% of CO_2 emissions as well as 90% of energy consumption [3]; these figures are reported despite complying with requirements such as the continuous improvement on the environmental product design, the use of environmentally friendly raw materials, the reuse of products, and the elimination of waste after a product's useful life. In fact, it is known that GM implementation leads to the improvement of manufacturing performance, but also requires the development of an organizational structure to establish a relationship between the GM implementation and the benefits obtained to encourage organizations to adopt a GM [7].

Nevertheless, declaring that a company applies GM practices implies that it complies with a series of industrial requirements, where it is possible to determine its level of implementation by evaluating a list of attributes of its production process. Likewise, the fulfillment of these attributes must guarantee the obtaining of benefits, reflected in the production processes, in the expansion of the market, and, as a consequence, driving the better economic performance of the company. Therefore, this article is focused on quantifying the existing relationship between those attributes that allow us to characterize a GM process, as well as analyzing how they impact the benefits obtained.

The attributes and benefits obtained from the implementation of a GM process are briefly described below.

1.1. Green Attributes in a Production Process

A GM addresses basic aspects during its manufacturing process, such as reduction, reuse, recycling, remanufacturing, preserving, managing waste, protecting the environment, complying with regulations, controlling pollution, and a variety of related issues [9]. The fulfillment of the previous criteria allows us to determine quickly and easily if the manufacturing processes can be considered a GM process.

As a matter of fact, currently in the industry some attributes are used to identify whether a manufacturing process fulfills the characteristics that confirm it is a real GM system. In addition, these attributes work as a measure in GM implementation systems or to assess how green a manufacturing process is at a certain moment. In addition, several authors have tried to identify and investigate some attributes to evaluate GM processes. In this sense, the majority of researchers have identified certain attributes as "mandatory" for any GM process, such as; reduction of emissions and waste towards the environment, energy, water, and resources preservation, environmental certification, clean production, green products generation, technologies implementation, and reverse logistics [10].

Additionally, other authors have labeled these attributes as "vital" in the environmental innovation adaptation based on the technological innovation, environmental monitoring systems, and the environmental customer collaboration [11]; however, attributes that are focused on personnel, management, suppliers, and government commitment are also included [12]. Finally, some authors have identified several attributes referred to as "key elements" on a GM process: green practices, green design, green purchasing or marketing, green packaging, ecological transportation, GSC management, reverse logistics, among others, as well as the total quality and life cycle management of a product [13].

Given that raw material suppliers play an important role in the sustainable performance of the company, a series of green attributes must be evaluated before the production process; the company does not have complete control of this, although they can be part of the supplier management program. On the other hand, there are attributes that occur during the production process over which the company has total control.

The attributes before production (ABP) are those attributes detected before starting the production process, and show the relationship between organizations and suppliers, as well as programs for the use and preservation of natural resources, green product design and processes, and the eco-business model. On the other hand, attributes during production process (ADP) are presented throughout the GM process; they usually include reduction of emissions, clean production process, use of green technologies, use of alternative or sustainable energies, green practices in productive processes, and the implementation of new technologies. In general, these attributes are essential and an opportunity for companies to maximize performance, quality, and profits from their production processes. Similarly, lean manufacturing tools, the use of total quality management philosophy (TQM), remanufacturing or reworking of products, green labeling, green practices in the distribution system, and social responsibility are also included in the ADP classification.

In a literature review, 24 attributes have been found that help determine the level of the GM implementation processes, including as follows:

Attributes before the process:

- There are programs for the use and preservation of natural resources [14,15]
- Green purchases [16,17]
- There is an environmental certification or ISO 14001 [18–20]
- Green process design [19,21]
- Environmental collaboration with suppliers [12,16]
- Eco-business models [12,17]
- Use of environmentally friendly raw materials [17]
- Selection of green suppliers [18,22,23]
- Green product design [8,24]
- Green practices in provisioning [19,25].

Attributes during the process:

- Reduction of emissions towards the environment [21,26]
- Lean manufacturing tools are implemented [7,17]
- There is a clean production process [10,12]
- Green technologies implementation [16,27]
- Use of alternative or sustainable energies [7,15]
- Damage towards the environment is reduced [16,28,29]
- Green practices in productive processes [12,25]
- New technologies implementation [8,24]
- TQM philosophy implementation [30,31]
- Green labeling or eco-labeling [32,33]
- Remanufacturing of products [16,17]
- Green practices in the distribution system [19,25]
- Social responsibility [12,20,24].

However, there are also trends to investigate which attributes are having an impact on production costs that may lead to an increase in the number of customers for companies, and consequently, companies may improve their income [34]. Therefore, some managers believe that the objective of generating wealth is lost when using ABP and ADP, which is the purpose the company was created for [35]. As a result, if those attributes represent a financial cost to the company, the question that must be asked is: What are the benefits that a company may obtain when implementing these GM practices? The following section presents an answer to that question.

1.2. Benefits of a Green Manufacturing Process

The search for environmental improvements is generally associated with rising costs at the beginning of a sustainable environmental manufacturing implementation [30]. It is a common belief that the costs of energy, products, inputs, and regulatory pressure will increase when ADP are monitored [36]. However, if there are changes of paradigms in the industrial structure towards an increasingly green future, it is possible to identify a series of benefits on different aspects.

Furthermore, researchers have shown that there are positive impacts from GM on environmental, commercial, economic, operational, and social performance that have led to the reduction of raw materials costs, energy, and labor, adding better value to the final product, improving production efficiency, increasing market fees, social corporation image, as well as minimizing waste and pollution [37]. Likewise, substantial improvements are made in the company's organization and technology, helping to reduce the use of resources, which suggests better choices in the use of alternative materials and energy [11], eliminating the generation of wastewater, CO_2 and heat emission, as well as residual sounds [38].

Finally, the most important benefit is that green growth may help to link sustainability with the economic performance [22], which is the reason why the manufacturing industry has confirmed that there is a significant relationship between a reduction in emissions and improvements in financial performance, thus generating short- and long-term economic gain [37]. A list of the benefits obtained from the GM implementing processes, which some authors have supported, is given below, structured into three categories:

Operating Benefits:

- Increase the quality of their processes [15,39]
- Product design improvement [18,40]
- Increase technological innovation [11,41]
- Greater competitiveness, productivity, and efficiency [18,19]
- Optimization in the use of available resources [12,42]
- Low product rework [8,43]
- Increase the quality of the final product [9,18].

Commercial Benefits:

- Local market expansion [18,33,44]
- Better customer service [3,45,46]
- Increase the number of products classified as green [47,48]
- Greater environmental certifications [9,33].

Economic Benefits:

- Increase in sales [16,49]
- Increase in economic gains [9,24]
- Reduction of marketing costs [3,12]
- Reduction of material waste [37,42]

- Reduction of production costs [10,50,51]
- Reduction of workforce for reprocessing [43,52]
- Cost reduction for guarantees [18,40].

1.3. Research Problem, Objective, and Contribution

Nowadays, a GM process is not just a competitive advantage but a necessity to mitigate the effects of the manufacturing industry on the environment. However, companies wonder: is it profitable to change from a traditional manufacturing process to a GM process? Is there any new, easy, and fast way to make the transformation to a GM process? and, of course, Should attributes be measured in a production process to know the level of implementation of GM?

Although the existing literature deals with the green attributes that must be monitored before and after the production process, as well as the benefits obtained, the relationship between them is unknown, which has been the reason many GM implementation processes were abandoned. Therefore, a study based on empirical data that relate those attributes present in the production processes with the benefits is needed.

In order to answer the previous questions and the problems posed, the objective of this research is to design a second-order structural equations model that relates the green attributes to the benefits obtained by implementing a GM process, which facilitates decision-making in industry when transforming a traditional manufacturing process into a GM process.

In the research into those relationships between attributes and benefits of a GM process, our contributions are as follows: (i) Two new classifications of green attributes are presented, the attributes before the process (ABP) and the attributes during the process (ADP); in addition, this classification of attributes allows us to demonstrate a close and significant relationship between the green attributes that a GM process must have, which have been identified as necessary, mandatory, or vital; the reason is that, when reviewing previous works, mention is only made of simple attributes or requirements, but they have not been related to each other; (ii) 18 benefits have been identified and classified, taken from previous works related to the implementation of a GM process, and classified into three categories: operational, commercial, and economic benefits; (iii) the relationship between the ABP and ADP green attributes and operational, commercial, and economic benefits is presented and quantified; (iv) an easy, fast, and novel way for organizations to monitor their GM implementation process is provided.

The article is divided into six main sections. After this introduction, Section 2 presents the research hypothesis and the literature review; Section 3 presents the materials and methods; Section 4 presents the results; Section 5 exposes the industrial and academic implications; and, finally, Section 6 presents the conclusions of the research.

2. Hypothesis and Literature Review

ABP and ADP in a GM process are used to evaluate the green implementation level; these attributes allow for the generation of a significant change in a corporation's operational, and environmental objectives, as well as maximizing the *Operating Benefits* (OB) for organizations. As a result, these will be reflected almost instantly in the innovation of green products and processes [53], such as product quality, technological innovations [28], competitiveness improvement, productivity and efficiency of operations [54], time reduction in a product cycle and reprocess, increase of the quality of processes, and final product [13].

Fortunately, several world-class companies have changed their traditional production processes to GM processes voluntarily, based on the ABP and ADP control [40]. In addition, they have been implemented without analyzing their importance, since these attributes directly assess the potential impacts of their factories, facilities, and operations in the environment. They also involve an active change in the social environment, generating a series of benefits that are reflected in their organizations. Therefore, the following hypothesis is proposed:

Hypothesis H1. *Attributes before and during a Green Manufacturing Process* have a direct and positive effect on the *Operating Benefits* obtained by implementing GM practices.

Ji, et al. [55] mention that supplier collaboration through ABP is one of the most important GM business practices, crucial for the efficiency of green practices, as it helps companies to achieve greater performance in terms of flexibility, delivery time, quality, and costs. Similarly, other authors indicate that ABP and ADP are useful through environmental certification or ISO 14001, eco-business models, use of alternative or sustainable energies, use of TQM philosophy, and eco-labeling, which can be implemented to evaluate GM processes and practices commercially [5], because they are focused on innovation and green business design to improve production efficiency by reducing costs while presenting a green reputation as well [56].

Theoretically, the improvement of traditional manufacturing processes with a GM process may be fundamental to decide when to charge a higher price for a product or ensure wider market coverage, as well as fulfill customers' or clients' requirements while maintaining environmental awareness [16]. In addition, it supports local and global market expansion, better customer service, and a better reputation with customers and competitors, which leads to an updated company, product image, and product itself [57]. Therefore, the following hypothesis is proposed:

Hypothesis H2. *Attributes before and during a Green Manufacturing Process* have a direct and positive effect on the *Commercial Benefits* obtained by implementing GM practices in the production process.

Furthermore, several empirical findings have confirmed a positive relationship between GM and improvement in the operational performance in an organization [35]. Additionally, the GM is closely related to environmental management and achievements from ecological operating objectives. GM also stimulates the generation of *Operating Benefits* since the use of ecologically innovative products and processes not only reduces the negative environmental impact, but also increases a company's economic and social performance through the reduction of waste and costs [28]. Finally, the achievement of these *Operating Benefits* promotes a better corporate image, market penetration, and strengthening of the brand to outperform the competition [58].

As a matter of fact, consumers realize that the improvement in the environmental characteristics of a product is associated with high quality, so it is assumed that it will have a higher price. However, the GM process improvements may encourage real quality improvements as well as provide affordable prices to customers [59]. In addition, it has been demonstrated that continuous improvement in capabilities that support environmental improvements generate positive effects in terms of product quality [35]. Therefore, the following hypothesis is proposed:

Hypothesis H3. *Operating Benefits* have a direct and positive effect on the *Commercial Benefits* obtained by implementing GM practices in the production lines.

GM processes are considered a competitive tool to achieve a positive corporate image, a reduction in costs, an improvement in market perception, improvements in the process, and a better product quality [55]. Moreover, it is crucial to address economic and environmental aspects simultaneously [58]. However, after investing money in environmental topics, companies expect to obtain *Economic Benefits* in a shorter period of time [60].

In the same way, *Economic Benefits* refer to a reduction of production costs, workforce for reprocessing aspects, and costs for guarantees, among others; However, the *Economic Benefits* go along with the *Operating Benefits* obtained in a GM process, since at the same time as improving the product, competitiveness, productivity, and efficiency of the processes, production will be improved as well and, consequently, less waste and pollution will be generated [38]. Therefore, the following hypothesis is proposed:

Hypothesis H4. *Operating Benefits* have a direct and positive effect on the *Economic Benefits* obtained by implementing a GM process.

Similarly, companies need to obtain greater income with the minimal environmental impact as well as use raw materials that have a minimally negative effect on the environment and society—but the product must still be economically viable [61]. Thus, the search for *Economic Benefits* will not be only focused on manufacturing operations, but also on the marketing and distribution of the product, as a strategy to encourage the consumption of green products among customers [62]. Likewise, by achieving greater *Commercial Benefits* with the increase in the number of products classified as green and environmentally friendly [33], *Economic Benefits* are generated and measured by the increase in sales and economic profits. In that case, the development of processes, products, and green distribution will be a competitive advantage and a way to maximize *Commercial Benefits* and *Economic Benefits* with GM [60]. Therefore, the following hypothesis is proposed:

Hypothesis H5. *Commercial Benefits* have a direct and positive effect on the *Economic Benefits* obtained by implementing a GM process.

The five hypotheses that have been presented are represented graphically in Figure 1, where the dependence of the variables (represented by ellipses) is illustrated by arrows.

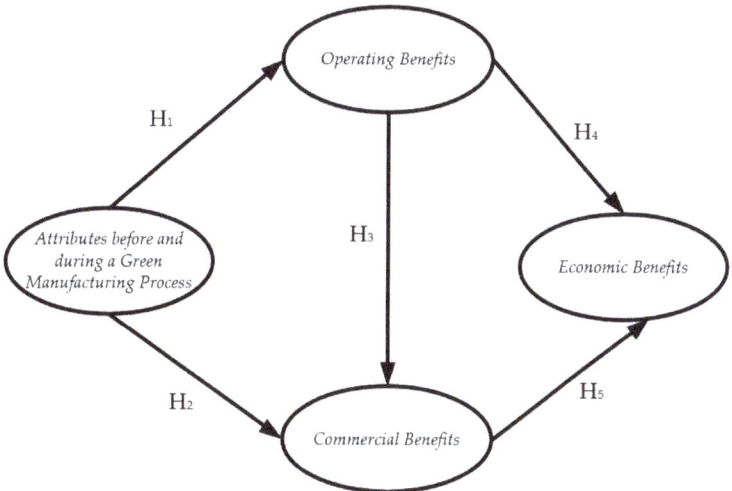

Figure 1. Proposed hypotheses.

3. Materials and Methods

In order to validate the five hypotheses that have been proposed, data from the industrial sector is required; therefore, for this purpose, a questionnaire is designed, data are collected, and the model is validated. The developed activities are shown below.

3.1. Literature Review

An in-depth review of the literature is carried out as a universal method in the development of research to collect, validate, and compare reliable and up-to-date information [1] or to identify aspects not discussed in previously published research [57]. The literature review was done for the period from 1995 to 2018 using more than 100 articles obtained from databases, such as ScienceDirect,

EBSCOhost, or Ingenta, among many others, and using keywords such as "green production process," "green attributes," "operational benefits," "commercial benefits," and "economic benefits."

This in-depth literature review was based on previous works conducted by authors such as Sarkis [26,63–65], Tseng [23,44], and Govindan [29,66], among others, who allowed us to identify the most used green attributes and the benefits that can be obtained when applying GM.

3.2. Survey Design

To collect information and validate the latent variable and model, we used a designed questionnaire, one of the most used methods to gather information easily and quickly [64]. The questionnaire had three sections: the first aimed to get sociodemographic information from participants and companies, the second included the green attributes (before and during the GM process), and the third described the benefits obtained from the GM implementation process. Similarly, a first draft was evaluated by academic judges and managers working in manufacturing companies to make a better adjustment to the geographical context; they must say whether the questionnaire is easy to understand and if there are linguistic issues or special words for a certain industrial sector. This review process by academics and industry personnel provides necessary validation for the questionnaire before it is applied to the manufacturing sector [67].

3.3. Data Acquisition

Each item in the questionnaire must be answered using a Likert scale with values between 1 and 5, where 1 indicates that the attribute is not present in the production process or that the benefit is not obtained, and 5 indicates that the attribute is always present or that the benefit is always obtained. In addition, the questionnaire is applied to personnel who have at least one year of experience in their job position, have working experience in manufacturing companies, have participated in GM implementation practices, and have knowledge about the results obtained from their projects. Consequently, the sampling is initially stratified and focused on personnel with previous manufacturing experience; however, the snowball technique is included as some respondents recommend that other colleagues answer the questionnaire as well.

Furthermore, possible participants are identified using information from AMAC (Maquiladora Association, A.C.) in Ciudad Juarez (Mexico). Therefore, each candidate was contacted by e-mail to arrange an appointment for the interview and answer the questionnaire in a face-to-face manner; if an appointment was cancelled for any reason, a new appointment was agreed, but after three missed appointments that case was discarded.

3.4. Statistical Debugging

The data collected through the applied surveys are registered in a database created in Software SPSS 24® software, because of its easy data analysis and integrated commands; each row represents a case or questionnaire, while a column represents an observed variable or item. In addition, the data are debugged and screened before performing any type of analysis, where the main activities are [68]:

- Identifying missing values that were not answered in the survey; if the percentage of missing values is under 10%, then it is replaced by the median of the item; however, if the percentage is higher, then that questionnaire is discarded.
- Identifying extreme values in each item in order to replace it with the median, since the values obtained are on a Likert scale.
- Identifying uncommitted respondents by estimating the standard deviation in every questionnaire; cases with a standard deviation under 0.35 are discarded.

3.5. Data Validation

Once the data have been debugged, they are validated through different indexes. In this case, the model in Figure 1 illustrates four latent variables, which are integrated by other variables called items or observed variables. Therefore, the validation and their measured indexes in each latent variable are:

- R^2 and adjusted R^2 are estimated to measure the predictive validity of the model, where values greater than 0.2 are required [60].
- Q^2 is estimated to measure the non-parametric predictive validity and values greater than zero and similar to R^2 are expected [13].
- The Cronbach's alpha and composite reliability index are used to measure the internal reliability, which requires values greater than 0.7 [64]; these indexes are obtained iteratively, since sometimes by eliminating any attributes or benefits, their values increased.
- The Average Variance Extracted (AVE) is used to measure the convergent validity, where values greater than 0.5 are required [28].

For the integration of the attributes and benefits in the latent variables analyzed in the model of Figure 1, a factorial analysis is carried out using the technique principal components technique with a PROMAX rotation, which is excited iteratively. The items that have a factorial load less than 0.5 are eliminated due to the low association with the latent factor or variable. In addition, Z values are estimated for the statistical test of the factorial load of the items (attributes and benefits) and their confidence interval is obtained for a confidence level of 95%.

3.6. Statistical Description of the Sample

The description of the sample is necessary in order to get relevant sociodemographic data from participants as well as the industrial sector where they are currently working, such as: years of experience, industrial sector, job position, and gender. Crosstabs are used to analyze the demographic characteristics of the sample to identify trends.

3.7. Development of the Structural Equation Modelling

In order to validate the hypotheses presented in Figure 1, the structural modeling equation (SEM) technique based on Partial Least Squares and integrated in the WarpPLS 6.0® software is used, because some latent variables have a double role as independent and dependent variables. Therefore, this technique is recommended when the data do not have a normal distribution, which appears in an ordinal scale or when there is a small sample size [69].

In addition, SEM allows researchers to evaluate or validate theoretical models, making it one of the most powerful tools for the study of causal relationships on non-experimental data when these relationships are linear. Furthermore, SEM is a novel technique and is currently used in different fields [70]; for instance, Farooq, et al. [71] analyze the impact of the quality service on customer satisfaction, Ojha, et al. [72] analyze the SC organizational learning, exploration, exploitation, and firm performance, and Qi, et al. [73] analyze the impact of operations and SC strategies on integration and performance, among many others researchers.

Partial least squares regression is estimated according to Equations (1) and (2) [74]:

$$X = TP^T + E \tag{1}$$

$$Y = UQ^T + F, \tag{2}$$

where X is an *nxm* matrix of predictors, Y is an matrix of responses; T and U are matrices that are, respectively, projections of X (the X *score*, *component* or *factor* matrix), and projections of Y (the Y *scores*); P and Q are, respectively, *mxl* and *pxl* orthogonal *loading* matrices; and E and F matrices are the error terms, which are assumed to be independent and identically distributed in randomly normal variables. Also, the decompositions of X and Y are made to maximize the covariance between T and U.

As a regression technique, the PLS idea is to generate a dependence measure between latent variables, which can be a simple regression or a multiple regression, depending on the number of independent latent variables that explain a dependent variable. The objective is to obtain a linear equation, as illustrated in Equation (3):

$$Y = \beta_0 + \beta_1 X_1 + \beta_2 X_2 + \ldots + \beta_p X_p, \tag{3}$$

where:

Y is a dependent latent variable

β_0 is the regression coefficient for the intercept

β_i values are the regression coefficients (for independent latent variables 1 to p) that have a direct effect on Y.

Specifically, WarpPLS 6.0® software reports and use standardized values for the β estimation, and then the $\beta_0 = 0$, simplifying Equation (3) into Equation (4):

$$Y = \beta_1 X_1 + \beta_2 X_2 + \ldots + \beta_p X_p + E, \tag{4}$$

where E represents an estimation error. The statistical validation for a hypothesis is carried out through the β value in direct effects, as appears in Figure 1, where the null hypothesis (H$_0$) is represented by Equation (5), and alternative hypothesis (H1) is represented by Equation (6):

$$H_0 : \beta i = 0 \tag{5}$$

$$H_1 : \beta i \neq 0. \tag{6}$$

The statistical test for β is done at a 95% confidence level. A hypothesis is accepted if the *p*-value associated is lower than 0.05, or the Z-value associated to the β significance test is bigger than 1.96, corresponding to a two-tailed test. In addition, a confidence interval for β value is estimated, looking to have intervals excluding the zero. The confidence intervals are estimated according to Equation (7) for the lower bound and Equation (8) for the upper bound:

$$\beta_{low} = \beta_i - 1.96 \text{ SE} \tag{7}$$

$$\beta_{up} = \beta_i + 1.96 \text{ SE}, \tag{8}$$

where:

β_i is the estimated value for a relationship between two variables;

1.96 is the Z value for a 95% confidence value for a two-tailed test;

SE is the standard error for β_i.

Three different types of effects are estimated in the structural equation model, and every relationship between latent variables has a β value linked. The effects measured in the model are: the direct effects that help to validate the hypotheses, which are represented by arrows in Figure 1, the total indirect effects that are presented when there are mediating variables, which require two or more segments, and the total effects that represent the sum of the direct and indirect effects.

In addition, it is important to mention that the variable called *Attributes before and during a Green Manufacturing Process* is a second-order variable, since there is a set of attributes that are present before the GM process, which represents a latent variable, while the attributes during the production process represent a second variable.

Six quality model indexes are obtained before interpreting the model [75]:

- Average path coefficient (APC), where *p*-associated values under 0.05 are required.
- Average R-squared (ARS) and Average Adjusted R-squared (AARS), which require p-associated values under 0.05.
- Average block variance inflation factor (AVIF) and Average full collinearity VIF (AFVIF), which require values under 5.
- Tenenhaus GoF Index (GoF) that estimates the data adjustment in the model, which requires values over 0.36.

In Figure 1 some dependent latent variables are explained by two or more latent variables. Therefore, in the present research the R^2 value is disintegrated according to the portion of variance that each independent variable explains, which is called effect size (ES) and allows for separating the essential variables from the trivial ones.

3.8. Sensitivity Analysis

A sensitivity analysis is reported in order to know the effect that a possible change in an independent variable has on a dependent variable. In partial least square the values of the latent variables are standardized, so it is possible to estimate the probabilities of occurrence among them. In this case, it is assumed that a standardized value greater than 1 represents a "high" probability of occurrence, while a value less than −1 represents a "low" probability of occurrence. Therefore, for each hypothesis an analysis is done regarding the four stages where the variables may be involved.

Specifically, this study analyzes the probabilities of occurrence simultaneously in each scenario is represented by "&", while the conditional probability is represented by "if".

4. Results

4.1. Demographic Data of the Sample

After four months of sampling, 559 valid questionnaires were obtained, once the statistical purification was applied; Table 1 presents descriptive information of the sample. In the first column the demographic data are presented, such as: gender, industrial sector and years of experience in their work position. In the second column the frequency of the answers of the participants is presented; for example it can be seen that 362 people of the total of the respondents were men, 190 women, and seven people did not specify their gender.

Table 1. Sample characterization.

Demographic Data		Frequency	%
Gender	Male	362	64,758
	Female	190	33,989
	*NOS	7	1252
		** T = 559	T = 100
Industrial Sector	Automotive	342	61,180
	Plastics	72	12,880
	Metal—mechanical	49	8766
	Medical	34	6082
	Electronic	30	5367
	Electric	19	3399
	Aeronautics	7	1252
	NOS	6	1073
		T = 559	T = 100
Years of experience in the work position	2–5	185	33,095
	6–10	128	22,898
	1–2	119	21,288
	10–20	97	17,352
	20–30	28	5009
	NOS	2	0.358
		T = 559	T = 100

* NOS = Not specified, ** T = Total.

In the third column the percentage response of the participants is presented; returning to the previous example, it can be seen that 64.76% of the participants were men, 33.99% women, and 1.25% did not specify their gender. See Table 1 for a complete summary of the other demographics mentioned above.

4.2. Latent Variables Validation

Table 2 shows a summary of the validation indexes for the latent variables integrated in Figure 1. In the first column the validity indices of the latent variables are presented, such as R^2, or adjusted R^2, among others. In the second to fifth columns the values of the respective coefficients for each of the latent variables are presented. Finally, the interpretation and validation of each of the coefficients presented in the first column of Table 2 is presented.

Table 2. Indexes for latent variables validation.

Latent Variable Coefficients	Attributes before and during a Green Manufacturing Process	Operating Benefits	Commercial Benefits	Economic Benefits
R^2		0.371	0.705	0.737
Adj. R^2		0.370	0.704	0.736
Composite Reliability	0.942	0.940	0.909	0.952
Cronbach's Alpha	0.877	0.925	0.866	0.941
AVE	0.890	0.690	0.714	0.737
Full Collinearity VIF	1.772	3.928	4.026	3.764
Q^2		0.372	0.705	0.738

The latent variables have parametric and non-parametric predictive validity with R^2 and adjusted R^2 values greater than 0.2; however, the Q^2 value is greater than zero, and very similar to R^2, which leads us to conclude that there is enough non-parametric predictive validity.

Additionally, it is observed that the AVE values are greater than 0.5, which indicates enough convergent validity. On the other hand, the latent variables have internal reliability, since the values for Cronbach's alpha index and the composite reliability are greater than 0.7. Finally, it is observed that the latent variables do not have problems of multicollinearity, since the VIF values are less than 5. According to previous index values, it is concluded that the latent variables are suitable to be integrated in the structural equation model.

In Appendix A appears a list of attributes and benefits that integrates every latent variable, the Z ratios for the statistical test, and the confidence interval for every factor loading regarding the factor analysis. Some attributes or benefits were eliminated from the analysis due to low factor loadings, and therefore the appendix only illustrates the final list.

4.3. Structural Equation Model

Table 3 illustrates the indexes of validation obtained from the structural equation model after it was introduced and analyzed in the Software WarpPLS 6.0®. Observe that the APC, ARS, and AARS indexes comply with their approval values since the *p*-value is lower than 0.001. Similarly, the AVIF and AFVIF indexes have values lower than 5, the maximum allowed value, which indicates that there are no multicollinearity problems. Also, the Tenenhaus GoF (Goodness of Fit) index suggests that the model has enough explanatory power, since it has a value greater than 0.36. According to the previous values in indexes, it is concluded that the defined model is valid and can be interpreted.

Table 3. Indexes for model validation.

Indexes for the Model Validation
Average path coefficient (APC) = 0.484, $p < 0.001$
Average R-squared (ARS) = 0.604, $p < 0.001$
Average adjusted R-squared (AARS) = 0.603, $p < 0.001$
Average block VIF (AVIF) = 2.379, acceptable if ≤ 5, ideally ≤ 3.3
Average full collinearity VIF (AFVIF) = 3.373, acceptable if ≤ 5, ideally ≤ 3.3
Tenenhaus GoF (GoF) = 0.677, small ≥ 0.1, medium ≥ 0.25, large ≥ 0.36

Figure 2 illustrates the evaluated structural equation model, where the β values are indicated for each relationship among latent variables (hypotheses) and the associated *p*-value in order to determine their statistical significance. Appendix B illustrates the Z ratios for β values statistical test indicating that every value is higher than 1.96 according to the stablished confidence level. Also, Appendix B illustrates the confidence intervals for every β in direct effects and, according to that information, every hypothesis is accepted.

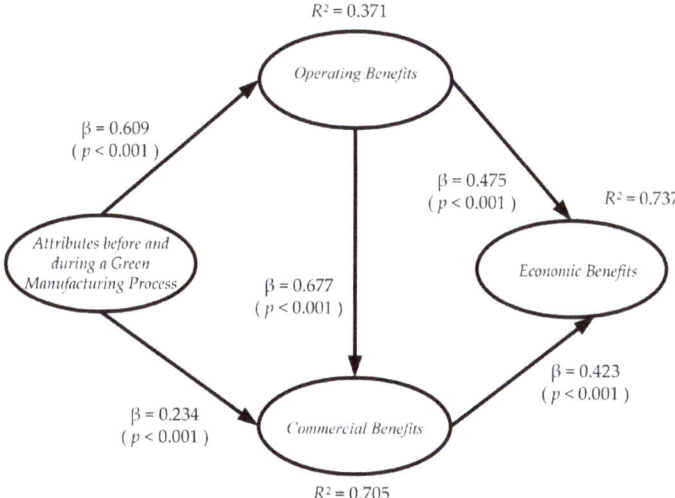

Figure 2. Evaluated model.

Considering the *p*-values and β values in each relationship among the latent variables and considering that the confidence level is 95%, the conclusions about the five hypotheses presented in this research are as follows:

H_1: There is enough statistical evidence to declare that *Attributes before and during a Green Manufacturing Process* have a direct and positive effect on the *Operating Benefits* obtained by implementing GM practices, because when the first latent variable increases its standard deviation in one unit, the second one goes up by 0.609 units.

H_2: There is enough statistical evidence to declare that *Attributes before and during a Green Manufacturing Process* have a direct and positive effect on the *Commercial Benefits* obtained by implementing GM practices in the production process, because when the first latent variable increases its standard deviation in one unit, the second one goes up by 0.234 units.

H_3: There is enough statistical evidence to declare that *Operating Benefits* have a direct and positive effect on the *Commercial Benefits* obtained by implementing GM practices in the production lines, because when the first latent variable increases its standard deviation in one unit, the second variable goes up by 0.677 units.

H_4: There is enough statistical evidence to declare that *Operating Benefits* have a direct and positive effect on the *Economic Benefits* obtained by implementing a GM process, because when the first latent variable increases its standard deviation in one unit, the second one goes up by 0.475 units.

H_5: There is enough statistical evidence to declare that *Commercial Benefits* have a direct and positive effect on the *Economic Benefits* obtained by implementing a GM process, because when the first latent variable increases its standard deviation in one unit, the second one goes up by 0.423 units.

According to the information in Figure 2, there are three dependent variables, and therefore the structural equations can be stated as follows:

$$Operating\ Benefits = 0.609\ Attributes\ before\ and\ during\ a\ Green\ Manufacturing\ Process + Error \quad (9)$$

$$Commercial\ Benefits = 0.234\ Attributes\ before\ and\ during\ a\ Green\ Manufacturing\ Process + 0.677\ Operating\ Benefits + Error \quad (10)$$

$$Economic\ Benefits = 0.475\ Operating\ Benefits + 0.423\ Commercial\ benefits + Error. \quad (11)$$

Table 4 presents in more detail the hypotheses and their results. In the first column the number of the hypothesis is presented; in the second and third columns the independent and dependent variables are presented, respectively; in the fourth and fifth columns the indices β and R^2 (used as part of the validation of the hypotheses) are presented, and in the sixth column the *p*-value is presented. The *p*-value is evaluated with a confidence level of 95%, to reject or accept the hypothesis. Finally, in the last column, the conclusion of rejection or acceptance of the hypothesis is determined.

Table 4. Hypothesis validation (direct effects).

Hypothesis	Independent Variable	Dependent Variable	β	R^2	*p*-Value	Conclusion
H_1	Attributes before and during a Green Manufacturing Process	Operating Benefits	0.609	0.371	$p < 0.001$	Accepted
H_2	Attributes before and during a Green Manufacturing Process	Commercial Benefits	0.234	0.150	$p < 0.001$	Accepted
H_3	Operating Benefits	Commercial Benefits	0.677	0.555	$p < 0.001$	Accepted
H_4	Operating Benefits	Economic Benefits	0.475	0.392	$p < 0.001$	Accepted
H_5	Commercial Benefits	Economic Benefits	0.423	0.345	$p < 0.001$	Accepted

Figure 2 shows that the dependent variables have an associated R^2 value as a measure of the variance explained, which is due to one or several independent variables. Table 5 decomposes the R^2 value according to the contribution of each independent variables. In the first column the dependent variables are presented, and the second, third, and fourth present the independent variables; finally, the last column contents the total value of R^2.

Table 5. R^2 value decomposition.

To	From			R^2
	Operating Benefits	Commercial Benefits	Attributes before and during a Green Manufacturing Process	
Operating Benefits			0.371	0.371
Commercial Benefits	0.555		0.150	0.705
Economic Benefits	0.392	0.345		0.737

For example, it is observed that the variable *Commercial Benefits* has a total value of $R^2 = 0.705$, which indicates that it is explained in a 70.5% by the variables *Attributes before and during a Green*

Manufacturing Process and *Operating Benefits*, but only 15.0% comes from the first variable while 55.5% comes from the second variable. In that sense, *Operating Benefits* has a greater explanatory power in the dependent variable (note that the sum of the contributions of each variable is the percentage of variance explained). In the case the variable *Economic Benefits*, which is explained in 73.7%, 39.2% is explained by the variable *Operating Benefits* and 34.5% by the variable *Commercial Benefits*.

Moreover, Table 6 portrays in its first column the type of effect, whether the sum of indirect effects or the total effects between variables. The last column presents the associated *p*-value and the ES as a measure of the explanatory power between the variables. These values were taken from the results obtained once the model was run in the WarpPLS 6.0 Software and are essential to interpret the results in Section 4.

Table 6. Sum of indirect and total effects.

Type Effect	From	To	
Indirect	Attributes before and during a Green Manufacturing Process	Commercial Benefits	0.413 ($p < 0.001$) ES = 0.265
	Attributes before and during a Green Manufacturing Process	Economic Benefits	0.563 ($p < 0.001$) ES = 0.310
	Operating Benefits	Economic Benefits	0.286 ($p < 0.001$) ES = 0.236
Total	Attributes before and during a Green Manufacturing Process	Operating Benefits	0.609 ($p < 0.001$) ES = 0.371
	Attributes before and during a Green Manufacturing Process	Commercial Benefits	0.647 ($p < 0.001$) ES = 0.416
	Attributes before and during a Green Manufacturing Process	Economic Benefits	0.563 ($p < 0.001$) ES = 0.310
	Operating Benefits	Commercial Benefits	0.677 ($p < 0.001$) ES = 0.555
	Operating Benefits	Economic Benefits	0.761 ($p < 0.001$) ES = 0.628
	Commercial Benefits	Economic Benefits	0.423 ($p < 0.001$) ES = 0.345

In addition, the only effect with three segments involves the *Attributes before and during to the Green Manufacturing Process* variable along with the *Economic Benefits* variable through the mediating variables *Operating Benefits* and *Commercial Benefits*, which are the highest and most significant indirect effects, since the *Attributes before and during a Green Manufacturing Process* variable does not have a direct effect on the *Economic Benefits* variable.

Likewise, this indirect effect is expected, since when a company makes the decision to update its traditional manufacturing process and implement a GM process with the use of these attributes, it promotes a series of savings associated with the use and consumption of resources, as well as raw materials, savings in the distribution of supply products and materials, waste reduction, and reprocessing.

It is important to mention that the indirect effect from *Attributes before and during to the Green Manufacturing Process* on *Commercial Benefits*, through the mediating *Operating Benefits* variable is higher than the direct effect that exists between them; this indirect effect is 0.413 while the direct effect is only 0.234, which means that when a company generates *Operating Benefits* and increases the quality in products and processes, companies are indirectly obtaining *Commercial Benefits*, because they are able to provide a better service to clients, which in the long term generates a better reputation and may lead to market expansion.

Similarly, Table 6 displays six total effects, which are statistically significant and indicate the importance of these aspects, as well as the magnitude of the effect that is used in the attributes to obtain operational, commercial, and *Economic Benefits*. This is crucial, since even nowadays it is still questioned if implementing a GM process will automatically bring a benefit, and these findings prove quantitatively and statistically that relationship.

In addition, the total effect of the *Operating Benefits* variable on the *Economic Benefits* variable can be remarked, since the highest total effect indicates that, when obtaining *Operational Benefits*, it is guaranteed that for any company that implements a GM process, essential *Economic Benefits* may be acquired. As a result, these processes and products have been improved by reducing material waste, production costs, workforce, reprocessing, warranty costs, marketing costs, etc.

In summary, the results presented in Table 6 may help companies have better confidence and initiative for implementing GM practices in their processes—a change that will significantly impact the environment. The data obtained show that companies that make the decision to adapt their traditional processes to GM processes will obtain a substantial series of operational, commercial, and *Economic Benefits*.

4.4. Sensitivity Analysis

Table 7 illustrates the sensitivity analysis from the relationships between the latent variables in the model when they have high and low levels independently, as well as a combination of levels (four stages for each relationship or hypothesis). In the first column (named "To") the dependent variables are presented; the second one presents the levels that can have the latent variables; the third column includes the value of P (i) (probability of occurrence in its high and low level of each of the latent variables); the other columns present the values of the probabilities of occurrence simultaneously in each scenario.

Table 7. Sensitivity analysis.

From			Attributes before and during a Green Manufacturing Process		Operating Benefits		Commercial Benefits	
To	Level		+	−	+	−	+	−
		P (i)	0.177	0.181	0.156	0.150	0.186	0.165
Operating Benefits	+	0.156	& 0.086 If 0.485	& 0.007 If 0.040				
	−	0.150	& 0.007 If 0.040	& 0.075 If 0.416				
Commercial Benefits	+	0.186	& 0.100 If 0.566	& 0.004 If 0.020	& 0.091 If 0.586	& 0.000 If 0.000		
	−	0.165	& 0.004 If 0.020	& 0.088 If 0.485	& 0.002 If 0.011	& 0.114 If 0.762		
Economic Benefits	+	0.161			& 0.081 If 0.517	& 0.000 If 0.00	& 0.095 If 0.510	& 0.004 If 0.022
	−	0.159			& 0.000 If 0.000	& 0.125 If 0.833	& 0.000 If 0.000	& 0.122 If 0.739

For example, it is observed that the probability that *Operating Benefits* are present independently at their low level is 0.156, while at their high level is 0.150. However, the probability of being at a high level simultaneously with the *Commercial Benefits* is only 0.091, but the conditional probability of having these *Commercial Benefits* at a high level due to high *Operating Benefits* is 0.586, which indicates that managers must focus on obtaining the second type of benefits, since there is a high probability of obtaining the first ones.

However, if there are simultaneously *Operating Benefits* and *Commercial Benefits* at their low levels, the probability of the two variables occurring together is only 0.114; therefore, the importance of the analysis becomes significant when analyzing the probability of occurrence for the first variable at its low level, since the second variable has occurred at that same level, because the probability of that event is 0.762. That information indicates that if a manager does not strive to achieve *Operating Benefits* with the GM implementation process, there is a high probability that *Commercial Benefits* will not be achieved.

The previous conclusion is verified when the *Operating Benefits* have a high level and the *Commercial Benefits* a low level, where it is observed that the probability of occurrence for that event simultaneously is only 0.002, which indicates that whenever the first benefits are obtained, the second variable will be present in that low scenario. As a result, the probability that the second variable occurs at its low level and the first at its high level is only 0.011—a very low probability, which indicates that *Operating Benefits* are always associated with *Commercial Benefits*.

5. Practical and Theoretical Implications

5.1. Theoretical Implications

The analysis shows that it is possible to acquire favorable results for the company when implementing GM, since it is possible to improve the use of resources, energy, and raw material, and therefore the environmental impact is reduced. On the other hand, the information presented in this research has great relevance as a GM evaluation or implementation tool, and allows for obtaining the following conclusions:

The variable *Attributes before and during a Green Manufacturing Process* and its items show their importance in GM, which can be verified as in Table 2. In addition, this relationship indicates that each GM process must have certain attributes, such as reduction of emissions and waste, preservation of natural resources, clean production, generation of green products, use of green technologies, and selection of green suppliers, as mentioned in Wang, Huscroft, Hazen and Zhang [21].

Furthermore, these attributes and their execution generate a series of *Operating Benefits* in the productive processes, which is validated statistically with H_1, since it has a direct, positive, and significant effect. Moreover, it indicates that when GM process is implemented, operational benefits will be generated in terms of competition, productivity, and efficiency [45]. According to the previous information, it is observed that a series of *Commercial Benefits* are obtained, a statement that is validated with H_2 and H_3; consequently, there is a direct, positive, and significant effect between those three latent variables. In addition, when implementing a GM process, there will be a better production process and better products, which will lead to new potential customers, market expansion, and a better reputation with clients and competitors [48].

Although the *Attributes before and during a Green Manufacturing Process* variable does not have a direct effect on the *Economic Benefits* variable (it has not been studied in this research), this variable does have a significant indirect effect. This fact is not surprising, since having a GM process in a company generates a series of *Economic Benefits* associated with the reduction of resources, supplies, and raw materials consumption, which agree with Roy and Khastagir [42]. In addition, the fact of having a green image, better quality, better customer service, and higher certifications will attract new customers, increasing sales and financial gains, as Sun, Miao and Yang [11] mention.

Furthermore, it is also relevant to review the direct, positive, and significant relationship that exists between operating, commercial, and *Economic Benefits*, which are validated through H_4 and H_5, confirming that, if a GM process is implemented with the use of attributes, benefits will be generated in terms of the processes, products, reputation, quality of the final product, and customer service, which, consequently, will be reflected in the savings, profits, and commercialization of the company.

Today, it is essential to emphasize that not having a GM process may place a company in a positive or negative situation, classifying it as obsolete. In addition, the GM is a valuable tool to promote the environmental awareness of clients, customers, suppliers, and manufacturers.

Finally, this research can be applied in different types of manufacturing as a new way to evaluate whether their manufacturing processes are green, with the use of the attributes mentioned in this work. Furthermore, using these attributes guarantees the obtaining of several operational, commercial, and *Economic Benefits* in the organization.

5.2. Practical Implications

It is important to emphasize that, although the research was carried out in the Mexican maquiladora industry context, it can also be applied in other countries with emerging economies that have similar conditions. Also, the results allow companies to recommend that maquiladoras develop green products and processes through proactive environmental management policies and environmental practices, which guarantees new green markets and customers who are more committed to their environment and its preservation.

From the data analysis in Table 7, the following conclusions and industrial inferences can be established for companies that implement GM in their production processes:

- The GM implementation is a continuous process that must be monitored throughout the production system; there are attributes that must be evaluated before and during the production process.
- The execution of activities that provide the *Attributes before and during a Green Manufacturing Process* helps to obtain *Operating Benefits*, since there is a probability that this will occur of 0.485 and *Commercial Benefits* with a probability of 0.566 to happen. In addition, the previous information indicates that managers must have a tracking system for GM practices in order to have control of them and make the necessary adjustments and guarantee the desired benefits, especially those of an operational type, since the commercial and economic benefits depend on them. Also, in the event of low levels of execution of the activities associated with the obtained attributes that characterize the GM process, there is also a risk of having low *Operating Benefits* (probability of 0.416) and *Commercial Benefits* (probability of 0.485).
- *Operating Benefits* at a high level guarantee the obtaining of high *Commercial Benefits* (probability of 0.586); therefore, the way that it is implemented should be a priority for managers when implementing GM. However, if these operating benefits are low for any reason, the risk of obtaining low *Commercial Benefits* is 0.762; if there is a very high value, since the implementation is associated with aspects related to the product quality and cost, it means these are not attractive to the customer, so the company loses market opportunities.
- According to the previous information, it is concluded that high levels of *Operating Benefits* bring *Commercial Benefits*, and these in turn bring *Economic Benefits*. In fact, it can be observed that it is not possible to have high economic benefits when there are low *Operating Benefits*, which again indicates that managers should focus on aspects associated with the cost, quality, and company image. Moreover, there is a high risk of having financial problems when *Operating Benefits* are not obtained, since when they have low levels, there is a high risk that the *Economic Benefits* are low (probability of 0.833).
- Companies must guarantee *Commercial Benefits* at high levels in order to obtain *Economic Benefits* at that same level (probability of 0.510), since, if there are low levels for the first variable, there is a high risk of also having low levels in *Economic Benefits* (probability 0.739).

5.3. Future Studies

The evaluation of a GM should also consider attributes after the manufacturing process since, in this study, only the attributes before and during the manufacturing process have been considered. In future studies, how the attributes would be related after the manufacturing process will be considered, by integrating them into the model proposed in this research or associating the three different types of attributes and knowing their relationship quantitatively by structural equation modeling. Likewise, the social benefits or some other possible type of benefits that can be obtained with the update of the traditional manufacturing process to a GM process could be associated.

6. Conclusions

Evaluating the performance of green production processes will continue to be of interest to academia and industry, which are committed to the environment in which they and their clients perform. Therefore, the identification of green attributes before and during the production process will be an indication of the level of implementation of GM processes and the possible benefits that will be obtained. The analysis of the attributes and benefits of GM processes leads to the following conclusions:

1. The monitoring of green attributes before and during the production process allows us to evaluate the company's GM process and facilitates the obtaining of operational benefits in the production line and commercial benefits to clients.
2. The operational benefits obtained from implementing a GM process help to improve the commercial and economic benefits to the companies.
3. *Commercial Benefits* obtained by implementing GM facilitate the increase of economic benefits for companies.

Author Contributions: All authors contributed in this paper. J.R.M.-F. collected the data and designed the structural equation model. J.L.G.-A. participated in the development of the methodology. J.B.-F. carried out the validation of the data. E.J.-M. carried out the data analysis and wrote the results section. J.R.D.-R. and J.B.-F. wrote the rest of the paper and did the final review of the paper as well as its modifications.

Funding: This research was funded by Mexican National Council for Science and Technology, grant number CONACYT-INS (REDES) 2018—293683 LAS.

Acknowledgments: All authors acknowledge the Mexican National Council for Science and Technology (CONACYT) and responders from industry.

Conflicts of Interest: All authors declare that there is no conflict of interest regarding the publication of this paper.

Appendix A. Latent variables validation

Confidence level used: 95%
Value for two-tailed tests: 1.960

Table A1. Z ratios for loadings.

Items	Operating Benefits	Commercial Benefits	Economic Benefits	ABP and ADP
Increasement in the quality of their processes	21,569			
Product design improvement	21,509			
Increasement in its technological innovation	20,976			
Optimization in the use of available resources	21,521			
Low product rework	21,465			
Greater competitiveness, productivity, and efficiency	22,469			
Increasement in the quality of the final product	21,682			
Local market expansion		22,164		
Better customer service		21,920		
Increasement in the number of products classified as green		22,032		
Greater environmental certifications		21,925		
Increasement in sales			21,939	
Increasement in economic gains			22,382	
Reduction of marketing costs			22,799	
Reduction of material waste			22,192	
Reduction of production costs			22,711	
Reduction of workforce for reprocessing			22,512	
Cost reduction for guarantees			22,318	
Attributes before the process				24,866
Attributes during the process				24,866

Table A2. Confidence intervals for loadings (Low and Up).

Items	Operating Benefits		Commercial Benefits		Economic Benefits		ABP and ADP	
Increasement in the quality of their processes	0.754	0.905						
Product design improvement	0.752	0.903						
Increasement in its technological innovation	0.733	0.884						
Optimization in the use of available resources	0.752	0.903						
Low product rework	0.750	0.901						
Greater competitiveness, productivity, and efficiency	0.786	0.936						
Increasement in the quality of the final product	0.758	0.909						
Local market expansion			0.775	0.925				
Better customer service			0.766	0.917				
Increasement in the number of products classified as green			0.770	0.921				
Greater environmental certifications			0.767	0.917				
Increasement in sales					0.767	0.918		
Increasement in economic gains					0.783	0.933		
Reduction of marketing costs					0.797	0.947		
Reduction of material waste					0.776	0.926		
Reduction of production costs					0.794	0.944		
Reduction of workforce for reprocessing					0.787	0.937		
Cost reduction for guarantees					0.780	0.931		
Attributes before the process							0.869	1.018
Attributes during the process							0.869	1.018

Appendix B. Z ratios and confidence intervals for β

Table A3. Z ratios for β values.

Latent variables	Operating Benefits	Commercial Benefits	ABP and ADP
Operating Benefits			15.45
Commercial Benefits	17.315		5.681
Economic Benefits	11.86	10.491	

Table A4. Confidence intervals for β values (Low and Up).

Latent variables	Operating Benefits		Commercial Benefits		ABP and ADP	
Operating Benefits					0.532	0.687
Commercial Benefits	0.601	0.754			0.153	0.315
Economic Benefits	0.396	0.553	0.344	0.502		

References

1. Tseng, M.-L.; Islam, M.S.; Karia, N.; Fauzi, F.A.; Afrin, S. A literature review on green supply chain management: Trends and future challenges. *Resour. Conserv. Recycl.* **2019**, *141*, 145–162. [CrossRef]
2. Singh, N.; Jain, S.; Sharma, P. Determinants of proactive environmental management practices in Indian firms: An empirical study. *J. Clean. Prod.* **2014**, *66*, 469–478. [CrossRef]
3. Kara, S.; Singh, A.; Philip, D.; Ramkumar, J. Quantifying green manufacturability of a unit production process using simulation. *Proc. CIRP* **2015**, *29*, 257–262. [CrossRef]
4. Denisa, M.; Zdenka, M. Perception of implementation processes of green logistics in SMEs in Slovakia. *Procedia Econ. Financ.* **2015**, *26*, 139–143. [CrossRef]
5. Maruthi, G.D.; Rashmi, R. Green Manufacturing: It's tools and techniques that can be implemented in manufacturing sectors. *Mater. Today Proc.* **2015**, *2*, 3350–3355. [CrossRef]

6. Paul, I.D.; Bhole, G.P.; Chaudhari, J.R. A review on green manufacturing: It's important, methodology and its application. *Procedia Mater. Sci.* **2014**, *6*, 1644–1649. [CrossRef]
7. Thanki, S.; Govindan, K.; Thakkar, J. An investigation on lean-green implementation practices in Indian SMEs using analytical hierarchy process (AHP) approach. *J. Clean. Prod.* **2016**, *135*, 284–298. [CrossRef]
8. Govindan, K.; Diabat, A.; Madan Shankar, K. Analyzing the drivers of green manufacturing with fuzzy approach. *J. Clean. Prod.* **2015**, *96*, 182–193. [CrossRef]
9. Rehman, M.A.A.; Shrivastava, R.L.; Shrivastava, R.R. Comparative analysis of two industries for validating green manufacturing (GM) framework: An Indian scenario. *J. Inst. Eng. India Ser. C* **2016**, *98*, 203–218. [CrossRef]
10. Charmondusit, K.; Gheewala, S.H.; Mungcharoen, T. Green and sustainable innovation for cleaner production in the Asia-Pacific region. *J. Clean. Prod.* **2016**, *134*, 443–446. [CrossRef]
11. Sun, L.-Y.; Miao, C.-L.; Yang, L. Ecological-economic efficiency evaluation of green technology innovation in strategic emerging industries based on entropy weighted TOPSIS method. *Ecol. Indic.* **2017**, *73*, 554–558. [CrossRef]
12. Woo, C.; Kim, M.G.; Chung, Y.; Rho, J.J. Suppliers' communication capability and external green integration for green and financial performance in Korean construction industry. *J. Clean. Prod.* **2016**, *112*, 483–493. [CrossRef]
13. Chan, H.K.; Yee, R.W.Y.; Dai, J.; Lim, M.K. The moderating effect of environmental dynamism on green product innovation and performance. *Int. J. Prod. Econ.* **2016**, *181*, 384–391. [CrossRef]
14. Oncel, S.S. Green energy engineering: Opening a green way for the future. *J. Clean. Prod.* **2017**, *142*, 3095–3100. [CrossRef]
15. Chen, S.-C.; Hung, C.-W. Elucidating the factors influencing the acceptance of green products: An extension of theory of planned behavior. *Technol. Forecast. Soc. Chang.* **2016**, *112*, 155–163. [CrossRef]
16. Grekova, K.; Calantone, R.J.; Bremmers, H.J.; Trienekens, J.H.; Omta, S.W.F. How environmental collaboration with suppliers and customers influences firm performance: Evidence from Dutch food and beverage processors. *J. Clean. Prod.* **2016**, *112*, 1861–1871. [CrossRef]
17. Rehman, M.A.; Seth, D.; Shrivastava, R.L. Impact of green manufacturing practices on organisational performance in Indian context: An empirical study. *J. Clean. Prod.* **2016**, *137*, 427–448. [CrossRef]
18. Teles, C.D.; Ribeiro, J.L.D.; Tinoco, M.A.C.; ten Caten, C.S. Characterization of the adoption of environmental management practices in large Brazilian companies. *J. Clean. Prod.* **2015**, *86*, 256–264. [CrossRef]
19. Ahi, P.; Searcy, C. An analysis of metrics used to measure performance in green and sustainable supply chains. *J. Clean. Prod.* **2015**, *86*, 360–377. [CrossRef]
20. Govindan, K.; Rajendran, S.; Sarkis, J.; Murugesan, P. Multi criteria decision making approaches for green supplier evaluation and selection: A literature review. *J. Clean. Prod.* **2015**, *98*, 66–83. [CrossRef]
21. Wang, Y.; Huscroft, J.R.; Hazen, B.T.; Zhang, M. Green information, green certification and consumer perceptions of remanufctured automobile parts. *Resour. Conserv. Recycl.* **2018**, *128*, 187–196. [CrossRef]
22. Neumüller, C.; Lasch, R.; Kellner, F. Integrating sustainability into strategic supplier portfolio selection. *Manag. Decis.* **2016**, *54*, 194–221. [CrossRef]
23. Lin, K.-P.; Tseng, M.-L.; Pai, P.-F. Sustainable supply chain management using approximate fuzzy DEMATEL method. *Resour. Conserv. Recycl.* **2018**, *128*, 134–142. [CrossRef]
24. Bai, C.; Sarkis, J.; Dou, Y. Corporate sustainability development in China: Review and analysis. *Ind. Manag. Data Syst.* **2015**, *115*, 5–40. [CrossRef]
25. Jayaram, J.; Avittathur, B. Green supply chains: A perspective from an emerging economy. *Int. J. Prod. Econ.* **2015**, *164*, 234–244. [CrossRef]
26. Chunguang, B.; Joseph, S. Determining and applying sustainable supplier key performance indicators. *Supply Chain Manag. Int. J.* **2014**, *19*, 275–291. [CrossRef]
27. Sáez-Martínez, F.J.; Lefebvre, G.; Hernández, J.J.; Clark, J.H. Drivers of sustainable cleaner production and sustainable energy options. *J. Clean. Prod.* **2016**, *138*, 1–7. [CrossRef]
28. Zailani, S.; Govindan, K.; Iranmanesh, M.; Shaharudin, M.R.; Sia Chong, Y. Green innovation adoption in automotive supply chain: The Malaysian case. *J. Clean. Prod.* **2015**, *108*, 1115–1122. [CrossRef]
29. Luthra, S.; Govindan, K.; Kannan, D.; Mangla, S.K.; Garg, C.P. An integrated framework for sustainable supplier selection and evaluation in supply chains. *J. Clean. Prod.* **2017**, *140*, 1686–1698. [CrossRef]

30. Dubey, R.; Gunasekaran, A.; Samar Ali, S. Exploring the relationship between leadership, operational practices, institutional pressures and environmental performance: A framework for green supply chain. *Int. J. Prod. Econ.* **2015**, *160*, 120–132. [CrossRef]
31. Qian, L.; Soopramanien, D. Incorporating heterogeneity to forecast the demand of new products in emerging markets: Green cars in China. *Technol. Forecast. Soc. Chang.* **2015**, *91*, 33–46. [CrossRef]
32. Kang, H.S.; Lee, J.Y.; Choi, S.; Kim, H.; Park, J.H.; Son, J.Y.; Kim, B.H.; Noh, S.D. Smart manufacturing: Past research, present findings, and future directions. *Int. J. Precis. Eng. Manuf. Green Technol.* **2016**, *3*, 111–128. [CrossRef]
33. Zhu, Q.; Sarkis, J. Green marketing and consumerism as social change in China: Analyzing the literature. *Int. J. Prod. Econ.* **2016**, *181 Pt B*, 289–302. [CrossRef]
34. Ma, P.; Zhang, C.; Hong, X.; Xu, H. Pricing decisions for substitutable products with green manufacturing in a competitive supply chain. *J. Clean. Prod.* **2018**, *183*, 618–640. [CrossRef]
35. Seth, D.; Rehman, M.A.A.; Shrivastava, R.L. Green manufacturing drivers and their relationships for small and medium(SME) and large industries. *J. Clean. Prod.* **2018**, *198*, 1381–1405. [CrossRef]
36. Lü, Y.-L.; Geng, J.; He, G.-Z. Industrial transformation and green production to reduce environmental emissions: Taking cement industry as a case. *Adv. Clim. Chang. Res.* **2015**, *6*, 202–209. [CrossRef]
37. Saufi, N.A.A.; Daud, S.; Hassan, H. Green growth and corporate sustainability performance. *Procedia Econ. Financ.* **2016**, *35*, 374–378. [CrossRef]
38. Pampanelli, A.B.; Found, P.; Bernardes, A.M. A lean & green model for a production cell. *J. Clean. Prod.* **2014**, *85*, 19–30. [CrossRef]
39. Kirezieva, K.; Jacxsens, L.; Hagelaar, G.J.L.F.; van Boekel, M.A.J.S.; Uyttendaele, M.; Luning, P.A. Exploring the influence of context on food safety management: Case studies of leafy greens production in Europe. *Food Policy* **2015**, *51*, 158–170. [CrossRef]
40. Luan, C.-J.; Tien, C.; Chen, W.-L. Which "green" is better? An empirical study of the impact of green activities on firm performance. *Asia Pac. Manag. Rev.* **2016**, *21*, 102–110. [CrossRef]
41. Soubihia, D.F.; Jabbour, C.J.C.; de Sousa Jabbour, A.B.L. Green manufacturing: Relationship between adoption of green operational practices and green performance of Brazilian ISO 9001-certified firms. *Int. J. Precis. Eng. Manuf. Green Technol.* **2015**, *2*, 95–98. [CrossRef]
42. Roy, M.; Khastagir, D. Exploring role of green management in enhancing organizational efficiency in petro-chemical industry in India. *J. Clean. Prod.* **2016**, *121*, 109–115. [CrossRef]
43. BR, R.K.; Agarwal, A.; Sharma, M.K. Lean management—A step towards sustainable green supply chain. *Compet. Rev.* **2016**, *26*, 311–331. [CrossRef]
44. Lim, M.K.; Tseng, M.-L.; Tan, K.H.; Bui, T.D. Knowledge management in sustainable supply chain management: Improving performance through an interpretive structural modelling approach. *J. Clean. Prod.* **2017**, *162*, 806–816. [CrossRef]
45. Lin, M.-H.; Hu, J.; Tseng, M.-L.; Chiu, A.S.F.; Lin, C. Sustainable development in technological and vocational higher education: Balanced scorecard measures with uncertainty. *J. Clean. Prod.* **2016**, *120*, 1–12. [CrossRef]
46. Rajeev, A.; Pati, R.K.; Padhi, S.S.; Govindan, K. Evolution of sustainability in supply chain management: A literature review. *J. Clean. Prod.* **2017**, *162*, 299–314. [CrossRef]
47. Yu, F.; Xue, L.; Sun, C.; Zhang, C. Product transportation distance based supplier selection in sustainable supply chain network. *J. Clean. Prod.* **2016**, *137*, 29–39. [CrossRef]
48. Kong, D.; Feng, Q.; Zhou, Y.; Xue, L. Local implementation for green-manufacturing technology diffusion policy in China: From the user firms' perspectives. *J. Clean. Prod.* **2016**, *129*, 113–124. [CrossRef]
49. Hursen, C.; Chun, S.-H.; Hwang, H.J.; Byun, Y.-H. Application to small and medium enterprises. *Proced. Soc. Behav. Sci.* **2015**, *186*, 862–867. [CrossRef]
50. Jaggernath, R.; Khan, Z. Green supply chain management. *World J. Entrep. Manag. Sustain. Dev.* **2015**, *11*, 37–47. [CrossRef]
51. Govindan, K. Sustainable consumption and production in the food supply chain: A conceptual framework. *Int. J. Prod. Econ.* **2018**, *195*, 419–431. [CrossRef]
52. Fercoq, A.; Lamouri, S.; Carbone, V. Lean/green integration focused on waste reduction techniques. *J. Clean. Prod.* **2016**, *137*, 567–578. [CrossRef]

53. Huang, X.-X.; Hu, Z.-P.; Liu, C.-S.; Yu, D.-J.; Yu, L.-F. The relationships between regulatory and customer pressure, green organizational responses, and green innovation performance. *J. Clean. Prod.* **2016**, *112*, 3423–3433. [CrossRef]
54. Zhu, W.; He, Y. Green product design in supply chains under competition. *Eur. J. Oper. Res.* **2017**, *258*, 165–180. [CrossRef]
55. Ji, P.; Ma, X.; Li, G. Developing green purchasing relationships for the manufacturing industry: An evolutionary game theory perspective. *Int. J. Prod. Econ.* **2015**, *166*, 155–162. [CrossRef]
56. Gorane, S.J.; Kant, R. Supply chain practices: An implementation status in Indian manufacturing organisations. *Benchmarking Int. J.* **2016**, *23*, 1076–1110. [CrossRef]
57. De Oliveira, U.R.; Espindola, L.S.; da Silva, I.R.; da Silva, I.N.; Rocha, H.M. A systematic literature review on green supply chain management: Research implications and future perspectives. *J. Clean. Prod.* **2018**, *187*, 537–561. [CrossRef]
58. Gandhi, N.S.; Thanki, S.J.; Thakkar, J.J. Ranking of drivers for integrated lean-green manufacturing for Indian manufacturing SMEs. *J. Clean. Prod.* **2018**, *171*, 675–689. [CrossRef]
59. Mittal, V.K.; Sindhwani, R.; Kapur, P.K. Two-way assessment of barriers to lean–green manufacturing system: Insights from India. *Int. J. Syst. Assur. Eng. Manag.* **2016**, *7*, 400–407. [CrossRef]
60. Sen, P.; Roy, M.; Pal, P. Exploring role of environmental proactivity in financial performance of manufacturing enterprises: A structural modelling approach. *J. Clean. Prod.* **2015**, *108*, 583–594. [CrossRef]
61. Mittal, V.K.; Sangwan, K.S. Prioritizing drivers for green manufacturing: Environmental, social and economic perspectives. *Proced. CIRP* **2014**, *15*, 135–140. [CrossRef]
62. McDonagh, P.; Prothero, A. Sustainability marketing research: Past, present and future. *J. Mark. Manag.* **2014**, *30*, 1186–1219. [CrossRef]
63. Bai, C.; Sarkis, J. Improving green flexibility through advanced manufacturing technology investment: Modeling the decision process. *Int. J. Prod. Econ.* **2017**, *188*, 86–104. [CrossRef]
64. Han, J.H.; Wang, Y.; Naim, M. Reconceptualization of information technology flexibility for supply chain management: An empirical study. *Int. J. Prod. Econ.* **2017**, *187*, 196–215. [CrossRef]
65. Rezaei, J.; Nispeling, T.; Sarkis, J.; Tavasszy, L. A supplier selection life cycle approach integrating traditional and environmental criteria using the best worst method. *J. Clean. Prod.* **2016**, *135*, 577–588. [CrossRef]
66. Awasthi, A.; Govindan, K.; Gold, S. Multi-tier sustainable global supplier selection using a fuzzy AHP-VIKOR based approach. *Int. J. Prod. Econ.* **2018**, *195*, 106–117. [CrossRef]
67. Gualandris, J.; Klassen, R.D.; Vachon, S.; Kalchschmidt, M. Sustainable evaluation and verification in supply chains: Aligning and leveraging accountability to stakeholders. *J. Oper. Manag.* **2015**, *38*, 1–13. [CrossRef]
68. Iacobucci, D.; Posavac, S.S.; Kardes, F.R.; Schneider, M.J.; Popovich, D.L. Toward a more nuanced understanding of the statistical properties of a median split. *J. Consum. Psychol.* **2015**, *25*, 652–665. [CrossRef]
69. Kock, N.; Verville, J.; Danesh-Pajou, A.; DeLuca, D. Communication flow orientation in business process modeling and its effect on redesign success: Results from a field study. *Decis. Support Syst.* **2009**, *46*, 562–575. [CrossRef]
70. Richter, N.F.; Cepeda, G.; Roldán, J.L.; Ringle, C.M. European management research using partial least squares structural equation modeling (PLS-SEM). *Eur. Manag. J.* **2016**, *34*, 589–597. [CrossRef]
71. Farooq, M.S.; Salam, M.; Fayolle, A.; Jaafar, N.; Ayupp, K. Impact of service quality on customer satisfaction in Malaysia airlines: A PLS-SEM approach. *J. Air Transp. Manag.* **2018**, *67*, 169–180. [CrossRef]
72. Ojha, D.; Struckell, E.; Acharya, C.; Patel, P.C. Supply chain organizational learning, exploration, exploitation, and firm performance: A creation-dispersion perspective. *Int. J. Prod. Econ.* **2018**, *204*, 70–82. [CrossRef]
73. Qi, Y.; Huo, B.; Wang, Z.; Yeung, H.Y.J. The impact of operations and supply chain strategies on integration and performance. *Int. J. Prod. Econ.* **2017**, *185*, 162–174. [CrossRef]
74. Hair, J.F.; Ringle, C.M.; Sarstedt, M. Editorial-partial least squares structural equation modeling: Rigorous applications, better results and higher acceptance. *Long Range Plan* **2013**, *46*, 1–12. [CrossRef]
75. Kock, N. *WarpPLS 6.0 User Manual*; ScriptWarp Systems: Laredo, TX, USA, 2018.

© 2019 by the authors. Licensee MDPI, Basel, Switzerland. This article is an open access article distributed under the terms and conditions of the Creative Commons Attribution (CC BY) license (http://creativecommons.org/licenses/by/4.0/).

Article

Supply Chain Innovation in Scientific Research Collaboration

Chih-Hung Yuan [1], Yenchun Jim Wu [2,*] and Kune-muh Tsai [3]

[1] College of Economic and Trade, University of Electronic Science and Technology of China, Zhongshan Institute, Zhongshan 528400, China; ialexyuan@gmail.com
[2] Graduate Institute of Global Business and Strategy, National Taiwan Normal University, Taipei 10645, Taiwan
[3] Department of Logistics Management, National Kaohsiung University of Science and Technology, Kaohsiung 824, Taiwan; kmtsai@nkfust.edu.tw
* Correspondence: wuyenchun@gmail.com; Tel.: +886-2-7734-3996

Received: 6 December 2018; Accepted: 29 January 2019; Published: 31 January 2019

Abstract: Innovations in supply chains and logistics, which help businesses reduce their costs and meet customer needs, have become increasingly vital. In this study, we first conducted a content analysis followed by a social network analysis to systematically review 104 research papers on supply chain innovation (SCI) that were published between 1987 and 2018. The results suggest that SCI research was originally concentrated in the United States and did not receive much attention in Europe and Asia, until more recently. An analysis of collaboration networks indicates that an SCI research community has just started to form, with the United Kingdom at the center of the international collaborative network. Implications of the study and directions for future research are summarized in detail, based on the systematic literature review.

Keywords: supply chain innovation; systematic review; social network analysis; collaboration

1. Introduction

In today's highly competitive global market, effective logistics innovation (LI) and supply chain innovation (SCI) play a key role in improving the organizational performance and competitive advantage of companies [1]. During the past few decades, new business models based on new logistics flows and supply chains have emerged [2]. SCI includes new production, marketing, or logistics processes that use technology and process innovations to generate information processing and new logistics services, improving operational efficiency and service effectiveness [3]. LI is seen as a new and innovative service related to logistics, whether basic or complex, which is particularly beneficial to specific stakeholders [4]. By comparing broader supply chain management with logistics management, we can deduce that the definition of SCI is more extensive than that of LI. Therefore, this study uses the term SCI to reflect this broader scope, including LI.

Competitive pressures and turbulent business environments are driving companies to innovate to ensure their survival, accelerate transformation, or support growth [5]. Among the activities in various sections of a supply chain, such as purchasing, material management, production, distribution, and marketing, one of the biggest challenges is meeting customer requirements while managing costs. Depending on the economy's level of development, the total supply chain cost accounts for, on average, approximately 5% to 10% of a company's annual income [6–8]. According to the GMA 2010 Logistics Benchmark Report, effectively reducing logistics costs has become a major goal for supply chains of US companies in 2008 and 2010 [7]. Cooperation allows supply chain participants to create complementary effects within innovation [9,10]. Therefore, SCI is regarded as a necessary tool.

LI and SCI are also becoming popular topics in the academic community [9]. An increasing numbers of scholars have begun reviewing supply chain innovation, such as supply chain collaboration [11], innovation process [2], network configuration [6,12], information technology [13,14], and sustainable development [15,16]. Kwak et al. [17] examined the effects of SCI on risk management capabilities based on 174 Korean manufacturers and logistics intermediaries; the results showed that SCI has a significantly positive effect on both the robustness and the resilience of risk management capabilities and, in turn, exerts a significant effect on the enhancement of competitive advantage. Abdelkafi and Pero [3] adopted a qualitative research tool to investigate how companies can use SCI to generate new business models; they found that SCI is primarily used to solve specific operational issues and often leads to incremental innovation of business models, but rarely results in radical innovation. Innovations can also lead to positive logistics business performance and customer satisfaction [18,19]. Wang et al. [20] pointed out that innovation performance may be negatively affected by certain supplier–client contracts. Lastly, in knowledge-intensive industries, value-creating activities are dispersed among the specialized companies in the supply chain, and the key companies serve as the controllers of the network and the knowledge integrators. Under this context, the difficulties faced by these key companies in knowledge integration are deemed obstacles to innovation [21].

Furthermore, the collaborative research team is a necessary and ideal component in the development of emerging research fields. Cooperation and networking are beneficial and useful strategies to increase productivity and influence research activities [22]. Interaction between researchers is expected to create new knowledge and expedite knowledge transfer. Cooperation between researchers is a laudable goal, one that has also been accepted and promoted by many policymakers [23]. Research projects funded by the European Union and China will include researcher cooperation as one of the necessary conditions to guarantee research funding.

In recent years, cooperation in academic research has become increasingly diversified across institutions, professions, practices, and countries. Adopting an interdisciplinary research strategy enhances the coherence and social relevance of the results that researchers produce [24]. Chen et al. [25] mentioned that scholars and practitioners have taken actions to cooperate through conferences, forums, and professional education courses. The research capabilities of major countries in Europe and Asia have been substantially improved and have had a major impact on the research environment that was previously dominated by American universities [26]. Advances in communications technology also enhance researcher cooperation, which may even be international [14,27].

The main purpose of this study is to analyze literature in the SCI field using content analysis and social network analysis (SNA) in a systematic review. The first contribution offered by this study is an overall picture of the field of SCI. Second, it allows readers to understand which topics researchers address in their discussions of SCI. Third, through academic collaboration relationships, it allows us to understand growth and changes in the SCI community. Finally, it reveals the finding that the field of SCI requires additional collaboration between academia and industry.

The rest of this paper is arranged as follows: Section 2 details the structured methodology used for literature review and evaluation in this study, Section 3 illustrates the analytical results, and Section 4 offers insights in terms of future research directions and conclusions.

2. Methodology

In the present study, we performed a systematic review to analyze SCI articles. The content analysis focused on published works rather than on data collection [28]. The steps of the systematic review process were (1) identification of suitable keywords and search queries; (2) compilation of a consideration set; (3) specification of eligibility criteria and assessment; (4) data processing and analysis; and (5) classification and typology of the results [29–31].

The study carried out independent searches using Scopus and the Web of Science to search the following keywords: "logistics innovation" and "supply chain innovation." Scopus is the world's largest abstract database that covers over 21,000 titles from peer-reviewed journals, conference proceedings, books, and business journals to ensure its multidisciplinary nature. The Web of Science is a citation index service developed by the Institute for Scientific Information; it grants access to literature and abstracts published in over 12,000 journals in the areas of science, engineering, medicine, agriculture, humanities, and social sciences.

Table 1 summarizes the articles retrieved and selected. To determine whether a study should be included, the inclusion and exclusion criteria were (1) the article is in English; (2) the article is from a peer-reviewed journal; (3) repeated, retracted, or articles published as conference papers, notes, comments, books, magazines, or trade publications, were excluded; (4) the article is relevant to the search terms defined in Section 1. The full texts of 142 articles were analyzed, and those not related to SCI were eliminated from the sample. In total, 104 SCI-related articles, published between 1987 and 2018, were selected.

Table 1. Summary of keyword searching.

Database	Keyword Search	English only	Full Paper Analysis	Final Analysis
Scopus				
SCI	125	119		
LI	71	69		
			142	104
Web of Science				
SCI	41	41		
LI	17	17		

Notes: SCI refers to supply chain innovation; LI refers to logistics innovation.

These SCI-related articles were reviewed systematically through content analysis, and each was examined twice to derive data of interest. All the articles were analyzed under the themes of descriptive statistics, collaboration, methodology, and theory. Data from descriptive statistics encompassed the following categories: distribution of the articles, country of data collection, and topic of research [9,16]. Subsequently, the collaborative relationships indicated in the articles were analyzed by author name, affiliation, and country [27]. Moreover, the theories used in the articles were summarized [16]. Finally, to elucidate trends in SCI research, we divided articles by publication period into two groups, 1987–2011 and 2012–2018, and compared them; this was because the number of SCI-related articles published, per year, during the 2012–2018 period was more than ten times that during the 1987–2011 period (see Figure 1).

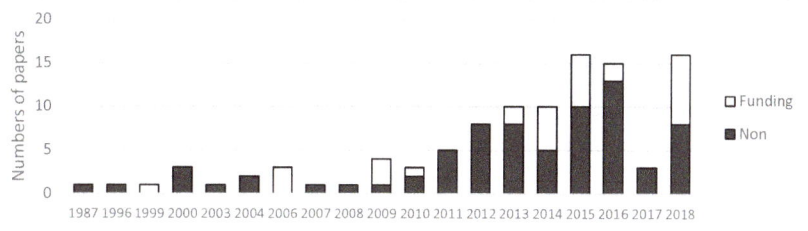

Figure 1. Distribution of reviewed papers by year.

3. Results

This section presents the results of the analysis of the SCI-related articles in terms of descriptive statistics, research collaboration, and theory analysis.

3.1. Descriptive Statistics

Figure 1 presents the number of funded and nonfunded SCI-related articles by year between 1987 and 2018. On average, the number of SCI-related articles published, from 1987—when the first

SCI-related article was published—to 2011, was fewer than two per year (standard deviation = 1.45). However, this figure is more than seven times higher in 2012, when eight SCI-related articles were published. No SCI-related articles were published in certain years prior to 2006. These results indicate that SCI research needed more attention during the 1987–2011 period. Moreover, the majority of published SCI-related studies were undertaken without funding from any organization (74.15%). In 1999, 2006, 2009, 2014, and 2015 only, over 30% of SCI-related studies were funded. This suggests that the number of published SCI-related studies with funding did not increase as the body of SCI research expanded. Benner and Sandström [32] noted that funding is a crucial mechanism underlying the reform of a regulatory system, because the reward framework in the system influences the research performance and appraisal. Additionally, in the United States and other countries, funding is one of the most crucial policies for fostering industrial development [33]. SCI can significantly shorten the cycle of new product introduction, lower inventory, increase the frequency of customer feedback, and improve decision-making processes [13]. Thus, more funds should be invested in promoting SCI research and application.

Table 2 categorizes SCI-related articles by the geographic region in which the studies were undertaken. Among the 104 SCI-related articles, 19 did not mention the region where the data were collected, while some covered either multiple countries ($n = 7$), or just the European Union ($n = 7$), or the whole world ($n=2$). Over the 1987–2018 period, most of the studies were conducted in the United States ($n = 20$), followed by Italy ($n = 8$), the European Union ($n = 7$), and India ($n = 6$). The top ten regions with the most SCI-related studies were largely developed countries. Furthermore, between 1987 and 2011, most SCI research was conducted in the United States. This trend continued into the 2012–2018 period, when more researchers began to focus on Italy, China, the European Union, India, Denmark, and Taiwan. Therefore, SCI research was concentrated in the United States and spread to Europe and Asia, suggesting that American academia plays a leading role in the SCI field.

Table 2. Countries of focus.

Collaboration Form	1987–2011	2012–2018	Total
United States	7	13	20
Italy	0	8	8
European Union	1	6	7
India	1	5	6
China	0	5	5
Republic of Korea	2	3	5
Taiwan	1	4	5
Australia	2	2	4
Denmark	0	4	4
United Kingdom	1	3	4
German	0	3	3
Malaysia	0	3	3
Netherlands	2	1	3
Sweden	0	3	3

Note: shown only if >2.

The dynamics of SCI research topics are shown in Table 3. During the research period, academics primarily investigated SCI from the perspectives of adoption, collaboration, and green/sustainable development. From 2012 to 2018, researchers focused on the topics of green/sustainable development, managerial function, managerial tools, and collaboration. The topics of green/sustainable development and managerial tools experienced the fastest growth, accounting for more than 8% of total growth. Organizational performance and adoption have remained topics of interest among scholars [34].

Table 3. Topic profile.

Topics	1987–2011		2012–2018		Total	
Green/sustainable development	1	(3.8)	18	(28.6)	19	(21.3)
Collaboration	7	(26.9)	10	(15.9)	17	(19.1)
Managerial function	4	(15.4)	13	(20.6)	17	(19.1)
Adoption	7	(26.9)	9	(14.3)	16	(18.0)
Managerial tools	2	(7.7)	13	(20.6)	15	(16.9)
Organizational performance	5	(19.2)	8	(12.7)	13	(14.6)
Key success factor	1	(3.8)	5	(7.9)	6	(6.7)
Network	1	(3.8)	2	(3.2)	3	(3.4)
Others	3	(11.5)	3	(4.8)	6	(6.7)
Total	31	(119.2)	81	(128.6)	112	(125.8)

Note: 1. The numbers in parentheses are percentage; 2. Of the 104 articles, seven articles dealt with multiple topics.

3.2. Collaboration Networks

Table 4 summarizes the patterns of SCI collaboration in academic communities. Compared with the period of 1987–2011, in the period of 2012–2018, the proportion of articles written by single authors declined, and the rate of collaboration increased. This is indicative of the fact that, in recent years, researchers in the SCI field have published articles through collaboration. With regard to team collaboration patterns, collaboration is primarily conducted within a single institution (37.5%), followed by co-authorship among academics at the international level (26.0%). The extent of cross-sector collaboration at national (4.8%) and international (2.9%) levels is low. Furthermore, there has been a downward trend in the ratio of cross-sector collaboration in recent years. The most important studies were the result of researchers' personal involvement in practical activities or face-to-face interactions [35].

Table 4. Collaboration analysis.

Collaboration Form	1987–2011		2012–2018		Total	
Single authored	4	(15.4)	7	(9.0)	11	(10.6)
Institutional level	7	(26.9)	32	(41.0)	39	(37.5)
National level						
Among academics	6	(23.1)	13	(16.7)	19	(18.3)
Cross-sector collaboration	4	(15.4)	1	(1.3)	5	(4.8)
International level						
Among academics	4	(15.4)	23	(29.5)	27	(26.0)
Cross-sector collaboration	1	(3.8)	2	(2.6)	3	(2.9)
Total	26	(100.0)	78	(100.0)	104	(100.0)

Note: The numbers in parentheses are percentage.

After the 11 single-authorship SCI-related articles were eliminated from the sample, an SNA was performed to visualize collaboration networks based on authors' personal information, providing an insight into the collaboration in terms of authors (Figure 2), institutions (Figure 3), and countries (Figure 4). The collaboration networks shown in Figures 2–4 comprised (1) nodes, which represented authors, institutions, or countries; and (2) links which denoted one or more relationships that existed between nodes. A thicker link indicated more frequent collaboration between authors, institutions, or countries. The weight of a link is represented by the strength of the link. NodeXL, Ucinet, and Netdraw were used to visualize the collaboration networks and calculate their weight.

Figure 2 illustrates the co-authorships on degree centrality based on publications over the past three decades. Degree centrality is defined as the number of direct ties that a node has with other nodes in the network graph. Five co-authorship networks were relatively strong; these authors published at least two papers. In particular, these networks led by Hazen B.T., Habidin N.F., Shazazli N.A., Santa R., and Storer M. involved more scholars.

Figure 2. Co-authorship networks. Notes: 93 articles among 253 authors.

However, Figure 2 suggests that the collaborative relationships in SCI research were fragmented, and no connections were formed between different or unconnected clusters. As future SCI papers are indexed by Scopus and Web of Science, the collaboration networks will solidify.

The affiliations of SCI authors were also linked to explore institutional collaboration in SCI research. The co-institutional research collaboration on degree centrality based on publications is shown in Figure 3. Figure 3 indicates that some institutions, such as the Zhejiang University, University of Waikato, Xiamen University, Cardiff University, Politecnico di Milano, Stanford University, and Alfaisal University, were in a "structural hole" and could link with different clusters. The structural hole refers to the optimal position where an organization—capable of linking two or more partners or collaboration groups without any connection—can gain information and control benefits over these partners or collaboration groups [36].

Figure 4 depicts international collaboration networks in SCI research. In the figure, only four node-to-node networks appear over the 1987–2011 period; this is because a) co-authored studies were mostly conducted in the same countries, and b) productive researchers tended to work with colleagues in the same institutions [37]. However, a cross-national collaboration network centering on the United Kingdom formed during the 2012–2018 period. This echoes the results in Table 5. As the core issues as well as the data acquisition, analysis, and interpretation in recent SCI studies have become increasingly complex, a wide range of skills and expertise is required. Cross-national collaboration may lead to debates and can thus generate new ideas and viewpoints [38].

Over the 1987–2018 period, the SCI research community were divided into 1) the core cluster led by the United Kingdom; 2) a peripheral cluster comprising Malaysia and Nigeria; and 3) a peripheral cluster comprising Canada and the United Arab Emirates. The core cluster led and influenced the direction of SCI research, and can effectively tap the knowledge resources of the global research community [39]. Over the 30-year period, researchers from India, Brazil, Greece, Japan, and Russia did not collaborate with their counterparts in other countries. Six co-authored SCI papers were published in India, forming an independent research community.

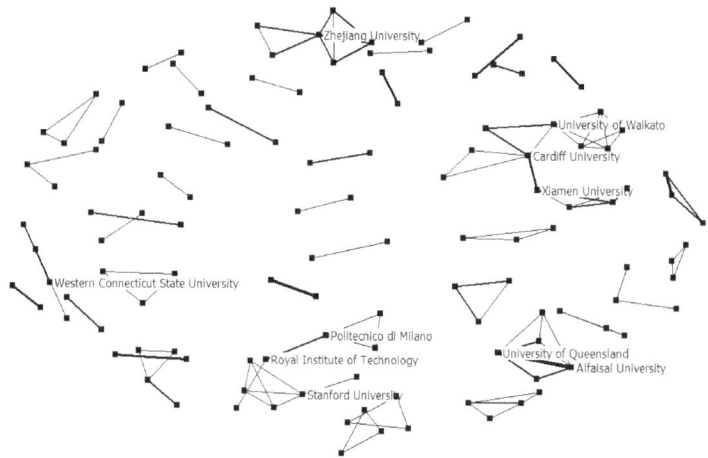

Figure 3. Institutional collaboration networks in supply chain innovation. Notes: 54 articles among 123 institutions.

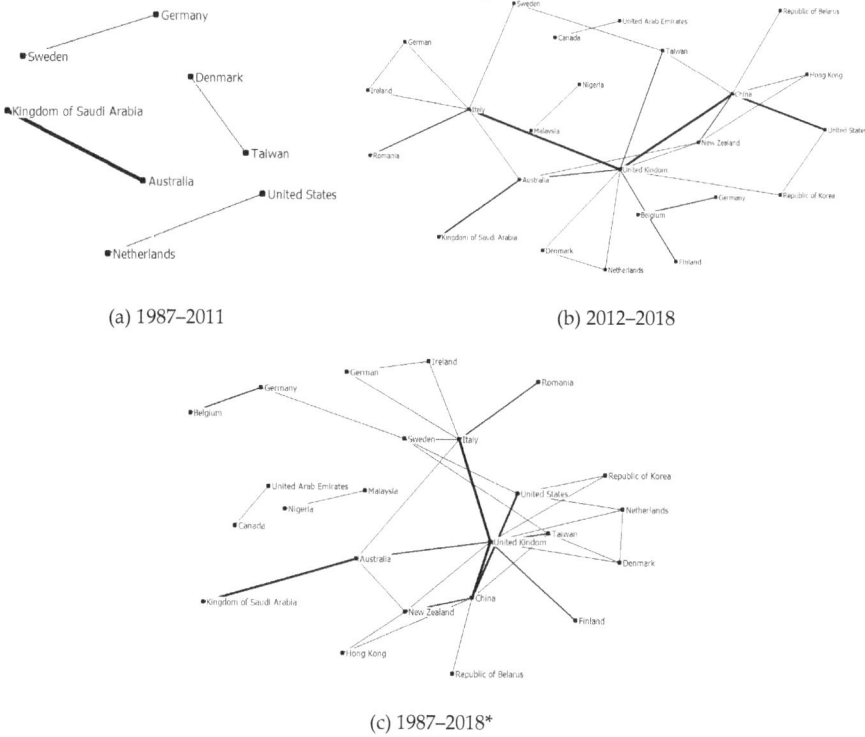

(a) 1987–2011

(b) 2012–2018

(c) 1987–2018*

Figure 4. International collaboration networks in supply chain innovation. * 30 articles among 25 countries during 1987–2018

3.3. Theory Analysis

Table 5 provides the frequency that various theories were used in the sampled SCI-related articles. Approximately 30% of the articles adopted theories to discuss SCI over the 30-year period. More than 80% did not use any theory over the 1987–2011 period; however, this figure dropped to approximately 64% over the 2012–2018 period. Throughout these 30 years, the most-used theories were dynamic capability ($n = 9$) and the resource-based view ($n = 9$), whereas the second most-used were the resource-advantage ($n = 3$) and transaction cost ($n = 3$) theories.

Table 5. Theory profile.

Theory	1987–2011	2012–2018	Total
Resource-based view	1	10	11
Dynamic capabilities theory	2	8	10
Resource dependence theory	2	1	3
Resource-advantage theories	0	3	3
Transaction-cost theory	1	2	3
Contingency theory	0	2	2
Game theory	1	1	2
Innovation diffusion theory	0	2	2
Social exchange theory	0	2	2
Others	0	9	11
Total	7	40	47

Note: The theories above were used 47 times in 33 articles.

4. Discussion and Conclusions

Considering the significance of innovation for businesses, supply chains, and the wider industry, this study sampled English-language research papers on SCI, published from 1987 to 2018, to identify trends in SCI research. We found that SCI-related articles were first published in 1987; then, the number of articles was low, which increased substantially to 8 in 2012. The systemic literature review in this study analyzed SCI-related studies from the perspectives of research funding, research topics, author's collaborative relationships, and theories. We note the following implications and limitations, as well as future research directions based on its findings.

4.1. Implications

The results of the descriptive statistics suggest that over 70% of sampled SCI-related studies were not funded. Research grants enable the production of new knowledge [40]. Performance-oriented ex ante review of research proposals can inhibit the development of studies, particularly those in emerging fields [32]. The same applies to SCI, an emerging topic of supply chain research. Although SCI has been discussed from various perspectives, more research is required to understand this topic. As a result, scholars focus primarily on topics like greenness, collaboration, and managerial function. Greenness and the environment have become the main challenges for regional and global sustainable development [9]. Generally speaking, companies are limited by their own capabilities and do not possess all the resources necessary for innovation. The main reason for collaboration is to obtain such resources, especially knowledge, from other companies [41].

SCI research is concentrated in the United States, and did not receive much attention in Europe and Asia until 2012. This is consistent with the findings of Gao, Xu, Ruan and Lu [9], who found that selection of the study site depends largely on the size of gross domestic product (GDP). GDP grows faster in India, China, South Korea, and Taiwan than in Western countries [8].

There was a shift in article publishing from single authors towards collaboration. However, the number of articles written through cross-sector collaboration declined. Some researchers have highlighted the necessity for increasing their influence on industry [42,43]. Academic research differs considerably from managerial practices because both have varying assumptions and beliefs [44]. Numerous critical studies are based on researchers' practical experiences and face-to-face interactions [35]. Such studies can be performed through collaboration between academia and industry. SCI is an emerging field of research with disproportionate collaboration between scholars who have yet to develop close connections.

This study revealed that some institutions in the SCI research community were situated in the position of a structural hole, which disseminates information and knowledge in a social network [45]. When these institutions collaborate with more partners to conduct SCI research, they gain more social capital. Additionally, although a majority of SCI studies were conducted or published in the United States, it was not the center of the core cluster. This suggested that the authors made more collaborative efforts with scholars outside the United States. Over the 2012–2018 period, the United Kingdom was at the core of the SCI research community, largely because British researchers conducted more cross-national research than their counterparts in other countries. This indicates a possibility that SCI-related articles from the United Kingdom would be cited more frequently.

A growing number of recent SCI studies have applied theories, and the adopted theories have also been increasingly diverse. This trend corresponds with the recommendation of Carter and Rogers [15] that theories should be connected to investigate SCI topics. Specifically, more theories should be developed for and applied in supply chain management research. Additionally, to extend supply chain management to social and environmental contexts necessitates the introduction of more perspectives to pave the foundation for developing creative solutions and new theories [46].

4.2. Limitations

This study has some limitations. First, only the exact keywords "supply chain innovation" and "logistics innovation" were used to retrieve articles from Scopus and Web of Science; therefore, SCI-related articles with other keywords might have been ignored. However, we argue that the sampled articles were adequately representative of SCI research. Second, this study collected research papers written in English and excluded conference papers, notes, commentaries, books, magazines, and trade publications. Third, many of the trends we identified in this study were based on the SCI-related articles that were rigorously selected. Lastly, the findings and their implications depended to a great extent on the experiences and educational background of the reviewers.

4.3. Future Directions of Research

The findings of this study indicate that SCI research still has a large scope for growth. Based on the literature review, we suggest the following directions for further SCI research:

1. SCI-related articles retrieved by this study appear in various fields, suggesting that SCI concerns diverse disciplines and industries. Accordingly, not only manufacturing, but also service, agricultural, and transport industries have improved the efficiency of their respective supply chains through innovation [9]. Moreover, the average number of authors per SCI article has increased. Thus, conducting SCI studies in a cross-disciplinary setting will be a trend in the foreseeable future, yielding insightful findings for industrial communities.
2. Only 15% of SCI-related studies sampled by this study focused on more than two countries. Most studies focused on a single country. In the competitive global market, it can be assumed that the raw materials of a product are produced in one country, whereas its production, assembly, transportation, and sales occur in various other countries. Therefore, future researchers should conduct SCI research at a cross-national level to analyze the generalization of SCI. This is because when innovation is integrated into the global supply chain, stakeholders can benefit.
3. SCI research is currently benefiting from cross-national collaboration and, thus, is developing rapidly. Future SCI studies should be conducted based on collaboration between academia and industries to narrow the academia–industry gap and develop new research topics.

4. The SCI research community is still developing. The country or institution at the center of a core cluster of collaboration networks for SCI research should promote and fund SCI research to acquire critical information and benefits. Both research and business organizations can benefit accordingly. This study analyzed the SCI research community on the scale of collaboration networks. Future studies can discuss such networks by examining relevant abstracts and keywords or using relevant methods.
5. This study found that "adoption," collaboration," and "green/sustainable development" have remained popular topics among SCI researchers. These topics can be investigated by using various stages or forms of SCI, or on a cross-national basis.
6. This study showed that dynamic capabilities and the resource-based view are the most used theories in SCI research. Future studies should analyze the role of both theories in SCI research and conduct a citation analysis of the theories to improve understanding of the development of SCI research.

Author Contributions: Conceptualization, C.-H.Y. and Y.J.W.; Methodology, C.-H.Y. and K.-M.T.; Formal Analysis: Y.J.W. and K.-M.T.; Writing, C.-H.Y., Y.J.W., and K.-M.T.

Funding: This research was funded by University of Electronic Science and Technology of China, Zhongshan Institute (No. 418YKQN05).

Conflicts of Interest: The authors declare no conflict of interest.

References

1. Wu, Y.J.; Tsai, K.-M. Making connections: Supply chain innovation research collaboration. *Transp. Res. Pt. e-Logist.* **2018**, *113*, 222–224. [CrossRef]
2. Su, S.I.I.; Britta, G.; Yang, S.L. Logistics innovation process revisited: Insights from a hospital case study. *Int. J. Phys. Distrib. Logist.* **2011**, *41*, 577–600.
3. Abdelkafi, N.; Pero, M. Supply chain innovation-driven business models: Exploratory analysis and implications for management. *Bus. Process. Manag. J.* **2018**, *24*, 589–608. [CrossRef]
4. Flint, D.J.; Larsson, E.; Gammelgaard, B. Exploring processes for customer value insights, supply chain learning and innovation: An international study. *J. Bus. Logist.* **2008**, *29*, 257–281. [CrossRef]
5. Wu, Y.J.; Liu, W.-J.; Yuan, C.-H. A mobile-based barrier-free service transportation platform for people with disabilities. *Comput. Hum. Behav.* **2018**, in press. [CrossRef]
6. Hoole, R. Five ways to simplify your supply chain. *Supply Chain Manag.* **2005**, *10*, 3–6. [CrossRef]
7. Butner, K.; Iglesias, J. *The GMA 2010 Logistics Benchmark Report: Performance Reaches All-Time High During Economic Depression*; GMA Logistics Committee: Washington, DC, USA, 2010.
8. Wu, Y.J.; Yuan, C.-H.; Goh, M.; Lu, Y.-H. Regional port productivity in APEC. *Sustainability* **2016**, *8*, 689. [CrossRef]
9. Gao, D.; Xu, Z.; Ruan, Y.Z.; Lu, H. From a systematic literature review to integrated definition for sustainable supply chain innovation (SSCI). *J. Clean Prod.* **2017**, *142*, 1518–1538. [CrossRef]
10. Huang, Y.-C.; Rahman, S.; Wu, Y.J.; Huang, C.-J. Salient task environment, reverse logistics and performance. *Int. J. Phys. Distrib. Logist.* **2015**, *45*, 979–1006. [CrossRef]
11. Liao, S.-H.; Kuo, F.-I. The study of relationships between the collaboration for supply chain, supply chain capabilities and firm performance: A case of the Taiwan's TFT-LCD industry. *Int. J. Prod. Econ.* **2014**, *156*, 295–304. [CrossRef]
12. Carnovale, S.; Yeniyurt, S. The role of ego network structure in facilitating ego network innovations. *J. Supply Chain Manag.* **2015**, *51*, 22–46. [CrossRef]
13. Tan, K.H.; Zhan, Y.; Ji, G.; Ye, F.; Chang, C. Harvesting big data to enhance supply chain innovation capabilities: An analytic infrastructure based on deduction graph. *Int. J. Prod. Econ.* **2015**, *165*, 223–233. [CrossRef]
14. Ding, W.W.; Ding, S.G.; Stephan, P.E.; Winkler, A.E. The impact of information technology on academic scientists' productivity and collaboration patterns. *Manag. Sci.* **2010**, *56*, 1439–1461. [CrossRef]

15. Carter, C.R.; Rogers, D.S. A framework of sustainable supply chain management: Moving toward new theory. *Int. J. Phys. Distrib. Logist.* **2008**, *38*, 360–387. [CrossRef]
16. Mansouri, S.A.; Lee, H.; Aluko, O. Multi-objective decision support to enhance environmental sustainability in maritime shipping: A review and future directions. *Transp. Res. Pt. e-Logist.* **2015**, *78*, 3–18. [CrossRef]
17. Kwak, D.-W.; Seo, Y.-J.; Mason, R. Investigating the relationship between supply chain innovation, risk management capabilities and competitive advantage in global supply chains. *Int. J. Oper. Prod. Manag.* **2018**, *38*, 2–21. [CrossRef]
18. Wu, Y.J. Assessment of technological innovations in patenting for 3rd party logistics providers. *J. Enterp. Inf. Manag.* **2006**, *19*, 504–524.
19. Wu, Y.J.; Chou, Y.-H. A new look at logistics business performance: Intellectual capital perspective. *Int. J. Logist. Manag.* **2007**, *18*, 41–63.
20. Wang, L.; Yeung, J.H.Y.; Zhang, M. The impact of trust and contract on innovation performance: The moderating role of environmental uncertainty. *Int. J. Prod. Econ.* **2011**, *134*, 114–122. [CrossRef]
21. Narasimhan, R.; Narayanan, S. Perspectives on supply network–enabled innovations. *J. Supply Chain Manag.* **2013**, *49*, 27–42. [CrossRef]
22. Adams, J.D.; Black, G.C.; Clemmons, J.R.; Stephan, P.E. Scientific teams and institutional collaborations: Evidence from U.S. universities, 1981–1999. *Res. Policy* **2005**, *34*, 259–285. [CrossRef]
23. Defazio, D.; Lockett, A.; Wright, M. Funding incentives, collaborative dynamics and scientific productivity: Evidence from the EU framework program. *Res. Policy* **2009**, *38*, 293–305. [CrossRef]
24. Hulme, D.; Toye, J. The case for cross-disciplinary social science research on poverty, inequality and well-being. *J. Dev. Stud.* **2006**, *42*, 1085–1107. [CrossRef]
25. Chen, C.Y.; Wu, Y.J.; Wu, W.H. A sustainable collaborative research dialogue between practitioners and academics. *Manag. Decis.* **2013**, *51*, 566–593. [CrossRef]
26. Kato, M.; Ando, A. National ties of international scientific collaboration and researcher mobility found in nature and science. *Scientometrics* **2017**, *110*, 673–694. [CrossRef]
27. Wu, Y.J.; Goh, M.; Yuan, C.-H.; Huang, S.-H. Logistics management research collaboration in Asia. *Int. J. Logist. Manag.* **2017**, *28*, 206–223. [CrossRef]
28. Seuring, S.A.; Müller, M. From a literature review to a conceptual framework for sustainable supply chain management. *J. Clean Prod.* **2008**, *16*, 1699–1710. [CrossRef]
29. Tranfield, D.; Denyer, D.; Smart, P. Towards a methodology for developing evidence-informed management knowledge by means of systematic review. *Brit. J. Manag.* **2003**, *14*, 207–222. [CrossRef]
30. Fahimnia, B.; Sarkis, J.; Davarzani, H. Green supply chain management: A review and bibliometric analysis. *Int. J. Prod. Econ.* **2015**, *162*, 101–114. [CrossRef]
31. Crossan, M.M.; Apaydin, M. A multi-dimensional framework of organizational innovation: A systematic review of the literature. *J. Manag. Stud.* **2010**, *47*, 1154–1191. [CrossRef]
32. Benner, M.; Sandström, U. Institutionalizing the triple helix: Research funding and norms in the academic system. *Res. Policy* **2000**, *29*, 291–301. [CrossRef]
33. Sabharwal, M.; Hu, Q. Participation in university-based research centers: Is it helping or hurting researchers? *Res. Policy* **2013**, *42*, 1301–1311. [CrossRef]
34. Lu, Y.-H.; Wang, S.-C.; Yuan, C.-H. Financial crisis and the relative productivity dynamics of the biotechnology industry: Evidence from the Asia-Pacific countries. *Agric. Econ.* **2017**, *63*, 65–79.
35. Mohrman, S.A.; Gibson, C.B.; Mohrman, A.M. Doing research that is useful to practice a model and empirical exploration. *Acad. Manag. J.* **2001**, *44*, 357–375.
36. Burt, R.S. *Structural Holes: The Social Structure of Competition*; Harvard University Press: Cambridge, MA, USA, 1992.
37. Michael, M.; Craig, R.C.; Lutz, K. Author affiliation in supply chain management and logistics journals: 2008–2010. *Int. J. Phys. Distrib. Logist.* **2012**, *42*, 83–101.
38. Katz, J.S.; Martin, B.R. What is research collaboration? *Res. Policy* **1997**, *26*, 1–18. [CrossRef]
39. Leydesdorff, L.; Wagner, C.S. International collaboration in science and the formation of a core group. *J. Inform.* **2008**, *2*, 317–325. [CrossRef]
40. Hicks, D. Performance-based university research funding systems. *Res. Policy* **2012**, *41*, 251–261. [CrossRef]
41. Zimmermann, R.; Ferreira, L.M.D.F.; Carrizo Moreira, A. The influence of supply chain on the innovation process: A systematic literature review. *Supply Chain Manag.* **2016**, *21*, 289–304. [CrossRef]

42. Van de Ven, A.H.; Johnson, P.E. Knowledge for theory and practice. *Acad. Manag. Rev.* **2006**, *31*, 802–821. [CrossRef]
43. Bartunek, J.M. Academic-practitioner collaboration need not require joint or relevant research: Toward a relational scholarship of integration. *Acad. Manag. J.* **2007**, *50*, 1323–1333. [CrossRef]
44. Rynes, S.L.; Bartunek, J.M.; Daft, R.L. Across the great divide: Knowledge creation and transfer between practitioners and academics. *Acad. Manag. J.* **2001**, *44*, 340–355. [CrossRef]
45. Cowan, R.; Jonard, N. Structural holes, innovation and the distribution of ideas. *J. Econ. Interact. Coord.* **2007**, *2*, 93–110. [CrossRef]
46. Anthony, A.; Helen, W.; Mohamed, N. Decision theory in sustainable supply chain management: A literature review. *Supply Chain Manag.* **2014**, *19*, 504–522.

© 2019 by the authors. Licensee MDPI, Basel, Switzerland. This article is an open access article distributed under the terms and conditions of the Creative Commons Attribution (CC BY) license (http://creativecommons.org/licenses/by/4.0/).

Article

Warranty Decision Model and Remanufacturing Coordination Mechanism in Closed-Loop Supply Chain: View from a Consumer Behavior Perspective

Xiaodong Zhu [1], Lingfei Yu [1,*], Ji Zhang [2], Chenliang Li [1] and Yizhao Zhao [1]

1. School of Management Science and Engineering, Nanjing University of Information Science and Technology, Nanjing 210044, China; zxd@nuist.edu.cn (X.Z.); 20151307047@nuist.edu.cn (C.L.); 20161307034@nuist.edu.cn (Y.Z.)
2. Faculty of Health, Engineering and Sciences, University of Southern Queensland, Toowoomba, QLD 4350, Australia; Ji.Zhang@usq.edu.au
* Correspondence: ylf@nuist.edu.cn; Tel.:+86-025-5823-5552

Received: 4 November 2018; Accepted: 9 December 2018; Published: 12 December 2018

Abstract: The remanufacturing warranty strategy has become an effective mechanism for reducing consumer risk and stimulating market demand in closed-loop supply chain management. Based on the characteristics of consumers' behavior of purchase decisions, this paper studies the warranty decision model of remanufacturing closed-loop supply chain under the Stackelberg game model. The present study discussed and compared the decision variables, including remanufacturing product pricing, extended warranty service pricing, warranty period and supply chain system profit. The research shows that consumers' decision-making significantly affirms the dual marginalization effect of the supply chain system while significantly affecting the supply chain warranty decision; the improved revenue sharing contract and the two charge contracts respectively coordinates the manufacturer-led and retail-oriented closed-loop supply chain system, which effectively implements the Pareto improvement of the closed-loop supply chain system with warranty services. In the present study, the model is verified and analyzed by numerical simulation.

Keywords: remanufacturing; closed-loop supply chain; warranty decision; contract coordination

1. Introduction

The recycling and reuse of electronic waste is currently a global concern, and remanufacturing provides an effective solution for used electronic products. However, due to the slow development of the market entities of remanufactured products in China, the development of the remanufacturing industry has been seriously hindered. Manufacturers (including original manufacturers and remanufacturers) and retailers have been using price leverage to attract customers and promote sales performance. However, in addition to price factors, consumer brand loyalty is increasingly built rather on the quality of the (service of) remanufactured products. The product warranty strategy has become a new round of competition hotspots besides traditional price competition. The warranty strategy refers to the obligation or warranty provided by the manufacturer or the retailer, distributor, etc., to the consumer in terms of technical performance, product effect and maintenance of the product during the sales process. The consumer can predict the product quality based on the product warranty (Boulding and Kirmani, 1993) [1]. Warranty has become an effective mechanism to reduce consumer risk and stimulate market demand, while also being linked to corporate social responsibility, which is a key feature of corporate sustainability and business sustainability (Wei et al. 2015 [2]; Ahi and Searcy, 2013 [3]). In closed loop supply chain management, the length of the remanufactured product warranty period (related to remanufactured failure rate) and extended warranty pricing play a key role in determining the total cost of the product (Shafiee and Chukova, 2013 [4]). A satisfactory warranty policy will increase consumers' willingness to purchase remanufactured products while contributing to production sustainability and resource efficiency (Song et al. 2018 [5]). However, the supply chain must balance between the warranty inputs and outputs to maximize benefits.

This paper is primarily related to the research in three streams: the operation of the warranty strategy (which mainly focusing on the warranty period and the product life cycle), the supply chain pricing strategies and the consumer behavior theory. In the following, we review the literature in these three streams.

At present, scholars have done preliminary research on the operation of warranty strategies, mainly focusing on the warranty period and the product life cycle. Li et al. (2016a) [6] studied the impact of the warranty period on the closed-loop supply chain system from the perspective of product warranty period. Lan et al. (2014) [7] explored the impact of product price and quality on the development of warranty strategies under the fuzzy supply chain based on three types of warranties. Du et al. (2016) [8] accurately estimated product warranty costs based on product endogenous variables (product loss rate, etc.) and gave cost calculation methods. Arabi et al. (2017) [9] determined the the best warranty period from the perspective of the manufacturer and consumer to minimize the total cost of use and increase the service life. Chen et al. (2017) [10] considered product warranty as an economic compensation for consumers in the event of product failure and compared the impact of the extended warranty provided by the manufacturer or distributor on the overall market. Li et al. (2016b) [11] discussed issues related to the manufacturer's pricing strategy in two supply chains, including one manufacturer and two competing retailers with warranty-related requirements. Mai et al. (2017) [12] analyzed the optimal warranty strategy given the original product warranty provided by the manufacturer and the additional product warranty provided by the seller. Sabbaghi and Behdad (2012) [13] studied the impact of product characteristics, sales environment and market size on consumer purchasing decisions in the context of manufacturers providing warranty services. Xu et al. (2018) [14] studied the issue of bundled pricing of durable consumer goods with warranty services under the two market structures of monopoly and duopoly. Based on the research above on warranty decision-making, it is found that there are few existing studies on the coordination of supply chain systems when strong retailers provide extended warranty services.

In terms of supply chain pricing strategies, Savaskan et al. (2004) [15] used manufacturers and distributors as recycling bearers to explore how to select recycling channels and determine pricing decisions for remanufactured products. Fleischmann et al. (1997) [16] emphasized the coordination of price decision in the reverse channel of the closed-loop supply chain through qualitative analysis. Xu et al. (2018) [17] studied the impact of the retailer's overconfidence on supply chain pricing and performance in the duopoly market environment of demand uncertainty and concluded that the overconfident retailers tend to give higher pricing. Gong and Jiang (2018) [18] studied the optimal decision of the supply chain system recycling model for the four mixed recycling models, and determined the optimal pricing strategies in different situations. Kaya et al. (2014) [19] did empirical analysis of pricing decisions and incentives for remanufacturing closed-loop supply chains in uncertain environments. Luo et al. (2017) [20] studied two different brand manufacturers selling products through unified distributors by building the pricing decision model under horizontal and vertical competition, and then analyzed the impact of different power structures on product pricing. Yoo and Kim (2016) [21] built a three-tier supply chain consisting of manufacturers, distributors and refurbished processors, by considering multiple supply chain process combinations, introducing five different supply chain structures, and comparing the models in pricing and performance of each product. Gan et al. (2017) [22] constructed a closed-loop supply chain pricing model for short-lived products consisting of manufacturers, distributors, and recyclers. In this model, new products are sold through traditional means, and remanufactured products are passed through manufacturers. The study found that the degree of acceptance and the direct sales channel preferences affect supply chain pricing. However, existing research on the coordinated pricing of closed-loop supply chains is still relatively rare.

In addition, the current study also investigates consumer behavior theory. Consumer behavior directly affects the remanufacturing closed-loop supply chain sales performance of warranty services. Chen et al. (2012) [23] provided a variety of optional warranty services based on consumer heterogeneity risk preferences and moral hazard decisions in different market periods. Gaur et al. (2017) [24] conducted research on the influencing factors of consumer purchase and perceived behavior, and analyzed the pricing decision problem. In terms of factors affecting consumer purchasing behavior, there is an uncertainty in the duration of consumers holding their own goods (Joshi and Rahman, 2015 [25]), loss aversion and other influencing factors. Gallego et al. (2014) [26] and Su and Wang (2016) [27] designed flexible warranty services and tapped flexible warranty strategies to create higher profits. Li et al. (2018) [28] conducted an exploratory study of consumer assurance perspectives and consumer risk preferences based on the consumer perception dimension. Zhu and Yu (2018) [29] studied the closed-loop supply chain pricing decision-making and coordination mechanism based on the differentiated willingness to pay of consumers. Zhou et al. (2017) [30] studied the pricing strategies and service cooperation of manufacturers under uncertain consumer demand. However, the study of warranty decisions for remanufacturing closed-loop supply chains has less introduced consumer behavior theory.

Different from previous research, by considering a re-manufacturing closed-loop supply chain system with warranty services, where market demand is disturbed by consumer behavior, this paper studies the retail as the warranty subject, the optimal decision of supply chain for warranty pricing and warranty period and the contract coordination problem of supply chain. In addition, our article focuses on three different game models, namely the retailer-led Stackelberg game, the retailer-led Stackelberg game, and the centralized decision model, and consider how manufacturers and retailers can optimize the cooperation contract under full information conditions.

The main contributions of this paper are as follows. First, establish five decentralized decision models by considering a remanufacturing closed-loop supply chain system of two-tier extended warranty services consisting of one manufacturer and one retailer, and improve the closed-loop supply chain coordination contract with warranty services, and supplemented the current related research; secondly, through the use of game theory, the equilibrium wholesale price, retail price, warranty price and warranty period are obtained and analyzed in five decentralized decision models; third, through numerical simulation, We have gained some valuable management insights. For example, firstly, the increase in consumer preference will increase the profit of the node enterprises and play a positive incentive role. Secondly, the increase of consumer preference has aggravated the loss of decision-making efficiency, and, at the same time, widened the gap between the total profit of supply chain system in centralized and decentralized decision-making, amplifying the effect of "double marginalization". Thirdly, the traditional revenue-sharing contract based on commodity sales revenues fails to coordinate the closed-loop supply chain system with extended warranty services. The cooperation between manufacturers and retailers can be improved by adopting improved revenue sharing contracts.

The rest of the paper is organized as follows: in the second section, we introduce the problem description and symbolic assumptions. The third section describes the constructed warranty decision model in detail, and gives the model analysis of the properties in the fourth section. The fifth section sets up the coordination contract to improve the double marginalization effect generated by the supply chain system. The sixth section presents and verifies the decision model in the form of numerical simulation. Finally, we summarize our results and propose implications for future research.

2. Problem Description and Assumptions

2.1. Problem Description

This paper considers a single-stage closed-loop supply chain system consisting of a single manufacturer and a single retailer. First, the consumer determines whether to purchase the remanufactured product and its extended warranty service. The manufacturer then determines the wholesale price of the remanufactured product. Finally, the retailer determines the product price and provides the extended warranty service for the market after buying remanufacturing products from the manufacturer via wholesale. This paper discusses three market-predictive decision models, as shown in Figure 1: centralized decision-making models (manufacturers and retailers see decision-making as a whole), manufacturer-led decentralized decision models (i.e., M-R models), and the retailer-led decentralized decision-making model (i.e., the R-M model) corresponding to the practical cases under the business models of Gree Group (Zhuhai, China), Apple Inc. (Cupertino, CA, USA) and Suning Tesco (Nanjing, China), respectively.

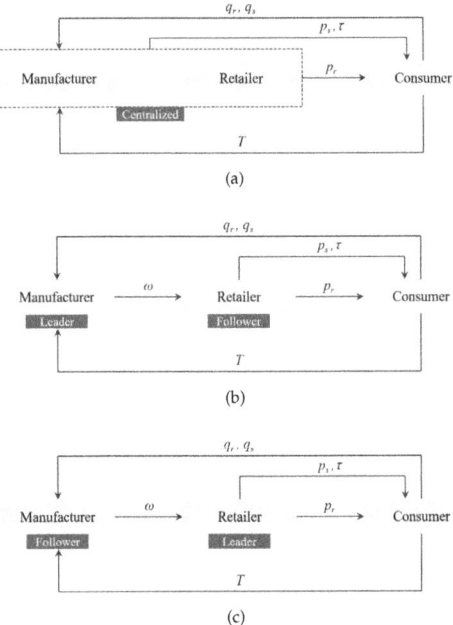

Figure 1. (a) centralized decision model; (b) M-R decision model; (c) R-M decision model.

2.2. Basic Assumptions and Parameters

Assumption 1: *Manufacturers and retailers in decentralized decision-making have a Stackelberg game relationship. They are risk-neutral and in the game of complete information.*

Assumption 2: *To be specific, we assume that there are only remanufactured products in the market, and the retail price of the product is p_r, the extended warranty price is p_s, the unit wholesale price of the remanufactured product is ω, the remanufactured unit cost is c_r, and the extended warranty service unit cost is c_s.*

Assumption 3: *The fixed cost of remanufacturing is T; τ is the warranty period of remanufactured products. According to the research results of Li et al. (2012) [31], the failure rate of remanufactured products during the warranty period is $\mu\tau^2$ (μ is a constant and $\mu > 0$), $\iota > 0$ is the average cost of the repair or replacement of the unit product that the manufacturer has failed, and the fixed service cost of the remanufactured product warranty.*

For $\frac{1}{2}\eta\tau^2$, where $\eta = \mu \iota > 0$ represents the final cost factor for the retailer to provide extended warranty service; the profit function is Π_j, where $j \in \{S, R, M\}$, and S represents the supply chain system, R the retailer, and M the manufacturer, respectively.

Assumption 4: *Let the production quantity (i.e., demand) of the remanufactured product be q_r. Let the market be Q the consumer's Willingness To Pay (WTP) is α and subject to a uniform distribution of $[0,1]$, and its distribution function is $F(\alpha) = \alpha$. If the consumer's WTP for the remanufactured product is α_r, the utility function of the consumer to purchase the remanufactured product is $\mu = \alpha_r - p_r$, and the consumer decides whether to purchase remanufacturing products by comparing the size of the consumer utility function. When the consumer utility function satisfies $\mu = \alpha_r - p_r > 0$, the consumer chooses to purchase the remanufactured product. At this time, under the influence of the consumer remanufactured product preference intensity, the remanufactured product demand function is represented as:*

$$q_r^N = \int_{p_r}^{Q} F'(\alpha_r)\, d\alpha_r + \theta q_r^e = \int_{p_r}^{Q} 1\, dF(\alpha_r) + \theta q_r^e, \tag{1}$$

where q_r^N represents the market demand function, q_r^e represents the user's expected consumer size for remanufactured products, and $\theta \in (0,1)$ represents the consumer preference over remanufactured products.

To simplify the discussion, assuming $q_r^N = q_r^e$, the remanufactured product demand function is converted to:

$$q_r = \frac{\int_{p_r}^{Q} 1\, dF'(\alpha_r)}{1 - \theta} = \frac{Q - p_r}{1 - \theta}. \tag{2}$$

Assumption 5: *Remanufactured product extended warranty service demand is q_s, remanufactured product extended warranty service demand is affected by extended warranty service price p_s, warranty period τ and product demand:*

$$q_s = q_r - \beta p_s + \gamma \tau. \tag{3}$$

Among them, β indicates the consumer's sensitivity to the price of the extended warranty service, γ is the coefficient of influence of the warranty period on the product demand, and $\beta > 0, \gamma > 0$. In particular, when $\beta p_s > \gamma\tau$, some consumers who purchase remanufactured products choose to purchase extended warranty services, hereinafter referred to as partial extensions.

Note: In the following, the superscripts "c", "d1" and "d2" in the variables represent centralized decision, decentralized M-R decision and decentralized R-M decision in closed-loop supply chain decision, respectively; superscript "r1", "r2" and "r3" respectively represent the revenue sharing contract under the coordination mechanism, the improved revenue sharing contract and the two charge contracts. The superscript "*" indicates the optimal decision result.

3. Closed-Loop Supply Chain Decision

3.1. Centralized Decision Making (Model C)

Under the centralized decision model, manufacturers and retailers form a unified joint decision to maximize the overall profit of the supply chain system (e.g., Gree's centralized business strategy), and its profit function is expressed as:

$$\max_{p_r, p_s, \tau} \Pi_S^c = q_r(p_r - c_r) + q_s(p_s - c_s) - \frac{1}{2}\eta\tau^2 - T \tag{4}$$

$$s.t \quad \beta p_s \geq \gamma\tau.$$

Sections 1 and 2 of Model (4) are the proceeds from the sale of remanufactured products and sales extension services, respectively, and the third part is the fixed fee required for the remanufacturing and sales process. When $\beta p_s > \gamma\tau$, i.e., $\beta > 1, \gamma^2 < \eta\beta, 0 < \theta < 1 - \dfrac{\eta}{2\eta\beta - \gamma^2}$, some consumers choose to purchase the remanufactured product extended warranty service. At this time, there is an optimal solution for the objective function. Combine Equation (2) and Formula (3) to solve model (4) and get Proposition 1.

Proposition 1: *Under the centralized decision, the optimal selling price of the remanufactured product, the optimal extended warranty price of the remanufactured product, the optimal warranty period, the remanufactured product output and the warranty service sales are respectively:*

$$p_r^{c*} = \dfrac{(\theta-1)\left(\beta\eta c_s - c_r\left(\gamma^2 - 2\beta\eta\right)\right) + Q\left(\eta(2\beta(\theta-1)+1) - \gamma^2(\theta-1)\right)}{\eta(4\beta(\theta-1)+1) - 2\gamma^2(\theta-1)}, \quad (5)$$

$$p_s^{c*} = \dfrac{\eta c_r + c_s\left(\eta(2\beta(\theta-1)+1) - 2\gamma^2(\theta-1)\right) + \eta(-Q)}{\eta(4\beta(\theta-1)+1) - 2\gamma^2(\theta-1)}, \quad (6)$$

$$\tau_s^{c*} = \dfrac{\gamma\left(c_r - 2\beta(\theta-1)c_s - Q\right)}{\eta(4\beta(\theta-1)+1) - 2\gamma^2(\theta-1)}, \quad (7)$$

$$q_r^{c*} = \dfrac{\left(\gamma^2 - 2\beta\eta\right)(Q - c_r) + \beta\eta c_s}{\eta(4\beta(\theta-1)+1) - 2\gamma^2(\theta-1)}, \quad (8)$$

$$q_s^{c*} = \dfrac{\beta\eta\left(c_r - 2\beta(\theta-1)c_s - Q\right)}{\eta(4\beta(\theta-1)+1) - 2\gamma^2(\theta-1)}. \quad (9)$$

At this point, under the centralized decision, the optimal profit of the supply chain system is:

$$\Pi_S^{c*} = -\dfrac{-\left(\gamma^2 - 2\beta\eta\right)(Q - c_r)^2 + 2\beta\eta c_s(c_r - Q) - 2\beta^2\eta(\theta-1)c_s^2}{2\left(\eta(4\beta(\theta-1)+1) - 2\gamma^2(\theta-1)\right)}. \quad (10)$$

Proof: To build a centralized decision Lagrangian function:

$$L^c(p_r, p_s, \tau, \lambda) = q_r(p_r - c_r) + q_s(p_s - c_s) - \dfrac{1}{2}\eta\tau^2 + \lambda(\beta p_s - \gamma\tau). \quad (11)$$

A centralized decision-making profit function for the Hessian Matrix of p_r, p_s, τ:

$$\mathbf{H}^c = \begin{bmatrix} \dfrac{2}{\theta-1} & \dfrac{1}{\theta-1} & 0 \\ \dfrac{1}{\theta-1} & -2\beta & \gamma \\ 0 & \gamma & -\eta \end{bmatrix}. \quad (12)$$

If and only if the order master subroutine $H_1 = -\dfrac{2}{1-\theta} < 0$, $H_2 = -\dfrac{4\beta(\theta-1)-1}{(1-\theta)^2} > 0$, $H_3 = \dfrac{2(2\eta\beta - \gamma^2)(1-\theta) - \eta}{(1-\theta)^2} < 0$, i.e., $\beta > 1, \gamma^2 < \eta\beta, 0 < \theta < 1 - \dfrac{\eta}{2\eta\beta - \gamma^2}$, the Hessian Matrix is negative, and the centralized decision profit function Π_S^{c*} is a strict concave function for p_r, p_s, τ so there is an optimal solution.

Solve the first-order partial derivative of p_r, p_s, τ for L^c and make it 0:

$$\dfrac{\partial L^c}{\partial p_r} = \dfrac{c_r + c_s - 2p_r - p_s + Q}{1-\theta} = 0, \quad (13)$$

$$\dfrac{\partial L^c}{\partial p_s} = \beta\lambda + \beta c_s + \gamma\tau + \dfrac{p_r}{\theta-1} - 2\beta p_s + \dfrac{Q}{1-\theta} = 0, \quad (14)$$

$$\dfrac{\partial L^c}{\partial \tau} = -\gamma\lambda - \gamma c_s - \eta\tau + \gamma p_s = 0. \quad (15)$$

The simultaneous decision equations can be obtained by substituting Equation (4), and Proposition 1 is proved. □

3.2. Decentralized Decision

3.2.1. M-R Decision (Model D1)

In the M-R decision-making(e.g., Apple Inc. and its affiliates), the manufacturer as the market leader first considers the retailer's optimal response function to determine the wholesale price. Then, the retailer determines the remanufactured product sales price and the extended warranty pricing decision according to the manufacturer's decision. The decision model is:

$$\max_{p_r,p_s,\tau} \Pi_M^{d1} = q_r (w - c_r) - T,$$

$$\text{s.t.} \quad \max_{p_r,p_s,\tau} \Pi_R^{d1} = q_r (p_r - w) + q_s (p_s - c_s) - \frac{1}{2}\eta\tau^2, \tag{16}$$

$$\beta p_s \geq \gamma\tau.$$

The first part of the objective function in model (16) is the gain from the manufacturer selling the remanufactured product, and the second part is the remanufactured fixed fee. The constraint is the retail price selected by the retailer under the maximization of the objective function and the price of the extended service. When $\beta p_s > \gamma\tau$, i.e., $\beta > 1, \gamma^2 < \eta\beta, 0 < \theta < 1 - \frac{\eta}{2\eta\beta - \gamma^2}$, some consumers choose to purchase the remanufactured product's extended warranty service. At this time, there is an optimal solution for the objective function. Combine Equation (2) and Formula (3) to solve the model (16). Proposition 2 is available.

Proposition 2: *Under the decentralized M-R decision, the optimal selling price of the remanufactured product, the optimal extended warranty price of the remanufactured product, the optimal warranty period, the remanufactured product output and the warranty service sales are respectively:*

$$p_r^{d1*} = -\frac{(\theta - 1)(c_r(\gamma^2 - 2\beta\eta) - \beta\eta c_s) + Q(3\gamma^2(\theta - 1) - 2\eta(3\beta(\theta - 1) + 1))}{2(\eta(4\beta(\theta - 1) + 1) - 2\gamma^2(\theta - 1))}, \tag{17}$$

$$p_s^{d1*} = -\frac{-\eta c_r + c_s(4\gamma^2(\theta - 1) - \eta(4\beta(\theta - 1) + \eta\phi\beta + 2)) + \eta Q}{2(\eta(4\beta(\theta - 1) + 1) - 2\gamma^2(\theta - 1))}, \tag{18}$$

$$\tau^{d1*} = \frac{\gamma(c_r + \beta c_s(-4\theta + \phi\eta + 4) - Q)}{2(\eta(4\beta(\theta - 1) + 1) - 2\gamma^2(\theta - 1))}, \tag{19}$$

$$w^{d1*} = -\frac{-(\gamma^2 - 2\beta\eta)(Q - c_r) - \beta\eta c_s}{2(\eta(4\beta(\theta - 1) + 1) - 2\gamma^2(\theta - 1))}, \tag{20}$$

$$q_r^{d1*} = \frac{(\gamma^2 - 2\beta\eta)(Q - c_r) + \beta\eta c_s}{\eta(4\beta(\theta - 1) + 1) - 2\gamma^2(\theta - 1)}, \tag{21}$$

$$q_s^{d1*} = \frac{\beta\eta\left((\gamma^2 - 2\beta\eta)(Q - c_r) - \beta c_s(\eta(8\beta(\theta - 1) + 1) - 4\gamma^2(\theta - 1))\right)}{2(\gamma^2 - 2\beta\eta)(\eta(-4\beta(\theta - 1) - 1) + 2\gamma^2(\theta - 1))}. \tag{22}$$

Note: To simplify the decision model, make $\phi = \frac{1}{\gamma^2 - 2\beta\eta}$.

At this point, under the decentralized M-R decision, the optimal profits of manufacturers and retailers are:

$$\Pi_M^{d1*} = -\frac{((\gamma^2 - 2\beta\eta)(Q - c_r) + \beta\eta c_s)^2}{4(\gamma^2 - 2\beta\eta)(\eta(-4\beta(\theta - 1) - 1) + 2\gamma^2(\theta - 1))}, \tag{23}$$

$$\Pi_R^{d1*} = \frac{-(\gamma^2 - 2\beta\eta)^2 (Q - c_r)^2 - 2\beta\eta c_s (\gamma^2 - 2\beta\eta)(Q - c_r) + \beta^2 \eta c_s^2 (\eta(16\beta(\theta-1)+3) - 8\gamma^2(\theta-1))}{8(\gamma^2 - 2\beta\eta)(\eta(-4\beta(\theta-1)-1) + 2\gamma^2(\theta-1))}. \quad (24)$$

Proof: The Lagrangian function to build retailer profits is:

$$L^{d1}(p_r, p_s, \tau, \lambda) = q_s(p_s - c_s) - \frac{1}{2}\eta\tau^2 + q_r(p_r - w) + \lambda(\beta p_s - \gamma\tau). \quad (25)$$

A decentralized M-R decision profit function for the Hessian Matrix of p_r, p_s, τ:

$$\mathbf{H}^{d1} = \begin{bmatrix} \frac{2}{\theta-1} & \frac{1}{\theta-1} & 0 \\ \frac{1}{\theta-1} & -2\beta & \gamma \\ 0 & \gamma & -\eta \end{bmatrix}. \quad (26)$$

If and only if the order master subroutine $H_1 = -\frac{2}{1-\theta} < 0$, $H_2 = -\frac{4\beta(\theta-1)-1}{(1-\theta)^2} > 0$, $H_3 = \frac{2(2\eta\beta - \gamma^2)(1-\theta) - \eta}{(1-\theta)^2} < 0$, i.e., $\beta > 1, \gamma^2 < \eta\beta, 0 < \theta < 1 - \frac{\eta}{2\eta\beta - \gamma^2}$ the Hessian Matrix is negative, and the decentralized M-R decision profit function Π_R^{d1*} is a strict concave function for p_r, p_s, τ, so there is an optimal solution.

Solve the first-order partial derivative of p_r, p_s, τ for L^{d1} and make it 0:

$$\frac{\partial L^d}{\partial p_r} = \frac{c_s - 2p_r - p_s + Q + w}{1 - \theta} = 0, \quad (27)$$

$$\frac{\partial L^d}{\partial p_s} = \beta\lambda + \beta c_s + \gamma\tau + \frac{p_r}{\theta-1} - 2\beta p_s + \frac{Q}{1-\theta} = 0, \quad (28)$$

$$\frac{\partial L^d}{\partial \tau} = -\gamma\lambda - \gamma c_s - \eta\tau + \gamma p_s = 0. \quad (29)$$

The simultaneous equations solve for $p_r^{d1}(w), p_s^{d1}(w), \tau(w)$ for w, and substitute Π_M^{d1} can get w^*. Finally, by substituting the above results into the model (16), we get Π_M^{d1*}, Π_R^{d1*}, so Proposition 2 is proved. □

3.2.2. R-M Decision (Model D2)

In the R-M decision-making (e.g., Wal-Mart), the retailer, as the market leader, prioritizes its own objective function and the manufacturer's optimal response function to determine the sales price. Then, the manufacturer determines the remanufactured product sales price and the extended warranty pricing decision based on the retailer's decision. The model is:

$$\max_{p_r, p_s, \tau} \Pi_R^d = q_r(p_r - w) + q_s(p_s - c_s) - \frac{1}{2}\eta\tau^2,$$

$$\text{s.t.} \quad \max_{p_r, p_s, \tau} \Pi_M^d = q_r(w - c_r) - T, \quad (30)$$

$$\beta p_s \geq \gamma\tau,$$

$$p_r = w + Z.$$

The first part of the model (30) objective function is the revenue earned by the retailer from remanufacturing the product, and the second part is the gain from the sales extension service. The constraint is the wholesale price chosen by the manufacturer under the maximization of the objective function. When $\beta p_s > \gamma \tau$, i.e., $\beta > 1, \gamma^2 < \eta\beta, 0 < \theta < 1 - \frac{\eta}{2\eta\beta - \gamma^2}$, some consumers choose to purchase remanufactured product's extended warranty service. At this time, there is an optimal solution for the objective function. Combine Equation (2) and Formula (3) to solve model (26). Proposition 3 is then available.

Proposition 3: *Under the decentralized R-M decision, the optimal selling price of the remanufactured product, the optimal extended warranty price of the remanufactured product, the optimal warranty period, the remanufactured product output, and the warranty service sales are respectively:*

$$p_r^{d2*} = \frac{(\theta-1)\left(\beta\eta c_s - c_r\left(\gamma^2 - 2\beta\eta\right)\right) + Q\left(\eta(6\beta(\theta-1)+1) - 3\gamma^2(\theta-1)\right)}{\eta(8\beta(\theta-1)+1) - 4\gamma^2(\theta-1)}, \tag{31}$$

$$p_s^{d2*} = \frac{\eta c_r + c_s\left(\eta(4\beta(\theta-1)+1) - 4\gamma^2(\theta-1)\right) - \eta Q}{\eta(8\beta(\theta-1)+1) - 4\gamma^2(\theta-1)}, \tag{32}$$

$$\tau^{d2*} = \frac{\gamma\left(-c_r + 4\beta(\theta-1)c_s + Q\right)}{\eta(-8\beta(\theta-1)-1) + 4\gamma^2(\theta-1)}, \tag{33}$$

$$q_r^{d2*} = \frac{\left(\gamma^2 - 2\beta\eta\right)(Q - c_r) + \beta\eta c_s}{\eta(8\beta(\theta-1)+1) - 4\gamma^2(\theta-1)}, \tag{34}$$

$$q_s^{d2*} = \frac{\beta\eta\left(c_r - 4\beta(\theta-1)c_s - Q\right)}{\eta(8\beta(\theta-1)+1) - 4\gamma^2(\theta-1)}. \tag{35}$$

At this point, under the decentralized R-M decision, the optimal profits of manufacturers and retailers are:

$$\Pi_M^{d2*} = -\frac{(\theta-1)\left(\left(\gamma^2 - 2\beta\eta\right)(Q - c_r) + \beta\eta c_s\right)^2}{\left(\eta(-8\beta(\theta-1)-1) + 4\gamma^2(\theta-1)\right)^2}, \tag{36}$$

$$\Pi_R^{d2*} = -\frac{-\left(\gamma^2 - 2\beta\eta\right)(Q - c_r)^2 + 2\beta\eta c_s(c_r - Q) - 4\beta^2\eta(\theta-1)c_s^2}{2\left(\eta(8\beta(\theta-1)+1) - 4\gamma^2(\theta-1)\right)}. \tag{37}$$

Proof: The Lagrangian function to build retailer profits is:

$$L^{d2}(p_r, p_s, \tau, \lambda) = q_s(p_s - c_s) - \frac{1}{2}\eta\tau^2 + q_r(p_r - \omega) + \lambda(\beta p_s - \gamma\tau). \tag{38}$$

A decentralized M-R decision profit function for the Hessian Matrix of p_r, p_s, τ:

$$\mathbf{H}^{d2} = \begin{bmatrix} \frac{2}{\theta-1} & \frac{1}{\theta-1} & 0 \\ \frac{1}{\theta-1} & -2\beta & \gamma \\ 0 & \gamma & -\eta \end{bmatrix}. \tag{39}$$

If and only if the order master subroutine $H_1 = -\frac{2}{1-\theta} < 0$, $H_2 = -\frac{4\beta(\theta-1)-1}{(1-\theta)^2} > 0$, $H_3 = \frac{2(2\eta\beta - \gamma^2)(1-\theta) - \eta}{(1-\theta)^2} < 0$, i.e., $\beta > 1, \gamma^2 < \eta\beta, 0 < \theta < 1 - \frac{\eta}{2\eta\beta - \gamma^2}$, the Hessian Matrix is negative, and the decentralized R-M decision profit function Π_M^{d2*} is a strict concave function for p_r, p_s, τ, so there is an optimal solution.

Solve the first-order partial derivative of ω for L^{d2}, introduce the artificial variable and make it 0:

$$\omega = c_r - p_r + Q. \tag{40}$$

Substituting Π_R^{d2} and solving the first-order partial derivatives of p_r, p_s, τ and making it 0, yields:

$$\frac{\partial L^d}{\partial p_r} = \frac{c_r + c_s - 4p_r - p_s + 3Q}{1 - \theta} = 0, \quad (41)$$

$$\frac{\partial L^d}{\partial p_s} = \beta c_s + \frac{\gamma(\theta - 1)\tau + p_r - Q}{\theta - 1} - 2\beta p_s = 0, \quad (42)$$

$$\frac{\partial L^d}{\partial \tau} = -\gamma c_s - \eta \tau + \gamma p_s = 0. \quad (43)$$

The simultaneous equations can be solved by deriving $p_r^{d2*}, p_s^{d2*}, \tau^{d2*}$, substituting Π_M^{d2}. Finally, the result is substituted into the model (30) to obtain Π_M^{d2*}, Π_R^{d2*}, and Proposition 3 is proved. □

In order to facilitate the analysis below, the optimal decision results of the closed-loop supply chain are collated, as shown in Table 1.

Table 1. Optimization decision summary table.

Model C	$q_r = \frac{(\gamma^2 - 2\beta\eta)(Q - c_r) + \beta\eta c_s}{\eta(4\beta(\theta - 1) + 1) - 2\gamma^2(\theta - 1)}$
	$q_s = \frac{\beta\eta(c_r - 2\beta(\theta - 1)c_s - Q)}{\eta(4\beta(\theta - 1) + 1) - 2\gamma^2(\theta - 1)}$
	$p_r = \frac{(\theta - 1)(\beta\eta c_s - c_r(\gamma^2 - 2\beta\eta)) + Q(\eta(2\beta(\theta - 1) + 1) - \gamma^2(\theta - 1))}{\eta(4\beta(\theta - 1) + 1) - 2\gamma^2(\theta - 1)}$
	$p_s = \frac{\eta c_r + c_s(\eta(2\beta(\theta - 1) + 1) - 2\gamma^2(\theta - 1)) + \eta(-Q)}{\eta(4\beta(\theta - 1) + 1) - 2\gamma^2(\theta - 1)}$
	$\tau = \frac{\gamma(c_r - 2\beta(\theta - 1)c_s - Q)}{\eta(4\beta(\theta - 1) + 1) - 2\gamma^2(\theta - 1)}$
	$\Pi_S^c = -\frac{-(\gamma^2 - 2\beta\eta)(Q - c_r)^2 + 2\beta\eta c_s(c_r - Q) - 2\beta^2\eta(\theta - 1)c_s^2}{2(\eta(4\beta(\theta - 1) + 1) - 2\gamma^2(\theta - 1))}$
Model D1	$q_r = \frac{(\gamma^2 - 2\beta\eta)(Q - c_r) + \beta\eta c_s}{\eta(4\beta(\theta - 1) + 1) - 2\gamma^2(\theta - 1)}$
	$q_s = \frac{\beta\eta((\gamma^2 - 2\beta\eta)(Q - c_r) - \beta c_s(\eta(8\beta(\theta - 1) + 1) - 4\gamma^2(\theta - 1)))}{2(\gamma^2 - 2\beta\eta)(\eta(-4\beta(\theta - 1) - 1) + 2\gamma^2(\theta - 1))}$
	$p_r = -\frac{(\theta - 1)(c_r(\gamma^2 - 2\beta\eta) - \beta\eta c_s) + Q(3\gamma^2(\theta - 1) - 2\eta(3\beta(\theta - 1) + 1))}{2(\eta(4\beta(\theta - 1) + 1) - 2\gamma^2(\theta - 1))}$
	$p_s = -\frac{-\eta c_r + c_s(4\gamma^2(\theta - 1) - \eta(4\beta(\theta - 1) + \eta\phi\beta + 2)) + \eta Q}{2(\eta(4\beta(\theta - 1) + 1) - 2\gamma^2(\theta - 1))}$
	$\tau = \frac{\gamma(c_r + \beta c_s(-4\theta + \phi\eta + 4) - Q)}{2(\eta(4\beta(\theta - 1) + 1) - 2\gamma^2(\theta - 1))}$
	$\Pi_M^{d1} = -\frac{((\gamma^2 - 2\beta\eta)(Q - c_r) + \beta\eta c_s)^2}{4(\gamma^2 - 2\beta\eta)(\eta(-4\beta(\theta - 1) - 1) + 2\gamma^2(\theta - 1))}$
	$\Pi_R^{d1} = \frac{-(\gamma^2 - 2\beta\eta)^2(Q - c_r)^2 - 2\beta\eta c_s(\gamma^2 - 2\beta\eta)(Q - c_r) + \beta^2\eta c_s^2(\eta(16\beta(\theta - 1) + 3) - 8\gamma^2(\theta - 1))}{8(\gamma^2 - 2\beta\eta)(\eta(-4\beta(\theta - 1) - 1) + 2\gamma^2(\theta - 1))}$

Table 1. Cont.

Model D2	$q_r = \dfrac{(\gamma^2 - 2\beta\eta)(Q - c_r) + \beta\eta c_s}{\eta(8\beta(\theta - 1) + 1) - 4\gamma^2(\theta - 1)}$	
	$q_s = \dfrac{\beta\eta(c_r - 4\beta(\theta - 1)c_s - Q)}{\eta(8\beta(\theta - 1) + 1) - 4\gamma^2(\theta - 1)}$	
	$p_r = \dfrac{(\theta - 1)(\beta\eta c_s - c_r(\gamma^2 - 2\beta\eta)) + Q(\eta(6\beta(\theta - 1) + 1) - 3\gamma^2(\theta - 1))}{\eta(8\beta(\theta - 1) + 1) - 4\gamma^2(\theta - 1)}$	
	$p_s = \dfrac{\eta c_r + c_s(\eta(4\beta(\theta - 1) + 1) - 4\gamma^2(\theta - 1)) - \eta Q}{\eta(8\beta(\theta - 1) + 1) - 4\gamma^2(\theta - 1)}$	
	$\tau = \dfrac{\gamma(-c_r + 4\beta(\theta - 1)c_s + Q)}{\eta(-8\beta(\theta - 1) - 1) + 4\gamma^2(\theta - 1)}$	
	$\Pi_M^{d2} = -\dfrac{(\theta - 1)\left((\gamma^2 - 2\beta\eta)(Q - c_r) + \beta\eta c_s\right)^2}{(\eta(-8\beta(\theta - 1) - 1) + 4\gamma^2(\theta - 1))^2}$	
	$\Pi_R^{d2} = -\dfrac{-(\gamma^2 - 2\beta\eta)(Q - c_r)^2 + 2\beta\eta c_s(c_r - Q) - 4\beta^2\eta(\theta - 1)c_s^2}{2(\eta(8\beta(\theta - 1) + 1) - 4\gamma^2(\theta - 1))}$	

4. Analysis of Decision-Making Properties of Closed-Loop Supply Chain

Property 1: *In model C, D1, D2 decisions, extended warranty service price p_s, warranty period τ, remanufactured product output q_r, extended warranty service sales q_s, and supply chain system profit Π increase as consumer preference θ increases, and remanufactured product sales price p_r decreases as consumer preference θ increases, indicating that consumers prefer in view of supply chain member companies that θ has a positive incentive for the revenue of member companies in the supply chain.*

Proof: Taking the decentralized MR decision as an example, when $\beta > 1, \gamma^2 < \eta\beta, 0 < \theta < 1 - \dfrac{\eta}{2\eta\beta - \gamma^2}$, the optimal solution is established:

$$\frac{\partial p_r^{d1*}}{\partial \theta} = \frac{\eta\left((\gamma^2 - 2\beta\eta)(Q - c_r) + \beta\eta c_s\right)}{2(\eta(4\beta(\theta - 1) + 1) - 2\gamma^2(\theta - 1))^2} < 0, \tag{44}$$

$$\frac{\partial p_s^{d1*}}{\partial \theta} = \frac{\eta\left(-(\gamma^2 - 2\beta\eta)(Q - c_r) - \beta\eta c_s\right)}{(\eta(-4\beta(\theta - 1) - 1) + 2\gamma^2(\theta - 1))^2} > 0, \tag{45}$$

$$\frac{\partial \tau^{d1*}}{\partial \theta} = \frac{\gamma(\gamma^2 - 2\beta\eta)(Q - c_r) + \beta\gamma\eta c_s}{(\eta(-4\beta(\theta - 1) - 1) + 2\gamma^2(\theta - 1))^2} > 0, \tag{46}$$

$$\frac{\partial q_r^{d1*}}{\partial \theta} = \frac{(\gamma^2 - 2\beta\eta)\left((\gamma^2 - 2\beta\eta)(Q - c_r) + \beta\eta c_s\right)}{(\eta(-4\beta(\theta - 1) - 1) + 2\gamma^2(\theta - 1))^2} > 0, \tag{47}$$

$$\frac{\partial q_s^{d1*}}{\partial \theta} = \frac{\beta\eta\left(-(\gamma^2 - 2\beta\eta)(Q - c_r) - \beta\eta c_s\right)}{(\eta(-4\beta(\theta - 1) - 1) + 2\gamma^2(\theta - 1))^2} > 0, \tag{48}$$

$$\frac{\partial \Pi_M^{d1*}}{\partial \theta} = \frac{\left((\gamma^2 - 2\beta\eta)(Q - c_r) + \beta\eta c_s\right)^2}{2(\eta(4\beta(\theta - 1) + 1) - 2\gamma^2(\theta - 1))^2} > 0, \tag{49}$$

$$\frac{\partial \Pi_R^{d1*}}{\partial \theta} = \frac{\left((\gamma^2 - 2\beta\eta)(Q - c_r) + \beta\eta c_s\right)^2}{4(\eta(4\beta(\theta - 1) + 1) - 2\gamma^2(\theta - 1))^2} > 0. \tag{50}$$

The same can be proved in the centralized decision-making and decentralized R-M decision. □

Property 2: Decentralized M-R and decentralized RM decision, retailers and manufacturers each make their own interests to maximize the goal of decision-making, resulting in double marginalization effect. The closed-loop supply chain system is not optimal, specifically in decentralized decision-making. The optimal selling price is greater than the optimal selling price under the centralized decision, and the optimal production quantity and the optimal warranty period are less than the optimal production quantity and warranty period under the centralized decision, namely:

1. $\tau^{c*} > \tau^{d1*}, \tau^{c*} > \tau^{d2*}$,
2. $p_r^{d1*} > p_r^{c*}, p_r^{d2*} > p_r^{c}; p_s^{d1*} > p_s^{c}, p_s^{d2*} > p_s^{c}$,
3. $q_r^{d1*} > q_r^{c}, q_r^{d2*} > q_r^{c}; q_s^{d1*} > q_s^{c}, q_s^{d2*} > q_s^{c}$.

In particular, as consumer preference for the degree of θ increases, the degree of differentiation between decentralized decision-making and centralized decision-making increases.

Proof: Take the decentralized M-R decision as an example, let:

$$Z_{pr} = p_r^{c*} - p_r^{d1*} = -\frac{(\theta-1)\left(-\left(\gamma^2-2\beta\eta\right)(Q-c_r)-\beta\eta c_s\right)}{2\left(\eta(4\beta(\theta-1)+1)-2\gamma^2(\theta-1)\right)}, \quad (51)$$

$$Z_{ps} = p_s^{c*} - p_s^{d1*} = \frac{\eta\left(\left(\gamma^2-2\beta\eta\right)(Q-c_r)+\beta\eta c_s\right)}{2\left(\gamma^2-2\beta\eta\right)\left(\eta(-4\beta(\theta-1)-1)+2\gamma^2(\theta-1)\right)}, \quad (52)$$

$$Z_\tau = \tau^{c*} - \tau^{d1*} = \frac{\gamma\left(\gamma^2-2\beta\eta\right)(Q-c_r)+\beta\gamma\eta c_s}{2\left(\gamma^2-2\beta\eta\right)\left(\eta(-4\beta(\theta-1)-1)+2\gamma^2(\theta-1)\right)}. \quad (53)$$

Solving a one-stage partial guide on θ,

$$\frac{\partial Z_{pr}}{\partial \theta} = -\frac{\eta\left(\left(\gamma^2-2\beta\eta\right)(Q-c_r)+\beta\eta c_s\right)}{2\left(\eta(4\beta(\theta-1)+1)-2\gamma^2(\theta-1)\right)^2}, \quad (54)$$

$$\frac{\partial Z_{ps}}{\partial \theta} = -\frac{\eta\left(\left(\gamma^2-2\beta\eta\right)(Q-c_r)+\beta\eta c_s\right)}{\left(\eta(-4\beta(\theta-1)-1)+2\gamma^2(\theta-1)\right)^2}, \quad (55)$$

$$\frac{\partial Z_\tau}{\partial \theta} = -\frac{\gamma\left(\gamma^2-2\beta\eta\right)(Q-c_r)+\beta\gamma\eta c_s}{\left(\eta(-4\beta(\theta-1)-1)+2\gamma^2(\theta-1)\right)^2}. \quad (56)$$

Since $\gamma^2 < \eta\beta$, i.e., $\gamma^2 - 2\beta\eta > 0$, $\frac{\partial Z_{pr}}{\partial \theta} > 0$, $\frac{\partial Z_{ps}}{\partial \theta} > 0$, $\frac{\partial Z_\tau}{\partial \theta} > 0$, Property 2 is certified. The same is true for re-centralized decision-making and decentralized RM decisions. □

Property 3: Under decentralized decision-making, the optimal total profit of the supply chain system is lower than the optimal profit of the centralized decision system, i.e., $\Pi_S^{c*} > \Pi_R^{d1*} + \Pi_M^{d1*}; \Pi_S^{c*} > \Pi_R^{d2*} + \Pi_M^{d2*}$. In addition, from the perspective of consumer behavior, the increase in consumer preference θ will be magnified under decentralized decision-making. The "dual marginal effect" exacerbates the loss of efficiency of decentralized decision-making systems.

Proof: Taking the decentralized MR decision as an example, when $\beta > 1, \gamma^2 < \eta\beta, 0 < \theta < 1 - \frac{\eta}{2\eta\beta - \gamma^2}$, the optimal solution is established:

$$Z_\Pi = \Pi_S^{c*} - (\Pi_M^{d1*} + \Pi_R^{d1*}) = -\frac{\left(\left(\gamma^2-2\beta\eta\right)(Q-c_r)+\beta\eta c_s\right)^2}{8\left(\gamma^2-2\beta\eta\right)\left(\eta(-4\beta(\theta-1)-1)+2\gamma^2(\theta-1)\right)} > 0 \quad (57)$$

$$\frac{\partial Z_\Pi}{\partial \theta} = \frac{\left(\left(\gamma^2-2\beta\eta\right)(Q-c_r)+\beta\eta c_s\right)^2}{4\left(\eta(4\beta(\theta-1)+1)-2\gamma^2(\theta-1)\right)^2} > 0 \quad (58)$$

Property 3 is proved. Similarly, the re-centralized decision-making and the decentralized R-M decision are also established. □

5. Coordination Mechanism Design

Because the decentralized M-R and R-M decisions have double marginal effects, the system is not optimal, and with the increase of consumer preference, the double marginalization effect is gradually expanded. Therefore, for the M-R model, the income sharing contract will be adopted. For the R-M model, two charge contracts will be adopted to improve the total system profit of the supply chain under the decentralized decision approach or to achieve the total system profit of centralized decision-making, and make each member company of the supply chain implement Pareto improvements.

5.1. M-R Decision: Revenue Sharing Contract (S-1 Model)

According to Cachon and Lariviere, (2005) [32], revenue sharing contracts can significantly increase the overall benefits of the supply chain: manufacturers give retailers lower wholesale prices, while retailers give manufacturers a certain amount of revenue. The sharing ratio is δ, and the corresponding share of the revenue earned by the manufacturer is $1 - \delta$, and the decision model is:

$$\max_{p_r, p_s, \tau} \Pi_M^{r1} = q_r(w - c_r) + (1 - \delta)p_r q_r - T,$$

$$\text{s.t.} \max_{p_r, p_s, \tau} \Pi_R^{r1} = q_r(\delta p_r - w) + q_s(p_s - c_s) - \frac{1}{2}\eta\tau^2, \tag{59}$$

$$\beta p_s \geq \gamma q_s.$$

The first part of the model (50) objective function is the gain from the manufacturer selling the remanufactured product, and the second part is the share-sharing share. The constraint is the retail price selected by the retailer under the maximization of the objective function and the price of the extended service. When $\beta p_s > \gamma\tau$, i.e., $\beta > 1, \gamma^2 < \eta\beta, 0 < \theta < 1 - \frac{\eta}{2\eta\beta - \gamma^2}$, some consumers choose to purchase the remanufactured product's extended warranty service. At this time, there is an optimal solution for the objective function. Combine Equation (2) and Formula (3) to solve model (50), and get Proposition 4.

Proposition 4: *The optimal retail price, optimal extended warranty price and optimal warranty period under the revenue sharing contract coordination M-R model are respectively:*

$$p_r^{r1*} = \frac{\eta\left(\eta(2\beta(\delta + 1)(\theta - 1) + 1) - \gamma^2(\delta + 1)(\theta - 1)\right)(2\beta(\theta - 1)c_s + Q)}{\eta - 2\delta(\theta - 1)(\gamma^2 - 2\beta\eta)} + (\theta - 1)c_r\left(\gamma^2 - 2\beta\eta\right), \tag{60}$$

$$p_s^{r1*} = \frac{c_r\left(\gamma^2 - 2\beta\eta\right)(2\delta(\theta - 1)(\gamma^2 - 2\beta\eta) - \eta)}{\eta - 2\delta(\theta - 1)(\gamma^2 - 2\beta\eta)} + \eta Q\left(\gamma^2 - 2\beta\eta\right)\left[2\delta\left(\delta(\theta - 1)\left(\gamma^2 - 2\beta\eta\right) - \eta\right)\right], \tag{61}$$

$$\tau^{r1*} = \frac{c_r\left(\gamma^2 - 2\beta\eta\right)(2\delta(\theta - 1)(\gamma^2 - 2\beta\eta) - \eta)}{2\delta(\theta - 1)(\gamma^2 - 2\beta\eta) - \eta} + Q\left(\gamma^2 - 2\beta\eta\right)\left[2\delta\left(\delta(\theta - 1)\left(\gamma^2 - 2\beta\eta\right) - \eta\right)\right]. \tag{62}$$

Proof: The Lagrangian function to build retailer profits is:

$$L^{r1}(p_r, p_s, \tau, \lambda) = q_s(p_s - c_s) + q_r(\delta p_r - w) - \frac{1}{2}\eta\tau^2 + \lambda(\beta p_s - \gamma\tau). \tag{63}$$

Under the revenue sharing contract coordination mechanism, the profit function about the Hessian Matrix of p_r, p_s, τ is:

$$H^{d1} = \begin{bmatrix} \frac{2}{\theta-1} & \frac{1}{\theta-1} & 0 \\ \frac{1}{\theta-1} & -2\beta & \gamma \\ 0 & \gamma & -\eta \end{bmatrix}. \tag{64}$$

If and only if the order master subroutine $H_1 = -\frac{2}{1-\theta} < 0$, $H_2 = -\frac{4\beta(\theta-1)-1}{(1-\theta)^2} > 0$, $H_3 = \frac{2(2\eta\beta - \gamma^2)(1-\theta) - \eta}{(1-\theta)^2} < 0$, i.e., $\beta > 1, \gamma^2 < \eta\beta, 0 < \theta < 1 - \frac{\eta}{2\eta\beta - \gamma^2}$, the Hessian Matrix is negative. At this time, the decentralized revenue sharing contract decision profit function Π_R^{r2*} is a strict concave function for p_r, p_s, τ, so there is an optimal solution.

Solve the first-order partial derivative of p_r, p_s, τ for L^{r1} and make it 0:

$$\frac{\partial L^{r1}}{\partial p_r} = \frac{c_s - 2\delta p_r - p_s + \delta Q + \omega}{1-\theta} = 0, \tag{65}$$

$$\frac{\partial L^{r1}}{\partial p_s} = \beta\lambda + \beta c_s + \gamma\tau + \frac{p_r}{\theta-1} - 2\beta p_s + \frac{Q}{1-\theta} = 0, \tag{66}$$

$$\frac{\partial L^{r1}}{\partial \tau} = -\gamma\lambda - \gamma c_s - \eta\tau + \gamma p_s = 0. \tag{67}$$

The simultaneous equations solve for $p_r^{r1}(\omega), p_s^{r1}(\omega), \tau(\omega)$ for ω, and by substituting Π_M^{r1} we can get ω^*. Finally, substituting the results above into the model (59) will result in Π_M^{r1*}, Π_R^{r1*}, and Proposition 4 will be proved. □

Property 4: *The traditional revenue sharing contract cannot coordinate the remanufacturing closed-loop supply chain system with warranty services. The specific implementation is as follows: under the coordination of revenue sharing contract, the overall benefit of the decentralized M-R decision system is improved, but the Pareto improvement is not achieved. The coordinated retailer's revenue is less than that before the coordination, i.e., $\Pi_R^{d1*} > \Pi_R^{r1*}$. At this time, the retailer refuses to accept the contract coordination, and the revenue sharing contract is invalid.*

Proof: When $\beta > 1, \gamma^2 < \eta\beta, 0 < \theta < 1 - \frac{\eta}{2\eta\beta - \gamma^2}$, the optimal solution is established. At this point, let $Z_\Pi^1 = \Pi_M^{d1*} + \Pi_R^{d1*}, Z_\Pi^2 = \Pi_M^{R1*} + \Pi_R^{r1*}$, then:

$$Z_\Pi^1 - Z_\Pi^2 =$$
$$\frac{(\delta-1)(\theta-1)\left(2\eta(\beta(3\delta+1)(\theta-1)+1) - \gamma^2(3\delta+1)(\theta-1)\right)\left((\gamma^2 - 2\beta\eta)(Q-c_r) + \beta\eta c_s\right)^2}{8\left(\eta(-4\beta(\theta-1)-1) + 2\gamma^2(\theta-1)\right)\left(\eta(-2\beta(\delta+1)(\theta-1)-1) + \gamma^2(\delta+1)(\theta-1)\right)^2} > 0, \tag{68}$$

$$\frac{\partial Z_\Pi^2}{\partial \theta} = \frac{\left[\gamma^2(\delta+1)(2\delta+1)(\theta-1) - \eta(2\beta(\delta+1)(2\delta+1))\right]\left((\gamma^2 - 2\beta\eta)(Q-c_r) + \beta\eta c_s\right)^2}{4\left(\eta(-2\beta(\delta+1)(\theta-1)-1) + \gamma^2(\delta+1)(\theta-1)\right)^3} > 0, \tag{69}$$

$$\Pi_R^{d1*} - \Pi_R^{r1*} = \frac{(\delta-1)^2(\theta-1)^2(2\beta\eta - \gamma^2)\left((\gamma^2 - 2\beta\eta)(Q-c_r) + \beta\eta c_s\right)^2}{8(\eta(-4\beta(\theta-1)-1) + 2\gamma^2(\theta-1))(\eta(-2\beta(\delta+1)(\theta-1)-1) + \gamma^2(\delta+1)(\theta-1))^2} > 0. \tag{70}$$

At this point, it can be judged that $Z_\Pi^1 > Z_\Pi^2, \Pi_R^{d1*} > \Pi_R^{r1*}$, and Property 4 is proved. □

5.2. M-R Decision: Improved Revenue Sharing Contract (S-2 Model)

Under the manufacturer-led Stackelberg game model environment, we assume that the manufacturer is trying to motivate the retailer to sell. It will give the retailer a lower wholesale price and allow the retailer to retain a portion of the F, while requiring a margin over the retained profit. Part of the profit sharing, with a retailer retention ratio of δ the manufacturer is divided into $1-\delta$, and its decision model is:

$$\max_{p_r,p_s,\tau} \Pi_M^{r2} = q_r(\omega - c_r) + (1-\delta)\left[q_r(p_r - \omega) + q_s(p_s - c_s) - \frac{\eta\tau^2}{2} - F\right],$$

$$s.t.\ (p_r, p_s, \tau) \in argmax \Pi_R^{r2} = \delta\left[q_r(p_r - \omega) + q_s(p_s - c_s) - \frac{\eta\tau^2}{2} - F\right] + F, \quad (71)$$

$$T \geqslant \Pi_R^{c*}.$$

The first part of the objective function in model (71) is the gain from the manufacturer selling the remanufactured product, and the second part is the share-sharing share obtained by the manufacturer. The constraints are the retailer's incentive compatibility constraints and the retailer's participation constraints. When $\beta p_s > \gamma\tau$, i.e., $\beta > 1, \gamma^2 < \eta\beta, 0 < \theta < 1 - \frac{\eta}{2\eta\beta - \gamma^2}$, some consumers choose to purchase the remanufactured product extended warranty service. At this time, there is an optimal solution for the objective function. Combine Equation (2) and Formula (3) to solve model (71), and get Proposition 5.

Proposition 5: *The optimal retail price, optimal extended warranty price, optimal wholesale price and optimal warranty period by improving the revenue sharing contract coordination under M-R are respectively:*

$$p_r^{r2*} = \frac{(\theta-1)c_r(\gamma^2 - 2\beta\eta) - \beta\eta(\theta-1)c_s + Q(\gamma^2(2\delta+1)(\theta-1) - \eta(2\beta(2\delta+1)(\theta-1) + \delta+1))}{(\delta+1)(\eta(-4\beta(\theta-1)-1)+2\gamma^2(\theta-1))}, \quad (72)$$

$$p_s^{r2*} = \frac{\eta(-c_r + \delta\eta\phi\beta c_s + \delta(-Q)) + c_s(\eta(-2\beta(\theta-1)-1)+2\gamma^2(\theta-1))+\eta Q}{\eta(-4\beta(\theta-1)-1)+2\gamma^2(\theta-1)}, \quad (73)$$

$$\omega^{r2*} = \frac{\delta c_r(\gamma^2 - 2\beta\eta) + \beta\delta^2\eta c_s + \delta^2 Q(\gamma^2 - 2\beta\eta)}{\delta(\delta+1)(\gamma^2 - 2\beta\eta)}, \quad (74)$$

$$\tau^{r2*} = \frac{\gamma(\delta(\eta\phi\beta c_s + Q) - c_r + 2\beta(\theta-1)c_s + Q)}{\eta(-4\beta(\theta-1)-1)+2\gamma^2(\theta-1)}. \quad (75)$$

Proof: The Lagrangian function to build retailer profits is:

$$L^{r2}(p_r, p_s, \tau, \lambda, A) = \delta\left[q_s(p_s - c_s) - F - \frac{\eta\tau\tau}{2} + q_r(p_r - \omega)\right] + F + \lambda(\beta p_s - \gamma\tau) + A(T - \Pi_R^{c*}). \quad (76)$$

Under the improved revenue sharing contract coordination mechanism, the profit function about the Hessian Matrix of p_r, p_s, τ is:

$$\mathbf{H}^{d1} = \begin{bmatrix} \frac{2}{\theta-1} & \frac{1}{\theta-1} & 0 \\ \frac{1}{\theta-1} & -2\beta & \gamma \\ 0 & \gamma & -\eta \end{bmatrix}. \quad (77)$$

If and only if the order master subroutine $H_1 = -\frac{2}{1-\theta} < 0$, $H_2 = -\frac{4\beta(\theta-1)-1}{(1-\theta)^2} > 0$, $H_3 = \frac{2(2\eta\beta - \gamma^2)(1-\theta) - \eta}{(1-\theta)^2} < 0$, i.e., $\beta > 1, \gamma^2 < \eta\beta, 0 < \theta < 1 - \frac{\eta}{2\eta\beta - \gamma^2}$, the Hessian Matrix is negative. At this time, the decentralized revenue sharing contract decision profit function Π_R^{r2*} is a strict concave function for p_r, p_s, τ, so there is an optimal solution.

Solve the first-order partial derivative of p_r, p_s, τ for L^{r1} and make it 0:

$$\frac{\partial L^{r2}}{\partial p_r} = -\frac{\delta(c_s - 2p_r - p_s + Q + \omega)}{\theta - 1} = 0, \quad (78)$$

$$\frac{\partial L^{r2}}{\partial p_s} = \beta\lambda + \delta\left[\beta c_s + \frac{\gamma(\theta-1)\tau + p_r - Q}{\theta-1} - 2\beta p_s\right] = 0, \tag{79}$$

$$\frac{\partial L^{r2}}{\partial \tau} = -\gamma\lambda + \gamma\delta(p_s - c_s) - \delta\eta\tau = 0. \tag{80}$$

The simultaneous equations solve for $p_r^{r2}(\omega), p_s^{r2}(\omega), \tau(\omega)$ for ω, and, by substituting Π_M^{r2}, we can get ω^*. Finally, substituting the results above into the model (59) will result in Π_M^{r2*}, Π_R^{r2*}, and Proposition 5 will be proved. □

Property 5: *The improved revenue sharing contract will effectively coordinate the remanufacturing closed-loop supply chain system with warranty services, which is embodied in the following: improving the total profit of the system under the coordination of the revenue sharing contract is greater than the total profit of the traditional revenue sharing contract supply chain system, and each member company of the supply chain has achieved a Pareto improvement, namely $Z_{\Pi}^1 > Z_{\Pi}^3 > Z_{\Pi}^2$. (Note: Let $Z_{\Pi}^3 = \Pi_M^{r2*} + \Pi_R^{r2*}$).*

Proof: When $\beta > 1, \gamma^2 < \eta\beta, 0 < \theta < 1 - \frac{\eta}{2\eta\beta - \gamma^2}$, the optimal solution is established, then:

$$Z_{\Pi}^3 - Z_{\Pi}^2 = \frac{(\delta-1)\eta\left[\delta\left(\eta(8\beta(\delta+1)(\theta-1)) + 4\gamma^2(\delta+1)(\theta-1)\right) + \eta\right]\left[(\gamma^2 - 2\beta\eta)(Q - c_r) + \beta\eta c_s\right]^2}{8(\delta+1)^2(\gamma^2 - 2\beta\eta)(\eta(-4\beta(\theta-1)) + 2\gamma^2(\theta-1))[\eta(-2\beta(\delta+1)(\theta-1)) + \gamma^2(\delta+1)(\theta-1)]^2} > 0, \tag{81}$$

$$\Pi_R^{d1*} - \Pi_R^{r2*} = 0. \tag{82}$$

It shows that the improved revenue sharing contract can realize the decision-making efficiency of the remanufacturing closed-loop supply chain system, and the two parties in the supply chain system receive the contract, improve the overall profit of the supply chain system and realize the Pareto improvement. At this point, how to determine the revenue sharing ratio of δ is particularly important, directly affecting the supply chain coordination effect. □

5.3. R-M Decision: Two Charging Contracts (T Model)

In the retail-led Stackelberg game model environment, in the market transaction process, the retailer will use the bargaining advantage to charge the manufacturer a certain channel fee of S. Therefore, this paper will use two charge contracts to coordinate the retailer-led closed-loop supply chain (Cachon, 2003 [33]). Its decision model is:

$$\max_{p_r, p_s, \tau} \Pi_R^{r3} = q_r(p_r - \omega) + q_s(p_s - c_s) - \frac{1}{2}\eta\tau^2 - S,$$

$$s.t. \ (p_r, p_s, \tau) \in argmax\Pi_M^{r3} = q_r(\omega - c_r) - T + S,$$

$$\Pi_M^{r3} > \Pi_M^{d3}, \tag{83}$$

$$\beta p_s \geq \gamma q_s,$$

$$p_r = \omega + Z.$$

In model (83), the first part of the objective function is the revenue obtained by the retailer for remanufacturing the product, the second part is the profit obtained by the retailer to sell the extended warranty service, and the third part is the channel fee paid by the retailer and the fixed service cost. The constraint is the manufacturer's participation constraint. When $\beta p_s > \gamma\tau$, i.e., $\beta > 1, \gamma^2 < \eta\beta, 0 < \theta < 1 - \frac{\eta}{2\eta\beta - \gamma^2}$, some consumers choose to purchase the remanufactured product's extended warranty service. At this time, there is an optimal solution for the objective function. Combine Equation (2) and Formula (3) to solve the model (83), and get Proposition 6.

Proposition 6: *The two charging contracts can coordinate the R-M decision model, and the parameters must satisfy* $\omega^{r3*} = p_r^{r3*} = p_r^{c*}, p_s^{r3*} = p_s^{c*}$,

$$S^* = -\frac{-(\gamma^2 - 2\beta\eta)(Q - c_r)^2 + 2\beta\eta c_s (c_r - Q) - 2\beta^2 \eta (\theta - 1) c_s^2}{2(\eta(4\beta(\theta - 1) + 1) - 2\gamma^2(\theta - 1))}. \tag{84}$$

Proof: For Π_M^{r3}, it is easy to know that $\Pi_M^{r3}(\omega, p_r)$ is a strict concave function for ω, p_r, and there is a unique optimal solution. If the two charging contract can coordinate the closed-loop supply chain, it must satisfy $p_r^{r3*} = p_r^{c*}, \omega^{r3*} = \omega^{c*}$, and substitute Π_R^{r3}; by deriving, we can see that $\Pi_R^{r3}(p_r, p_s, \tau)$ is a strict concave function on p_r, p_s, τ, and there is a unique optimal solution, and the value of S^* can be obtained with the two constraints and the retailer's target profit function. The verification is completed. □

6. Numerical Simulation

In order to reflect the conclusions obtained more clearly, the results of the models above are numerically simulated below. Referring to the parameter setting of Zhou et al. (2017) [34], let $Q = 1000, c_r = 100, c_s = 10, \eta = 1.3, \beta = 4, \gamma = 2, T = 50,000$. According to Zhu and Yu, (2018) [29], when the revenue sharing ratio $\delta \in [0.249, 0.512]$, the supply chain realized the income optimization. This paper assumes that $\delta = 0.5$, make model C, model D1 Model D2, S-1 model, S-2 model, and T model's remanufactured product's optimal retail price, optimal extended warranty service price, optimal warranty period, remanufacturing output, warranty service sales, and the profit of member companies of the supply chain and supply chain system's total profit comparison charts, respectively, and coordinate optimization of the model.

6.1. Remanufacturing Closed-Loop Supply Chain Warranty Decision Model

(1) Under the decision of each game model, consider the comparison of consumer preference θ for remanufactured product retail price p_r and output q_r.

From Figure 2, under each decision model, the retail price p_r decreases as the consumer preference θ increases, and the remanufactured output q_r increases with the increase of θ. It shows that, with the increase of consumers' preference, retailers can increase the sales volume of products by lowering the price; that is, the "small profits but quick turnover" model, and, at the same time, the number of potential consumers who can expand the subsequent warranty services.

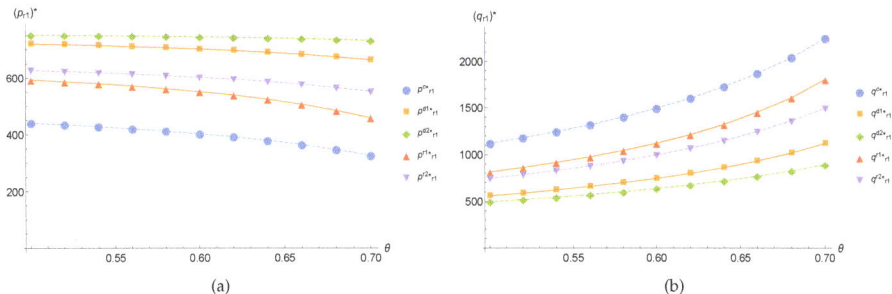

Figure 2. (a) the optimal retail price; (b) optimal yield.

(2) Under the decision of each game model, consider the comparison of consumer preference θ for remanufactured product extended service price p_s, warranty period τ and output q_s.

From Figure 3, under each decision model, the remanufactured product extension service price p_s, warranty period τ and production q_s all increase with the increase in consumer preference θ. It shows that, with the increase of consumer preference, retailers can increase the performance of extended service by attracting more consumers to choose the remanufactured product warranty service by increasing the service policy of the extended warranty period.

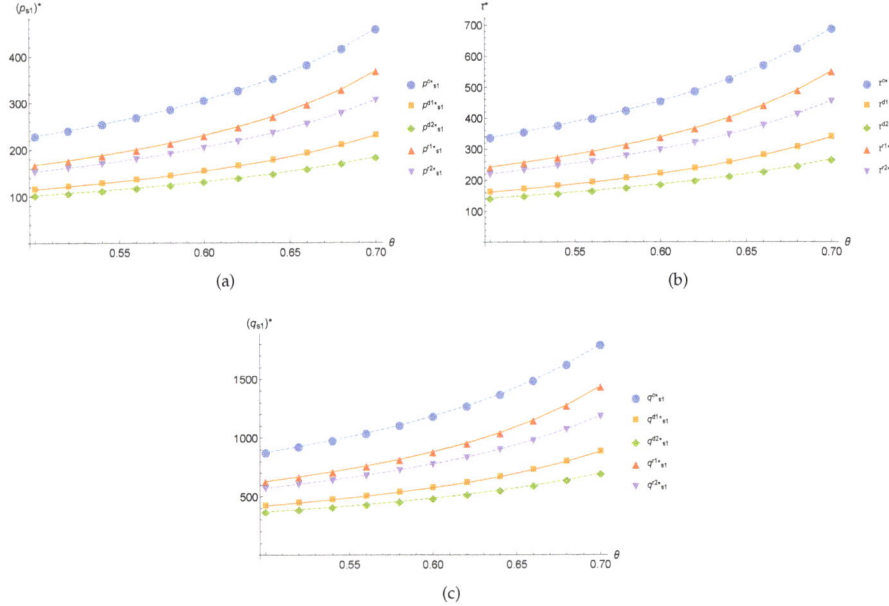

Figure 3. (a) optimal extended warranty price; (b) optimal warranty period; (c) optimal extended warranty sales.

(3) Under the decision of each game model, consider the consumer preference θ for the comparison of the income of each member company in the supply chain and the total profit of the supply chain system.

As shown in Table 2, firstly, as the consumer's preference θ increases, the profits of manufacturers, retailers, and supply chain systems increase. Secondly, the retailers under the R-M decision have the best profit, and, under the M-R decision, the manufacturers have the best profit, indicating that the leaders of the supply chain will gain greater profits with greater bargaining power. The system profit of M-R decision-making is greater than the system profit of R-M decision-making, indicating that the double marginal benefit of the manufacturer-led extended service supply chain is less than the retailer-oriented system. Moreover, under the R-M decision, the optimal retail price, optimal extended warranty price, optimal warranty period and production volume of remanufactured products are less than the optimal level under the M-R decision. Finally, as the degree of consumer preference increases, the gap in decision-making between models increases.

Table 2. The impact of the change in the parameter θ on the total profit of different decision models, manufacturers, retailers and closed-loop supply chains.

θ	Centralized Decision	Decentralized M-R Decision			Decentralized R-M Decision		
	Π_S^{c*}	Π_M^{d1*}	Π_R^{d1*}	$\Pi_M^{d1*}+\Pi_R^{d1*}$	Π_M^{d2*}	Π_R^{d2*}	$\Pi_M^{d2*}+\Pi_R^{d2*}$
0.3	332,462	166,150	83,237.3	249,387.3	82,566.5	153,316	235,882.5
0.4	399,130	199,484	99,904.3	299,388.3	98,889.1	181,202	280,091.1
0.5	499,263	249,550	124,938	374,488	123,181	221,503	344,684
0.6	666,501	333,169	166,747	499,916	163,062	284,884	447,946

6.2. Remanufacturing Closed-Loop Supply Chain Coordination Mechanism

Substituting the simulation data into the optimal profit results of the coordination mechanism model can reveal the optimal profit changes of manufacturers, retailers and supply chain systems under different game models. For the M-R decision model (see Table 3), the traditional revenue sharing contract cannot coordinate the supply chain system with warranty services, namely $\Pi_M^{r1*} + \Pi_R^{r1*} > \Pi_M^{d1*} + \Pi_R^{d1*}, \Pi_M^{r1*} > \Pi_M^{d1*}, \Pi_R^{r1*} < \Pi_R^{d1*}$, and through an improved revenue sharing contract the system can get Pareto improvement, i.e., $\Pi_M^{r3*} + \Pi_R^{r3*} > \Pi_M^{d1*} + \Pi_R^{d1*}, \Pi_M^{r3*} > \Pi_M^{d1*}, \Pi_R^{r3*} = \Pi_R^{d1*}$, effectively implementing supply chain coordination. For the R-M decision model (see Table 4) combined with decentralized M-R decision data, it is shown that $\Pi_M^{r3*} = \Pi_M^{d2*}, \Pi_R^{r3*} > \Pi_R^{d2*}$, indicating that the retailer is the market leader in the R-M decision model. The entire excess profit after coordination, and the manufacturer still retains the optimal profit before coordination.

Table 3. Profit comparison of the supply chain system of the M-R decision model under revenue sharing contract.

Decentralized M-R Decision			Revenue Sharing Contract			Revenue Sharing Contract (Improved)		
Π_M^{d1*}	Π_R^{d1*}	$\Pi_M^{d1*} + \Pi_R^{d1*}$	Π_M^{r1*}	Π_R^{r1*}	$\Pi_M^{r1*} + \Pi_R^{r1*}$	Π_M^{r3*}	Π_R^{r2*}	$\Pi_M^{r2*} + \Pi_R^{r2*}$
166,150	83,237.3	249,387.3	212,817	69,047.8	281,864.8	182,334	83,237.3	265,571.3
199,484	99,904.3	299,388.3	263,360	81,419.7	344,779.7	226,779	99,904.3	326,683.3
249,550	124,938	374,488	341,630	98,862.1	440,492.1	293,535	124,938	418,473
333,169	166,747	499,916	479,065	124,442	603,507	405,027	166,747	571,774

Table 4. Profit comparison of the R-M decision model supply chain system under two charge contracts.

Decentralized M-R Decision			Two Charge Contracts		
Π_M^{d2*}	Π_R^{d2*}	$\Pi_R^{d2*} + \Pi_M^{d2*}$	Π_M^{r3*}	Π_R^{r3*}	$\Pi_M^{r3*} + \Pi_R^{r3*}$
82,566.5	153,316	235,882.5	82,566.5	248,955	331,521.5
98,889.1	181,202	280,091.1	98,889.1	290,762	389,651.1
123,181	221,503	344,684	123,181	271,168	394,349
163,062	284,884	447,946	163,062	378,543	541,605

7. Conclusions

This paper considers the consumer's purchasing behavior, and analyzes the warranty decision model and coordination problem of remanufactured products under the three types of decisions: centralized, decentralized manufacturer-oriented and retailer-led. Based on the consumer behavior decision theory, the remanufactured product pricing, extended warranty service pricing, warranty period and system profit of each member of the supply chain are discussed. With strong bargaining power, the dominant players tend to earn more than their followers. For the dual marginal effects of decentralized manufacturer-led and retailer-led decision-making models, the first is to use commodity-based revenue sharing. The contract and the improved revenue sharing contract coordinate the M-R decision model, and then use the two toll system contracts to optimize the supply chain for the R-M decision model. Finally, the numerical simulation is used for further analysis.

The results show that: firstly, the increase of consumer preference can effectively increase the output of remanufactured products (i.e., demand), the extended service sales volume and the profit of each member of the supply chain. According to the different value ranges of consumer preferences, this paper derives five different optimal dynamic pricing and production strategies for manufacturers and retailers, which can provide practical guidance for corporate warranty decision-making. Secondly, the increase in consumer preference magnifies the dual margins of supply chain members. The effect of the process has aggravated the loss of decision-making efficiency. Finally, the improved revenue sharing contract and the two charging system contracts can effectively coordinate the closed-loop supply chain system with warranty services, so that the system can be improved by Pareto, which provides a decision-making reference for cooperation among member companies in the supply chain.

Several extensions to this article are possible. Firstly, this paper only considers the secondary supply chain consisting of a single manufacturer and a single retailer, and does not consider the supply chain warranty system under the market competition environment. Secondly, the three decision models only consider the supply chain. The system's warranty service is provided by the retailer, but, as the market capacity of the extended warranty service continues to expand, many large manufacturers and third-party guarantors (e.g., Safeware (Dublin, Ireland), etc.) are beginning to enter the warranty market, and the future can be studied from the perspective of the entire supply chain.

Finally, in addition to the design of the warranty entity combination model, in the actual operation, the guaranteed closed-loop supply chain system also has key warranty factors such as warranty quality, government regulation, sales efforts, etc., which will affect consumers' purchasing behavior decision-making. This will also become a further research direction of the article.

Author Contributions: Conceptualization, X.Z. and L.Y. Methodology, L.Y. Software, L.Y. Formal Analysis, L.Y. Resources, X.Z.; Validation, L.Y.; Visualization, C.L. and J.Z.; Project Administration, Y.Z.

Funding: This research was funded by the Priority Academic Program Development of Jiangsu Higher Education Institutions, Top-notch Academic, the Programs Project of Jiangsu Higher Education Institutions, Grant No. PPZY2015A072; the Practice Innovation Training Program of College Students in Jiangsu Province, Grant No. 201810300025Z.

Conflicts of Interest: The authors declare no conflict of interest.

References

1. Boulding, W.; Kirmani, A. A consumer-side experimental examination of signaling theory: do consumers perceive warranties as signals of quality? *J. Consum. Res.* **1993**, *20*, 111–123. [CrossRef]
2. Wei, J.; Zhao, J.; Li, Y. Price and warranty period decisions for complementary products with horizontal firms' cooperation/noncooperation strategies. *J. Clean. Prod.* **2015**, *105*, 86–102. [CrossRef]
3. Ahi, P.; Searcy, C. A comparative literature analysis of definitions for green and sustainable supply chain management. *J. Clean. Prod.* **2013**, *52*, 329–341. [CrossRef]
4. Shafiee, M.; Chukova, S. Maintenance models in warranty: A literature review. *Eur. J. Oper. Res.* **2013**, *229*, 561–572. [CrossRef]
5. Song, M.; Wang, S.; Sun, J. Environmental regulations, staff quality, green technology, R&D efficiency, and profit in manufacturing. *Technol. Forecast. Soc. Chang.* **2018**, *133*, 1–14.
6. Li, K.; Wang, L.; Chhajed, D.; Mallik, S. The Impact of Quality Perception and Consumer Valuation Change on Manufacturer's Optimal Warranty, Pricing, and Market Coverage Strategies. *Decis. Sci.* **2018**. [CrossRef]
7. Lan, Y.; Zhao, R.; Tang, W. A fuzzy supply chain contract problem with pricing and warranty. *J. Intell. Fuzzy Syst.* **2014**, *26*, 1527–1538.
8. Du, S.; Hu, L.; Song, M. Production optimization considering environmental performance and preference in the cap-and-trade system. *J. Clean. Prod.* **2016**, *112*, 1600–1607. [CrossRef]
9. Arabi, M.; Mansour, S.; Shokouhyar, S. Optimizing a warranty–based sustainable product service system using game theory. *Int. J. Sustain. Eng.* **2018**. [CrossRef]
10. Chen, C.K.; Lo, C.C.; Weng, T.C. Optimal production run length and warranty period for an imperfect production system under selling price dependent on warranty period. *Eur. J. Oper. Res.* **2017**, *259*, 401–412. [CrossRef]
11. Li, N.; Li, B. A Closed-loop Supply Chain Coordination Strategy Based on Product Warranty Period. *Syst. Eng.* **2016**, *34*, 90–96.
12. Mai, D.T.; Liu, T.; Morris, M.D.S.; Sun, S. Quality coordination with extended warranty for store-brand products. *Eur. J. Oper. Res.* **2017**, *256*, 524–532. [CrossRef]
13. Sabbaghi, M.; Behdad, S. Consumer decisions to repair mobile phones and manufacturer pricing policies: The concept of value leakage. *Resour. Conserv. Recycl.* **2018**, *133*, 101–111. [CrossRef]

14. Xu, L.; Shi, X.; Du, P.; Govindan, K.; Zhang, Z. Optimization on pricing and overconfidence problem in a duopolistic supply chain. *Comput. Oper. Res.* **2018**, *101*, 162–172. [CrossRef]
15. Savaskan, R.C.; Bhattacharya, S.; Van Wassenhove, L.N. Closed-loop supply chain models with product remanufacturing. *Manag. Sci.* **2004**, *50*, 239–252. [CrossRef]
16. Fleischmann, M.; Bloemhof-Ruwaard, J.M.; Dekker, R.; van der Laan, E.; van Nunen, J.A.E.E.; Van Wassenhove, L.N. Quantitative models for reverse logistics: A review. *Eur. J. Oper. Res.* **1997**, *103*, 1–17. [CrossRef]
17. Xu, Z.; Peng, Z.; Yang, L.; Chen, X. An Improved Shapley Value Method for a Green Supply Chain Income Distribution Mechanism. *Int. J. Environ. Res. Public Health* **2018**, *15*, 1976. [CrossRef] [PubMed]
18. Gong, Y.; Jiang, Y. Study on Pricing and Channel Selection of Closed-loop Supply Chain Hybrid Recovery Mode. *Soft Sci.* **2018**, *32*, 127–131+144.
19. Kaya, O.; Bagci, F.; Turkay, M. Planning of capacity, production and inventory decisions in a generic reverse supply chain under uncertain demand and returns. *Int. J. Prod. Res.* **2014**, *52*, 270–282. [CrossRef]
20. Luo, Z.; Chen, X.; Chen, J.; Wang, X. Optimal pricing policies for differentiated brands under different supply chain power structures. *Eur. J. Oper. Res.* **2017**, *259*, 437–451. [CrossRef]
21. Yoo, S.H.; Kim, B.C. Joint pricing of new and refurbished items: A comparison of closed-loop supply chain models. *Int. J. Prod. Econ.* **2016**, *182*, 132–143. [CrossRef]
22. Gan, S.S.; Pujawan, I.N.; Widodo, B. Pricing decision for new and remanufactured product in a closed-loop supply chain with separate sales-channel. *Int. J. Prod. Econ.* **2017**, *190*, 120–132. [CrossRef]
23. Chen, X.; Li, L.; Zhou, M. Manufacturer's pricing strategy for supply chain with warranty period-dependent demand. *Omega* **2012**, *40*, 807–816. [CrossRef]
24. Gaur, J.; Subramoniam, R.; Govindan, K.; Huisingh, D. Closed-loop supply chain management: From conceptual to an action oriented framework on core acquisition. *J. Clean. Prod.* **2017**, *167*, 1415–1424. [CrossRef]
25. Joshi, Y.; Rahman, Z. Factors affecting green purchase behaviour and future research directions. *Int. Strateg. Manag. Rev.* **2015**, *3*, 128–143. [CrossRef]
26. Gallego, G.; Wang, R.; Ward, J.; Hu, M.; Beltran, J.L. Flexible-duration extended warranties with dynamic reliability learning. *Prod. Oper. Manag.* **2014**, *23*, 645–659. [CrossRef]
27. Su, C.; Wang, X. Modeling flexible two-dimensional warranty contracts for used products considering reliability improvement actions. *Proc. Inst. Mech. Eng. Part O J. Risk Reliab.* **2016**, *230*, 237–247. [CrossRef]
28. Li, N.; Li, B.; Liu, Z. Research on the guaranty subject and warranty efficiency of remanufactured products. *Oper. Res. Manag.* **2016**, *25*, 249–257.
29. Zhu, X.; Yu, L. Differential Pricing Decision and Coordination of Green Electronic Products from the Perspective of Service Heterogeneity. *Appl. Sci.* **2018**, *8*, 1207. [CrossRef]
30. Zhou, L.; Naim, M.M.; Disney, S.M. The impact of product returns and remanufacturing uncertainties on the dynamic performance of a multi-echelon closed-loop supply chain. *Int. J. Prod. Econ.* **2017**, *183*, 487–502. [CrossRef]
31. Li, K.; Mallik, S.; Chhajed, D. Design of extended warranties in supply chains under additive demand. *Prod. Oper. Manag.* **2012**, *21*, 730–746. [CrossRef]
32. Cachon, G.P.; Lariviere, M.A. Supply chain coordination with revenue-sharing contracts: strengths and limitations. *Manag. Sci.* **2005**, *51*, 30–44. [CrossRef]
33. Cachon, G.P. Supply chain coordination with contracts. In *Handbooks in Operations Research and Management Science*; Elsevier: Amsterdam, The Netherlands, 2003; Volume 11, pp. 227–339.
34. Zhou, W.; Han, X.; Shen, Y. The Closed-loop Supply Chain Pricing and Service Level Decision and Coordination Considering Consumer Behavior. *Comput. Integr. Manuf. Syst.* **2017**, *23*, 2241–2250.

 © 2018 by the authors. Licensee MDPI, Basel, Switzerland. This article is an open access article distributed under the terms and conditions of the Creative Commons Attribution (CC BY) license (http://creativecommons.org/licenses/by/4.0/).

Article

Reverse Logistic Strategy for the Management of Tire Waste in Mexico and Russia: Review and Conceptual Model

Maria-Lizbeth Uriarte-Miranda [1,2,†], Santiago-Omar Caballero-Morales [1,*,†], Jose-Luis Martinez-Flores [1,†], Patricia Cano-Olivos [1,†] and Anastasia-Alexandrovna Akulova [2,†]

1. Postgraduate Department of Logistics and Supply Chain Management, Universidad Popular Autonoma del Estado de Puebla A.C.—UPAEP A.C., 72410 Puebla, Mexico; lizuriartem@gmail.com (M.-L.U.-M.); joseluis.martinez01@upaep.mx (J.-L.M.-F.); patricia.cano@upaep.mx (P.C.-O.)
2. Institute of New Materials and Technology, Ural Federal University—UrFU, 620002 Sverdlovsk, Russia; aa.akulova@urfu.ru
* Correspondence: santiagoomar.caballero@upaep.mx
† These authors contributed equally to this work.

Received: 14 August 2018; Accepted: 10 September 2018; Published: 25 September 2018

Abstract: Management of tire waste is an important aspect of sustainable development due to its environmental, economical and social impacts. Key aspects of Reverse Logistics (RL) and Green Logistics (GL), such as recycling, re-manufacturing and reusable packaging, can improve the management of tire waste and support sustainability. Although these processes have been performed with a high degree of efficiency in other countries such as Japan, Spain and Germany, the application in Mexico and Russia has faced setbacks due to the absence of guidelines regarding legislation, RL processes, and social responsibility. Within this context, the present work aims to develop an integrated RL model to improve on these processes by considering the RL models from Russia and Mexico. For this, a review focused on RL in Mexico, Russia, Japan and the European Union (EU) was performed. Hence, the integrated model considers regulations and policies performed in each country to assign responsibilities regarding RL processes for the management of tire waste. As discussed, the implementation of efficient RL processes for the management of tire waste depends of different social entities such as the user (customer), private and public companies, and manufacturing and state-of-the-art approaches to transform waste into different products (diversification) to consider the RL scheme as a total economic system.

Keywords: tire waste management; reverse logistics; green logistics; Mexico; Russia

1. Introduction

Logistics is defined as the element of the supply chain process that plans, implements and controls the efficient and effective flow and storage of goods, services and related information from the point of origin to the point of consumption to meet the needs of the client [1,2]. In recent years, Logistics has grown in complexity by opening its doors to a greater number of factors such as society, economics, social responsibility, sustainability and the environment. In this regard, concerns about environmental care have become an important factor, not only for the business sector, but also for society, government and social organizations.

This has paved the way for the concept of Green Logistics and Reverse Logistics:

- Reverse Logistics (RL) encompasses all the logistic activities from used products which are no longer required by the users to products again usable in a market. Within the environmental context, RL has been successfully applied for recovery, recycling and reuse of end-of-life electrical

and electronic equipment [3–5]. Figure 1 presents the basic activities or processes in an RL system [6].
- Green Logistics (GL) is focused on restricting damage to environment during the process of Logistics. It is based on the global environment maintenance and sustainable development [7–10].

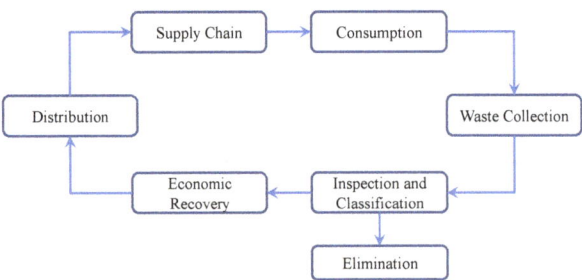

Figure 1. Basic activities in a Reverse Logistics (RL) system [6].

At present, the term GL is often used interchangeably with RL. However, in contrast to RL, GL resumes logistic activities that are primarily motivated by environmental considerations [11]. The most significant difference is that RL focuses on saving money and increasing value by reusing or reselling materials to recover lost profits and reduce operating costs; in turn, GL focuses on the transportation [8]. GL looks for alternatives so that the transport factor is favorable and the related costs are reduced, in addition to providing a green image for the company [12]. The activities considered are designed to measure environmental impacts on transportation, reduce energy consumption and reduce the use of materials. As presented in Figure 2, recycling, remanufacturing, and reusable packaging are the common processes that contribute to GL and RL. As discussed in [13], these processes have the most significant negative impacts on sustainability.

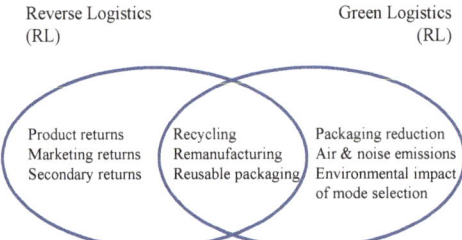

Figure 2. Comparison between Green Logistics (GL) and Reverse Logistics (RL) [8].

Hence, efforts to achieve sustainability within the supply chain should be focused on these processes. Particularly, tire waste management involves RL processes than can be improved to achieve sustainability.

The massive manufacture of tires and the difficulties to proper handling, once they have been used, constitute one of the most serious environmental problems of recent years at a worldwide level. This is because a tire needs large amounts of energy to be manufactured, and, in order to prevent it from being part of clandestine dumps, it requires a specialized recycling process after the end of its useful life [14].

Hence, management of tire waste is an important aspect of sustainable development due to the following environmental aspects [15,16]:

- as transport is one of the main logistic activities, there is an increasing demand for new tires and generation of scrap tires (end-of-life tires). Globally, an estimated one billion tires reach the end of their useful lives every year;
- scrap tires are usually shredded and disposed of in landfills, or stockpiled whole. Stockpiling leads to two significant hazards: it creates an ideal breeding ground for disease-carrier fauna, and fires;
- the void space of tires in landfills capture explosive methane gas which can represent a fire hazard, contaminate local water systems, and damage the landfill liners;
- chemical components such as stabilizers and flame retardants added to tires can kill advantageous bacteria in the soil.

Commonly, car owners take their worn tires to the tire shop, where they are replaced by new ones. Then, after the store has accumulated a certain number of tires, they are taken out of the store. Depending on the economic conditions, worn tires can be restored, transformed into energy, transformed into a new product, or buried [17]. Figure 3 presents the various stages of the life of a tire, from the acquisition of the raw material to its manufacture, use, and disposal [18].

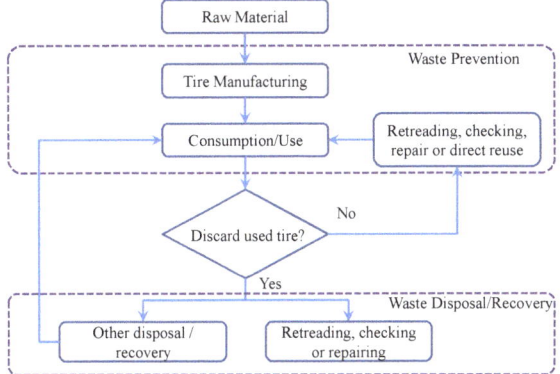

Figure 3. Stages of the life cycle of a tire [18].

However, RL processes in all industries are not widely performed. This is due to not enough knowledge about methodologies and tools to perform integrated Green-Reverse Logistics. In order to be sustainable, GL must consider environmental factors such as pollution, noise and climate change, which must keep a balance with economic and social factors. This can be achieved through RL as it influences the reduction of environmental impacts and the recovery of economic value. Particularly, within the manufacturing and transportation industries, RL and GL have become very important to reduce contaminants (chemical waste, CO_2 emissions, etc.).

In the specialized literature, different processes and strategies have been proposed to improve on sustainable practices to increase recycling rates of tire waste and reduce landfilling. Table 1 presents a review of some of the most significant proposals.

In general, most of the research has been performed on technical aspects of recycling processes and not on the management of RL processes. In this regard, recycling is one of the elements within a waste management system and an effective waste management system is crucial to address the problems caused by used tires.

Table 1. Works developed on sustainable processes and strategies for tire waste management (own work).

Work	Year	Journal	Type of Contribution	Focus
[19]	2018	Procedia CIRP (College International pour la Recherche en Productique)	Strategy	Determination of differences between tire waste recycling practices in South Africa and the European Union to improve on Reverse Logistics (RL) management.
[20]	2018	Resources, Conservation & Recycling	Process	Processing strategy of tire waste into activated carbon.
[21]	2018	Journal of Cleaner Production	Strategy	Study on the implementation of the extended producer responsibility scheme for tire waste in Colombia to improve on RL management.
[22,23]	2017–2018	Resources, Conservation & Recycling—Waste Management	Process	Study on the processing of recycled tire waste to be used as polymer modifier to improve strength of epoxy based composites.
[24]	2017	Coke and Chemistry	Process	Study on the use of tire waste in coke production.
[25]	2017	Journal of Cleaner Production	Process	Study on the application of recycled tire crumbs as insulator in lightweight cellular concrete.
[26]	2017	Journal of Cleaner Production	Process	Study on the processing of polymer-rubber composites from tire waste to obtain environmentally friendly materials.
[27]	2017	Environmental Research	Process	Processing strategy of tire waste into biofilm for wastewater treatment systems.
[28]	2017	Renewable and Sustainable Energy Reviews	Process	Processing strategy of tire waste into electric energy.
[29]	2017	Journal of Material Cycles and Waste Management	Strategy	Study on the management of tire waste recycling in Taiwan.
[30]	2015	Waste Management	Strategy	Study on the management of tire waste recycling in Italy and Romania.
[31]	2006	International Journal of Environmental Technology and Management	Strategy	Study on the management of RL processes for tire waste recycling in the United States of America.

As reported in [32], there are five key factors highlighted by the United Nations that must be present for the establishment of waste management systems:

- policies and regulations;
- supporting institutions;
- proper financial mechanisms;
- stake holder participation; and
- supporting technologies.

As reviewed in Table 1, there have been works on the analysis and development of strategies for tire waste management considering these factors in specific countries [19,21,29–31]. Thus, regional legislation plays an important role in the successful implementation of these strategies.

In this context, the present work is focused on developing a conceptual RL model for the tire waste management systems of Mexico and Russia due to an absence of works and consensus regarding RL strategies in these countries. While finding a better strategy cannot be assured due to different regional conditions of financial mechanisms, regulations, involving entities and technologies in each country, the contribution of the proposed RL model consists of the following:

- review of the factors (i.e., policies and regulations, supporting institutions, financial mechanisms) regarding the current tire waste management strategies in the EU, Japan, Mexico and Russia;
- analysis regarding the most important RL processes in these strategies associated with economic and sustainable benefits;
- development of a conceptual integrated RL model with these processes. The proposed RL model is focused on re-manufacturing or retreading and diversification to make it economically accessible and sustainable.

The advances of the present work are structured as follows: in Section 2, a review of the organization aspects of RL is presented. Then, in Section 3, the background of the RL processes and regulations in Russia, Mexico, the European Union (EU) and Japan are reviewed and discussed. The proposed conceptual RL model, which is focused on re-manufacturing and diversification, is presented in Section 4. State-of-the-art diversification techniques which can also be considered by general RL schemes are also presented. Finally, in Sections 5 and 6, the outcomes of the proposed model and conclusions are discussed.

2. Organizational and Managerial Aspects of Reverse Logistics in Companies

The RL process within companies requires the design, development and efficient control of a system to collect the out-of-use product and deliver it to the recovery entity that will apply the most appropriate management option for its optimal use. For this purpose, RL management must be supported by strategic, tactical and operational decisions (see Table 2). However, the application of these decisions and the overall development of RL will depend of the available reuse, re-manufacturing, or recycling systems for the out-of-use products. Thus, the development of RL can depend on [6]:

- The existence or absence of a traditional logistic system which can allow the reverse function.
- The type of product of interest in terms of its technology, ease of recovery, level of standardization, technical characteristics, etc.
- The recovery option (recycling, re-manufacturing or reuse) that will be applied to the returned product.
- The purpose of the RL system.
- The size of the company and its business objectives.
- The structure of the distribution channel.

Table 2. Strategic, tactical and operational decisions during the RL process [6].

Activities	Strategic Decisions	Tactical Decisions	Operational Decisions
Waste Collection	(a) location, quantity and capacity of collection facilities; (b) design of technologies for collection	(a) transportation of waste for collection centers; (b) management of collected waste inventories; (c) means of transportation	(a) collection routes; (b) collection lots; (c) load configuration
Inspection and Classification	(a) location, quantity and capacity of facilities for classification and inspection; (b) training of personnel	(a) inventory management of recoverable products; (b) task assignments; (c) sequencing of tasks: disassembly, cleaning, repairing	Option 3-R to be applied: reuse, re-manufacturing, recycling.
Economic Recovery	(a) technology; (b) effects on the long-term Production Plan	(a) effects on the aggregate Production Plan; (b) recovery lots; (c) management of inventories of recovered products	(a) effects on the Master Production Program; (b) Bill-of-Materials
Distribution	(a) distribution channels; (b) target markets	(a) assignment of products to markets; (b) means of transportation	(a) distribution routes; (b) distribution lots
Elimination	(a) removal systems; (b) target products to be eliminated	(a) management of inventories of non-recoverable products; (b) means of transportation	handling of waste

An important aspect of RL is the determination of its developing entity and scheme [6]. As presented in Table 3, RL can be developed on diverse schemes by the company itself or by a third-party company. Depending of the developing entity, the management of RL can be performed as follows:

- Company. In this case, the company itself designs, manages and controls the recovery and reuse of its out-of-use products. Companies that develop their own RL systems are often characterized by being leaders in their respective markets in which the identification between company and product is very high. They are generally manufacturers of complex and technologically advanced products, designed to recover some of the added value that they incorporated (e.g., Design for the

Environment—DFE, Design for Dismantling—DFDA). Although the ultimate responsibility for the system is the company itself, it is usual for some activities to be carried out by third-parties outside the company (e.g., the collection of products and their transport to the recovery center). The productive process considered to recover the added value of the out-of-use product is usually a complex process, with multiple tasks, in which there is an intensive use of labor. The logistic network that is developed to recover these products is characterized by being a complex, multi-link network, generally decentralized in which the recovered product is reintroduced into the original closed-loop Supply Chain (SC).

- Third-Party. In this case, the company responsible for the introduction of the product in the market does not directly manage the recovery process and this function is performed, for the most part, by third-parties outside the company. In this way the company can either choose to participate in an *Integrated Management System* or hire the services of *Logistics Professionals* specialized in the realization of RL services (e.g., organizations that promote and manage the recovery of out-of-use products for its subsequent treatment or proper disposal, generally these are constituted by suppliers, manufacturers, and distributors, who are the financiers of the scheme).

Table 3. Management schemes of RL systems according to the developing entity [6].

Scheme	Company	Third-Party: Integrated Management System	Third-Party: Logistics Professionals
Business	(a) Market leader; (b) Environmental strategy; (c) Dominant SC position	(a) Small and medium-sized enterprises; (b) Collaboration with other members of the SC	(a) Subcontracted direct logistics flow; (b) Development of the scheme for operational reasons: returns, toxic or hazardous waste
Product	(a) Very differentiated; (b) High added value; (c) Advanced technology; (d) Complex structure	(a) Little differentiated; (b) Low added value and residual; (c) Low technology; (d) Design for Recycling (DFR)	(a) Diversity of products; (b) Obsolete, defective, damaged, toxic or dangerous
Process	(a) Multiple tasks; (b) Intensive labor; (c) Very relevant transport	(a) Complex process; (b) Advanced technology; (c) High initial investment	(a) Simple process; (b) Few tasks; (c) Intensive labor
Market for Recovered Products	Same market as the originals	Different market than originals	(a) Share market in reuse; (b) Distinct market in returns
Network Design	(a) Integration of direct and reverse flows; (b) Decentralized and complex; (c) Closed-Loop; (d) Subcontracted activities	(a) Open Loop; (b) Centralized; (c) Simple with few levels; (d) Significant transport	(a) Open-loop on returns and closed-loop on reuse; (b) Simple and decentralized; (c) Significant transport
Reverse Scheme Goal	Recover elements of high added value	Regulatory compliance on waste	Regulatory compliance on waste and guarantees of consumption.
Management Options	Manufacturing	Recycling	Reuse and Returns
Examples	Xerox, IBM, Hewlett-Packard	Eco-glass, Eco-batteries, Eco-tires	Genco, UPS, GATX Logistics

The schemes presented in Table 3 can be considered starting points to design the appropriate scheme for each company. In order to accomplish this, the organizational schemes in RL require the vital participation of the government, the company, and the society (teamwork). However, in practice, there are deficiencies in the design of these schemes due to the following situations [33]:

- RL is not recognized as a factor that can generate a competitive advantage;
- belief that once the products are delivered, the responsibility of the company ends (to solve it, it is necessary to take into account a life cycle approach linked to the final distribution);
- failure to bridge the internal, external and associated processes in the E-Commerce and the return aspect of products in the SC (associated with process mapping, to understand its scope and complexity);
- assumption that part-time efforts are sufficient to deal with RL activities (RL is not recognized as a complex action that must have its own resources);

- belief that order-time cycles for returned products may be larger and more variable than those associated with the sale or distribution of new products;
- assumption that returned products and recycling and re-use of packaging will take care of themselves if enough time is given;
- difficult separation of products because commonly these arrive mixed at the distribution centers;
- belief that returned products are relatively unimportant in terms of costs, asset valuation, and potential revenues (returns tend to stay longer than new products on direct channels, resulting in high inventory, transportation, and storage costs, and at the same time, revenues decrease due to costs associated with obsolescence and degradation).

3. Reverse Logistics and Regulations in Russia, Mexico, the EU and Japan

3.1. Russian Context

Currently, Russia is one of the countries which can benefit from improvements in their logistic processes to increase its competitiveness within the global context. This is due to difficulties to solve main logistical problems such as infrastructure and environmental constraints in the collection of reusable waste [34]. In addition, there are few scientific publications regarding the return-flows in the Russian literature. Hence, it remains a little studied subject [35].

One of the main difficulties of developing RL in the SC in Russian companies is that the theoretical foundations are not widely studied. Nevertheless, many national organizations are eager to learn from the experience of foreign companies in improving logistic processes to reduce the costs of managing flows in SCs and to seek ways to optimize the movement of goods and materials within logistic systems at different levels [36].

However, this desire is not shared by many Russian companies as business implementations regarding RL are very slow. This is due to the following reasons [37]:

- underestimation of the importance of logistic processes in business activities;
- lack of a unified approach to understand RL in Russian literature (which causes misunderstandings and discussions, as well as making the study and implementation of the best practices of RL in organizations more difficult);
- insufficient explanation in the scientific literature of practical recommendations for the implementation of RL in the activities of Russian companies and evaluation of their effectiveness.

Adding to these reasons, there is a belief that GL and RL lead companies to higher logistical costs [38]. However, within the total logistic costs, the cost of RL is 4.0–6.0% where most of the product returns were initiated by consumers. By analyzing data from diverse global sources, the average percentage of product returns was estimated at 7.0% [39].

In recent years, the rapid development of the logistic market in Russia has given national companies the opportunity to create and design SCs that take into account reverse flows, which is generally an important step towards normalizing the sustainability of SCs [36]. This is an important advance on previous strategies where the recovery of secondary resources was practically deregulated by the government and environmental protection was focused on corrective measures to repair environmental impacts and not on preventive measures to minimize them [35]. Now, the environmental legislation in Russia does not only establish sanctions to the violation of waste collection and use of natural resources laws but establishes a production tax promotion, or subsidy, for companies using innovative technologies and environmentally friendly use of packaging materials in production, especially when recycling is impossible or difficult [38].

Another issue to consider is that most Russian companies do not create special storage facilities for product returns. Thus, it is advisable to allocate space for the implementation of RL processes and establish effective flow records of product returns. Special reception is most convenient for large and medium-sized companies creating their own base point for the collection and separation of returned products. This reception can take place through the creation of a base center, with the participation of several associated companies working in segments of related markets [35]. Another option is to transfer these logistic activities to outsourced companies, which is widely used by Russian companies since handling of returned products is not one of their core competencies.

Regulations for Tire Waste

In Russia, each of its regions has established its own municipal operator to manage the collection of waste, sorting the tires and sending them for recycling, as well as a regional model of waste management. However, not all regions have succeeded as regional officials have failed to make this system economically feasible. In addition, the government was unwilling to create additional financial sanctions in the form of recycling rates for its citizens. In the case of used tires, it is expected that, for any car owner or transportation company, it is best to throw the used tires away, instead of paying money for their proper recycling. As a result, state money remains a crucial part of waste management programs at the regional level [40]. Table 4 presents the main regulations established in Russia regarding the issue of environmental care and waste disposal.

Table 4. Regulations and normativities in Russia (own work).

Normativity	Characteristics
Basel Convention	Regulates the movement of hazardous wastes and recyclables and promotes their environmental management.
Federal Law 89-FZ	Regulates the production and consumption of waste and waste disposal and imposes obligations to recycle or pay an environmental duty.
Federal Law 458-FZ	States that manufacturers and importers of goods must provide the recycling, salvage, reclamation, and disposal of waste generated from the use of such goods including packaging which are no longer of value to consumers.
Federal Law on Production and Consumption of Waste	States that each region has to establish the means for waste processing such as sorting, recycling, sending to a landfill, or incinerating.
Decree 1886-r	States a list of finished goods, including packaging, that must be recycled at the end of their usable life.
GOST-8407-89	This standard applies to worn out tires and chambers that are unsuitable for further use and recovery, as well as tires and chambers rejected by inspection results (hereinafter referred to as secondary rubber raw materials).

In general, the subject of tire collection and re-manufacturing in Russia is not a priority. Just in 2010, over 35.0% of all types of tires were produced and sold compared to the same period of the previous year. Although sales were growing in the budget and the middle segment, in the premium segment, sales declined. This is because, if the car market and the tire market are compared, tire sales depend more on the existing fleet than on new car sales [41].

Experts give the following statistics: more than 700 thousand tons of waste are handled in Russia, and about 70 thousand tons are of old tires that are stored every year. The elimination of these takes place in 2.0–15.0% of the cases while the rest falls on the side of the road, landfills and general waste (which is prohibited by law). According to the accepted classification of rubber products, these products are classified as Risk Class IV of waste. Just in 2015, in Moscow the number of vehicles were expected to exceed 5 million cars, which represents a very pressing problem [42].

Until the 2000s, the recycling of tires and other rubber waste in Russia was a very underdeveloped area. However, in the last decade, the Russian market started to import equipment for the processing of tires from abroad [43] potentially developing the market for the processing of these wastes and making it economically more profitable. Figure 4 shows the RL processes for used tires in Russia [17].

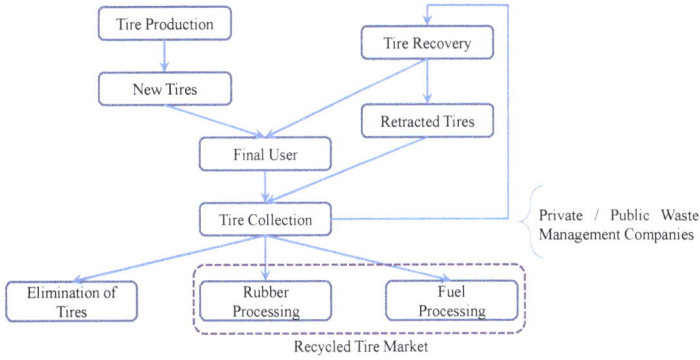

Figure 4. RL processes for used tires in Russia [17].

The current operating capacity of tire processing companies is around 150 thousand tons per year, but, according to the association *Shinecology*, its actual workload rarely exceeds 10.0–30.0%. Additionally, it is estimated (to date) that the amount generated of used tires is about 850 thousand tons per year. The estimated volume of the mechanical treatment of tires in Russia does not exceed 17.0% of the total annual waste of tires. Another 20.0% of used tires are burned. This represents the remaining volume for burial. In this case, for the year 2015, the volume produced annually by the waste of Russian tires reached approximately 935,000 tons per year [44]. The estimation of the total volume of tires to be recycled in Russia in 2010–2015 is presented in Figure 5.

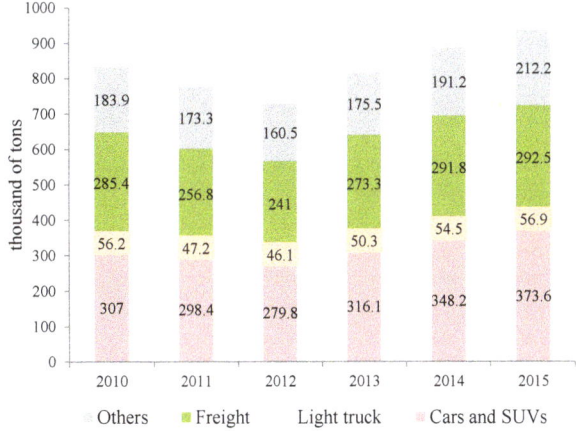

Figure 5. Tires recycled in Russia 2010–2015 (thousands of tons) [44].

The disposal of used tires is carried out only for 2.0–15.0% of this waste and the rest falls on the side of the road, landfills, or general waste (which is prohibited by law). Because in most Russian cities there are no places that are reserved for the permanent or temporary placement of this type of waste, the used tires are often simply thrown on the road on the outskirts and adjacent to the road area [45]. One of the main problems in large cities is that the reception centers for waste tires and the infrastructure intended for it are not commonly seen. As a result, in 2013, only 70 thousand tons of products derived from waste tires (of the more than 700 thousand tons generated) were recycled.

Although there is a public company that is dedicated to the collection of waste throughout the country, according to the tire specialist *Cordiant*, approximately 60.0% of tire recycling in Russia corresponds to just four cities: Volgograd, Moscow, Smolensk and Vladimir. Particularly, in cities with a small population, the collection system of used tires is absent, which means that the ecological situation is unfavorable for these regions.

On average, used tires account for around 80.0% of the previous year's consumption volume [45]. One of the main reasons for this situation is the absence of a developed market for the sale of recycled tires. In addition, Russia does not have an efficient system for collecting used tires for processing. Companies processing raw materials often deal with raw materials which do not guarantee the utilization of capacity [45].

3.2. Mexican Context

Globalization has ensured that the concept of logistic efficiency is an important factor in the competitiveness of countries worldwide—mainly because it gives the direction to follow, assess, prioritize and control the different elements of supply and distribution activities that affect customer satisfaction, both in costs and benefits [46].

In developing countries such as Mexico, with an incipient and unconsolidated recycling industry, it is necessary to improve its structure and activities in order to face the challenges and opportunities arising from the growing concern about environmental problems and the management of products at the end of their useful life [47]. By 2009, there were few companies in Mexico that had capitalized on RL as an area of opportunity to reduce operating costs, increase profits and increase their customers to be more competitive [48].

At present, there has been an increase in the concern of organizations to take full advantage of RL, and thus to be able to minimize costs. However, for small and medium-sized enterprises, this has not been of significant importance when considering the removal of waste, recyclables, perishables or materials. On the other hand, most of the large companies have directed their efforts in the reuse of some materials, designing products bearing in mind the future of recycling, as well as the possibility to cover other markets to classify the returns (e.g., those that can be discarded, reused and repaired, as well as sold throughout other channels or returned directly to the market) [48].

Regulations for Tire Waste

In Mexico, as in many other countries, regulation is the main instrument with which the government seeks to control and influence waste management practices [49]. At the level of federal entities, the normative framework has a moderate delay in relation to the national one. Approximately 40.0% of the Mexican states have defined standards in relation to the general law and the corresponding regulations, while 63.0% have already developed the state programs for the prevention and integral management of wastes [50]. Table 5 presents the main regulations adopted by Mexico regarding the issue of environmental care and waste.

As of 2011, in Mexico, approximately 40 million of used tires were disposed annually and only 12.0% was recycled, which represented about 5 million tires [51]. Mostly, recycling has been linked to the rubber industry [52] and the amount of used tires is additionally incremented by the uncontrolled entry of second-hand vehicles from neighbor countries.

Derived from this, management plans were established for this type of waste. However, as presented in Figure 6, since 2010, the implementation of this plan presented a decrease of 99.89% in the collection of used tires for recycling [14]. This was caused by the following aspects:

- there are no policies to make the collection and transportation of tires economically feasible;
- there are no policies to motivate or regulate the participation of companies specialized into recovery (collection, cleaning and re-conditioning) of this type of waste;
- the management of tire collection has been driven by market conditions; unfortunately, this collection chain is disaggregated from the parties such as the automotive industry and the

manufacturing industry of tires. The main reason for this is the lack of already consolidated networks that add value to the end of the tires' useful life [50].

Based on information provided by the Secretariat of the Environment (Secretaría del Medio Ambiente, SEDEMA) [14] and the Secretariat of the Environment and Natural Resources (Secretaría del Medio Ambiente y Recursos Naturales, SEMARNAT) [50] from the federal government, Figure 7 shows the RL processes for used tires in Mexico.

Table 5. Regulations and normativities in Mexico (own work).

Normativity	Characteristics
Secretariat of the Environment and Natural Resources (SEMARNAT)	Attention to natural resources, biodiversity, hazardous waste, and industrial urban environmental problems.
General Law on Ecological Equilibrium and Environmental Protection (LGEEPA, 1988)	Environmental Impact Assessment, Hazardous Waste, Prevention and Control of Atmospheric Pollution.
Political Constitution of the Mexican United States (24 February 2017)	Article 115° III (Responsibility to collect waste), Transients 17° (Protection and care of the environment) and 19° (Regulation and supervision of integral waste control)
General Law for the Prevention and Integral Management of Waste	Protection of the environment in the field of prevention and integrated management of waste in the national territory with the aim of guaranteeing a healthy environment and promoting sustainable development. This through the prevention and integral management of hazardous waste, urban solid residues and special handling.
NOM-161-SEMARNAT-2011	Establishes the criteria for classifying waste and determining which are subject to management plans, the procedure for their inclusion or exclusion, and the elements and procedures for the formulation of management plans.
NOM-027-SCT2/2009	Land transport of hazardous materials and waste, special and additional specifications for containers, packaging, intermediate bulk containers, portable tanks and transport of hazardous substances and waste materials.
NOM-052-SEMARNAT-2005	Establishes the characteristics, procedure for identification, classification and lists of hazardous waste (Diario Oficial de la Federación (Official Journal of the Federation), 23 June 2006).
ISO-14001	Environmental Management
ISO-26000	Social Responsibility
Basel Convention	Regulates the movement of hazardous wastes and recyclables and promotes their environmental management.
Rotterdam Convention	Promotes shared responsibility in international trade for certain hazardous chemicals and promote their efficient management in order to protect human health and the environment.
Stockholm Convention	Protects human health and the environment from the adverse effects of Persistent Organic Pollutants.

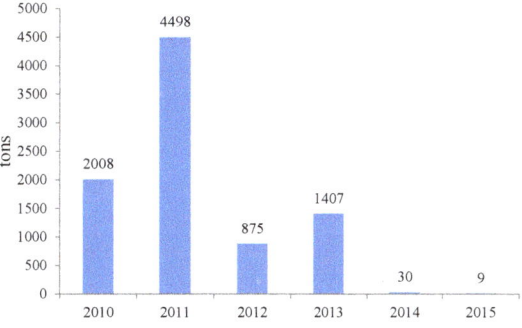

Figure 6. Tires recycled in Mexico 2010–2015 [14].

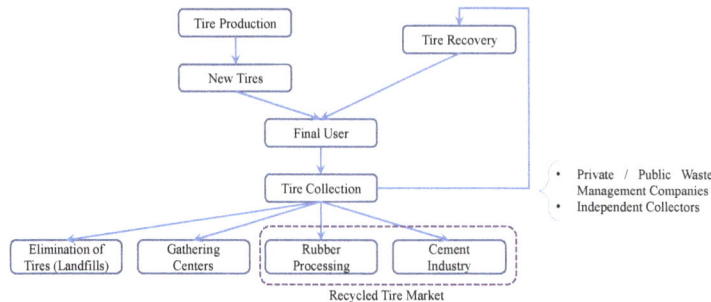

Figure 7. RL processes for used tires in Mexico (own work).

3.3. Context of Japan and the European Union

The United States of America and Japan are some of the first recyclers in the world. In the case of Japan, tire recycling rates were the highest of any country in 2003, averaging 85.0–90.0% with about 855,000 tons of discarded tires [53]. Just in 2017, the total recycled volume increased to 965,000 tons with a recycling rate of 93.0% [54]. Table 6 presents the main regulations adopted by Japan regarding the issue of tire waste.

Table 6. Regulations and normativities in Japan (own work).

Normativity	Characteristics
Basel Convention	Regulates the movement of hazardous wastes and recyclables and promotes their environmental management.
The Japanese End-of-Life Vehicle Recycling Law	Establishment of the characteristics to be met for the disposal of end-of-life vehicles and their components.

In general, scrap tire recycling is addressed as part of solid waste recycling. In addition, cooperative programs that gather tire manufacturers, government agencies and other industrial players, are integrated for scrap tire recycling.

Just as Mexico and Russia, Japan has its own RL processes. As presented in Figure 8, tire waste is obtained from two sources: (a) from users that discard their old tires to replace them by new tires, and (b) from end-of-life vehicles that are disposed completely. Under this scheme, users often have old tires near dealers including tire and car dealers, service stations and auto repair shops. These entities receive the old tires and deliver them to distributors which in turn deliver them to recycling or processing plants. Companies that own service cars, trucks and buses often dispose their scrap tires directly to processing plants as volume is higher. On the other hand, tires from end-of-life vehicles are discarded directly by scrap companies that deliver them to processing plants [53].

In recent years, regulation in the European Union for the tire and rubber industry has changed significantly due to the introduction of stricter requirements for safety, health and environmental purposes, and improved transparency of information to consumers. The complexity of EU legislation is increasing and focuses especially on the area of environment, features, and components that were not so legislated in the past.

The EU is the second largest generator of scrap tires after the United States of America. The EU landfill and end-of-life vehicles (ELV) directives have obliged the respective national governments to address the recycling of scrap tires. Table 7 shows the most important regulations adopted by the EU on the subject of environmental care and waste [19].

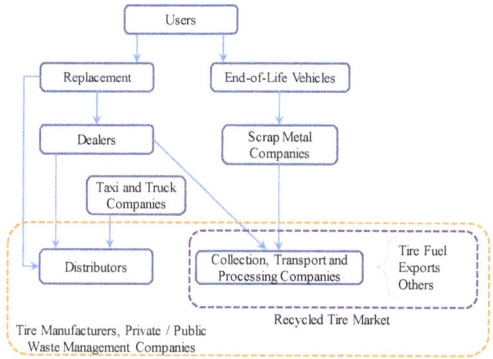

Figure 8. RL processes for used tires in Japan (adapted from [53]).

Table 7. Regulations and normativities in the European Union (own work).

Normativity	Characteristics
Basel Convention	Regulates the movement of hazardous wastes and recyclables and promotes their environmental management.
UNEP/CHW.10/6/Add.1	Technical guidelines for the environmental management of used tires and waste tires.
Royal Decree 1619/2005	Defines the importance of improving manufacturing techniques for tires so that they take longer to wear out, last longer, etc.
Royal Decree 1619/2005 on the Management of Out-of-Service Tires	Protects human health and the environment from the adverse effects of Persistent Organic Pollutants.
Directive 2008/98/EC	This directive establishes measures to protect the environment and human health by preventing or reducing the adverse impacts of waste generation and management, reducing the overall impacts of resource use and the effectiveness of such use.
Directive 2008/98/EC on Waste	This directive constitutes the current regulatory framework for the production and management of waste in the European Union.
Directive 2000/53/EC	This directive regulates the take-back responsibility for car-manufacturers because. Without take-back obligations for end-of-life products, there is almost no necessity for the manufacturer to coordinate the process chain beyond the point of sale.
Directive 2000/53/EC on End-of-Life Vehicles	This directive states that end-of-life vehicles have to be recovered and their tires have to be removed before they are scrapped.
Directive 1999/31/EC on the Landfill of Waste	This directive bans the disposal and stockpiling of tires on landfills.
Directive 2000/76/EC on Incineration of Waste	This directive prohibits combustion of end-of-life tires in old cement kilns.

Although the EU has left the method of implementing the directives at the discretion of the member countries, and with the deadlines for the implementation of the rapidly approaching directives, recycling rates are expected to increase in establishing the viable industry [53]. Within the EU, countries such as Spain, Sweden and France have achieved 100.0% recycling rates since 2011 [19].

In the EU, the tire industry operates with two schemes: (1) the *Industrial Liability Scheme*, where producers work together with other industry stakeholders to assume responsibility for the recycling and management of waste tires; and (2) the *Liability Scheme of the Producer*, where producers are responsible for the recycling of used tires. In both cases, consumers are responsible for the collection and recycling costs [53].

All member countries of the EU have to be in compliance with EU legislation in transposing directives into local legislation. They are free to set up national initiatives to achieve EU objectives. Tire manufacturers also face increasing environmental pressure from the general public and other stakeholders on dumping stocks [55].

For these reasons, it is in the interest of the tire industry in the EU to remain proactive and assume collective responsibility for the management of out-of-use tires. To date, there are three different schemes for the final management of these wastes. These are described in Table 8.

Table 8. Systems for the final management of out-of-use tires.

Extended Producer Responsibility	Liberal System (Free Market)	Government Responsibility
Systems with take-back obligations	The legislation sets the objectives to be met but does not designate those responsible.	Government responsibility financed through a tax.
The original manufacturer has a duty of care to ensure that the waste from the products it has created is disposed of responsibly, in an environmentally sound manner.	All the operators in the recovery chain contract under free market conditions and act in compliance with the legislation.	Each country is responsible for the management of tire waste.
	This may be backed up by voluntary cooperation between companies to promote best practice.	It is financed by a tax levied on tire producers and subsequently passed on to the consumer.

In this regard, the countries that have achieved a 100.0% recycling rate (i.e., Spain, Sweden and France) have operated under the Extended Producer Responsibility system [19]. This system uses economic incentives to encourage manufacturers to design environmentally friendly products by making them responsible for the costs associated with their disposal at the end of their useful life. Under this system, recycling costs are integrated within the product price. Figure 9 presents the general RL processes for used tires in the EU [53].

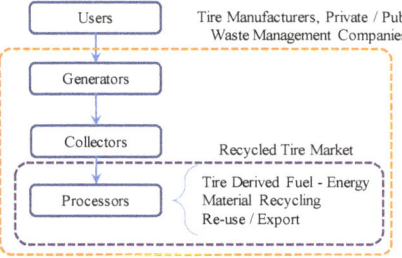

Figure 9. RL processes for used tires in the EU (adapted from [53]).

3.4. Entities Involved within RL Processes for Used Tires

As discussed in Section 2, the main RL processes consist of recycling, re-manufacturing, and reuse. Within the context of tire waste, the recovery systems associated with these processes consist of the following networks [6,56]:

- Networks for Recycling of Tires. They are usually simple structures, with few centralized links that are characterized by requiring, for an efficient management, a high volume of inputs (tires of waste) generally of little unit value. The high transformation costs determine the need for high utilization rates of these networks and the search for economies of scale.
- Networks for the Re-manufacturing of Tires. Its main objective is the recovery of parts and components of tires with high added value. In these systems, the original manufacturers usually play a very important role, sometimes being solely responsible for the design and management of RL systems. The design of the network responds to a multi-level, decentralized typology, for which synergies are usually sought with the direct channel.
- Reusable Tire Networks. In these systems, the recovered tires are reintroduced in the SC once the necessary cleaning and maintenance operations have been carried out. They tend to be decentralized structures where both original and reused tires circulate simultaneously and in which the cost of transport appears as the most significant.

Based on these networks, Tables 9–11 present the main entities which are considered to be involved in the main processes of collection, recycling and reuse of used tires [52].

Table 9. Entities involved in the collection of tires.

Entity	Objectives	Activities
Federal Government	Development of the collection market	(a) Facilitate and make attractive the collection and gathering business; (b) Develop economic stimuli and incentives; (c) Establish a list of recyclers and collectors.
	Establishment of collection processes	(a) Define the rules of the States and Municipalities concerning the collection by their public services; (b) Supervise compliance with the provisions.
	Establishment of facilities for gathering	(a) Define the rules of the State and Municipalities concerning the collection centers; (b) Monitor compliance with the provisions.
State and Municipal Government	Installation of a collection system	(a) Define the rules of the State and Municipalities concerning collection centers; (b) Monitor compliance with provisions
	Installation of storage centers	(a) Adapt the area and request permits; (b) Define rules and charges; (c) Allocate building resources; (d) Develop economic incentives for private initiative.
Government (user)	Adequate provision of tires	Direct used tires to pickers and collection centers.
Large Users	Adequate provision of tires	Direct used tires to pickers and collection centers.
Manufacturers, Importers and Traders	Dissemination of the collection centers for this purpose.	Reinforce with distributors the collection of used tires to authorized sites.

Table 10. Entities involved in the recycling of tires.

Entity	Objectives	Activities
Government	Development of the recycling market	(a) Facilitate and make attractive the business of recycling; (b) Develop economic stimuli and incentives; (c) Promote and fund research programs and grant support to educational institutions and students; (d) Regulate other recycling schemes; (e) Make associations of recyclers and promote conventions; (f) Government procurement of products under recycling specifications; (g) Give preference to new developments for the use of recycled used tires and recycle content of by-products of the used tires
Manufacturers, Importers	Cooperation agreements	(a) Establish agreements between Technical Committees to reach consensus with the Chambers and Associations of the branch; (b) Target the collection of used waste tires to meet the needs of recyclers and grant technical support; (c) Grant support to educational institutions and students; (d) Conduct research programs for the recycling of waste tire components
Recyclers	Involvement	Participate in activities promoted by government and private enterprises

Table 11. Entities involved in the reuse of tires.

Entity	Objectives	Activities
Government	Development of the reuse market	(a) Facilitate and make attractive the business of reuse; (b) Develop economic stimuli and incentives; (c) Make a list of potential companies for reuse of used waste tire; (d) Promote exhibition meetings or conventions; (e) Government procurement of products under reuse specifications; (f) Give preference to new developments for used products
(a) Potential companies for reuse; (b) Same manufacturing company	Involvement	Participate in activities promoted by the government and private enterprises

4. Proposed RL Strategy for Tire Waste

As previously discussed, tires at the end of their useful life are one of the biggest concerns with regard to the environment due to the pollution it causes. In addition, they have a practically unlimited degradation time due to the reticulated structure of rubbers and the presence of stabilizers [57].

Through the tire recycling process, the main materials that make up the tire can be used, such as: 70.0% rubber powder, 25.0% steel, and 5.0% textile fibers. According to the importance of this waste in terms of its composition material and its role in society, its recycling becomes vital for the care of the environment. In order to do this, and due to the impact that it generates all over the world, to a lesser or greater degree, mainly in countries like Russia and Mexico, it is indispensable to visualize it not only as a matter of recovery of recyclable material but as a total economic system.

The implementation of RL involves environmental and economic benefits in the different areas of implementation. In particular, the economic benefits are dependent on the degree of implementation in the process of transformation and distribution of companies. As for the recycling of products that have some degree of contamination such as tires, the economic benefits are obtained as follows:

- If efficient processes of re-use or re-manufacturing of goods (whether fast and inexpensive) are established, the cost of raw material and production of new goods can be reduced (instead of re-making a tire, re-conditioning of a used tire can be cheaper).
- The collection of used tires that are candidates for re-conditioning must be efficient (e.g., it must be of minimum time). Therefore, the optimization of collection routes can provide faster and cheaper re-utilization processes that can substitute a percentage of the purchase cost of raw material or manufacturing (from scratch) of a new product.
- Improvement in the environmental aspect is a consequence of these measures, and better government support (investment funds) can be accessed if these practices are encouraged.
- If a fixed area of re-conditioning of tires is established, it will be possible to have an area that receives raw material (in this case, used tires) that can give sustenance to inventory in case of breakdown. This also involves savings and service level improvement.
- On the other hand, tires that have reached the end of their useful life (it is not possible to re-condition them) support other industries (generate income).

For this reason, two strategies focused on re-manufacturing (or re-treading) and diversification are considered.

4.1. Re-Manufacturing or Retreading Strategy

The tire is mostly discarded due to wear of the tread, which has between 10.0% and 20.0% of all the material and the energy contained in it [58]. The process of retreading involves placing a layer of non-vulcanized rubber on a worn tire that will cover the tire's tread and shoulders. The process is then completed by placing the tire inside a mold so that, by means of pressure and temperature, the new drawing is inserted therein.

The re-manufacturing of this waste is a strategy that takes care of the environment, as it recovers the value of the components of the tire waste, which would otherwise end up in landfills. There is no doubt that the tire retreading process leads to significant savings in energy demand by 66.0% in production capacity and materials in the manufacturing process due to the minimization of raw material requirements [59].

The costs of retreaded tires are 30.0–50.0% less than the cost of a new tire. This makes it attractive to consumers, such as truck fleet operators who travel long distances and demand higher tire replacement rates. This sector represents the biggest retreading industry due to the following aspects:

- Maintenance and replacement of tires is the third highest cost for fleet operators, after labor and fuel.
- The renewal rate for tire replacement is much higher for heavy truck fleets.
- Tire retreading is desirable from the point of view of economic and material saving.

According to the Michelin Factbook (2001), retreaded tires account for 44.0% of the total tire replacement market for heavy truck tires. Retreading has an environmental impact in terms of processing of raw materials, manufacture and use. Among these impacts, the following can be mentioned:

- Reduction of energy savings in the field of transient technological changes in tires;

- Reconditioning of energy-saving tires in the field of inefficiency of degradation of retreaded tires compared to new equivalent tires;
- Energy saving of retreading of tires in the scope of the variations of the product.

According to the Tire Retread and Repair Information Bureau (TRIB), retreaded tires can last 75.0% to 100.0% of the life of a new tire, based on the quality of the retreading process. Retreading has its advantages and disadvantages, which are shown in Table 12.

Table 12. Advantages and disadvantages of waste tire retreading [60].

Advantages	Disadvantages
(a) Extends the useful life of the tire; (b) It uses many of the original materials and much of the original structure; (c) The net result is a decrease in materials compared to the manufacture of new tires; (d) Rubber extracted from used tires prior to retreading is often sold as crushed rubber for other purposes; (e) The energy used is lower compared to the original manufacture: the energy used to retread a tire is 400 MJ while the manufacture of a new tire requires 970 MJ	(a) Concern for the volatile organic components emanating from solvents, adhesive agents and rubber compounds during vulcanization; (b) Smell can also be a problem in some places; (c) The process generates a large amount of waste

This strategy can use the criterion of creating a network for the re-manufacturing and tire re-use network. The first network is aimed for the recovery of the tires and their latter re-manufacture. In this type of network, the original tire manufacturers usually play a very important role, being sometimes solely responsible for the design and management of RL systems, whose design responds to multilevel, decentralized characteristics, for which synergies are sought with the direct channel. The second network is aimed for re-introducing the recovered tires into the SC (vulcanization) once the necessary cleaning and maintenance activities are carried out. In these networks, there are decentralized structures whereby original and re-used products are simultaneously circulated and in which the cost of transport appears as the most significant [6].

4.2. Diversification Strategy

To produce recycled products of equivalent quality and price, the industry must invest heavily in new technologies. Doing so would be taking advantage of a cheap resource, which allows them to generate secondary products for sale. This investment creates an opportunity to close the recycling cycle and therefore to close the SC, paying attention not only to production and processing, but also to the collection of the waste generated to be used as raw material for other processes within the company (or to open a channel with new clients).

One of the materials with greater presence in tires is the dust generated from them, as well as the steel and the textile fibers. The market for these materials is very broad and their use can be seen in family parks, sports courts, roads, among other elements, which are presented in Table 13.

Table 13. Uses of tire components [60].

Products	Use	Potential Clients
Tire Dust	(a) Parks for children, sports tracks, asphalt; (b) Artificial grass; (c) New tires, rubber and energy source; (d) Home insulation, waterproofing; (e) Security plates; (f) Shoe soles, carpets & rugs; (g) Conveyor belts	(a) Construction equipment; (b) Suppliers of building materials; (c) Plastic Products; (d) Insulation; (e) Government; (f) Cemeteries; (g) Steel companies
Steel	Blast furnaces, asphalt fillers	(a) Construction equipment; (b) Suppliers of building materials; (c) Steel companies
Textile fibers	(a) Artificial grass; (b) Shoe soles, carpets & rugs; (c) Clothing	(a) Suppliers of building materials; (b) Plastic Products; (c) Insulation

Diversification has its advantages and disadvantages, which are shown in Table 14.

Table 14. Advantages and disadvantages of waste tire retreading [60].

Advantages	Disadvantages
It increases the demand based on supply. However if the company focuses on offering a single product, a change in the market may affect its performance. For this reason it is important that the company offers the consumer several purchase alternatives.	(a) Lack of knowledge of the market: by offering new products and services, there is the possibility of not having enough knowledge and experience of the market to which it is intended, affecting the financial resources of the company; (b) Cost: it must have a reliable economic stability so that the production of new products will not affect the company.

For this strategy, a recycling network could be used. This type of network usually has simple structures, few links and it is centralized. These networks are characterized by requiring, for an efficient management of the same, a high volume of recovered products which generally are of little unit value. The high transformation costs determine the need for high utilization rates of these networks and the search for economies of scale [6].

4.3. Integrated RL Scheme

Both strategies show the advantages and disadvantages of its implementation within the framework of RL. However, it is necessary to know what the interests of the company are in order to take the best decision to generate greater benefits, but how does RL can influence the tire manufacturing process? The proposed adaptation of the RL process is described in Figure 10.

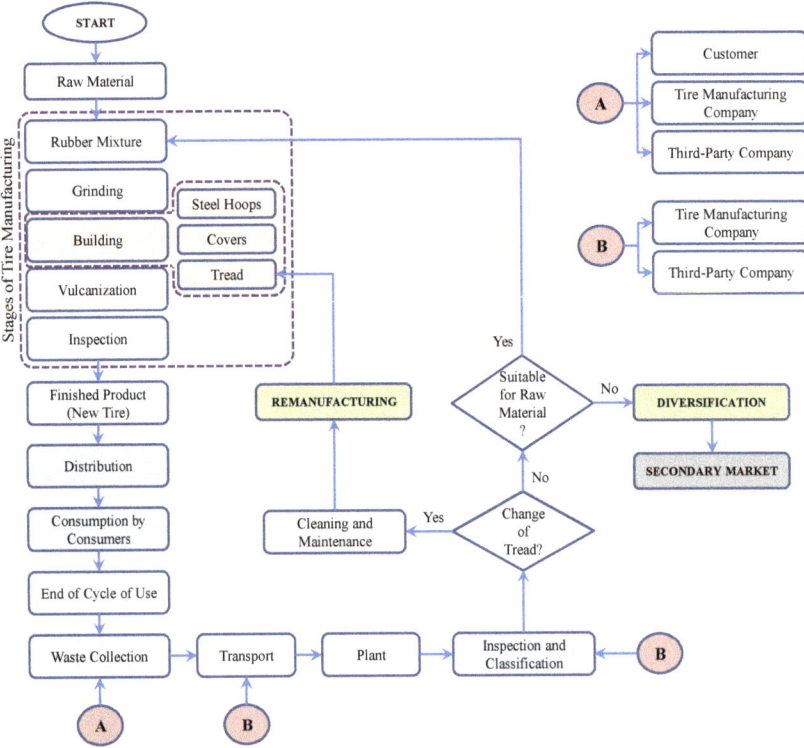

Figure 10. RL structure for tires (own work).

As presented in Figure 10, after the customers use the tires and they reach the end of their cycle of use, RL becomes a vital part of the proposed waste management strategy. In addition, it is important to observe that this RL structure has the purpose of minimizing the generation of non-recoverable waste, thus reducing the source of contaminants. This is the reason for the elimination process being absent in Figure 10.

In general, the management of the RL structure can be performed by the following entities:

- a third-party (public or private) company can perform the whole process of collection, classification, and re-manufacturing of the out-of-use tire;
- the manufacturing company itself can extend its participation within the RL process assuming some responsibilities.

While companies (either the manufacturing company, or third party companies) are considered for most of the RL processes, the users or customers are considered as a vital part of the waste collection process. This is proposed to support a culture of sustainable waste management as achieved in Japan and the EU. Nevertheless, it is important to point out that, as in the EU, good law legislation is the key for the success of the RL processes [26]. In Mexico, tax incentives can improve the collaboration of users and private entities in the collection process. Particularly, the adaptation of the *Extended Producer Responsibility* system, which has been successfully implemented in the EU and Colombia, can be of significant benefit in Mexico and Russia. In this aspect, while Russia is not a member of the EU, since 2015, there have been amendments to the Russian 1998 *Federal Law on Waste from Production and Consumption* to implement legislation similar to those of the EU for extended take-back obligations (i.e., collection and recycling) on producers and importers of certain products and packaging.

In addition, the proposed stages for waste collection can be extended with different logistic networks for recovery. As an example, Figure 11 presents the logistic chains proposed in [61] for tire waste management where:

- The first logistic chain considers that the product is recovered by the same company that manufactures it (thus, the company only recovers the tires that were manufactured by it).
- The second logistic chain considers that the company recovers (1) its own tires, and (2) tires manufactured by its competition (e.g., other companies), leading to reach a greater volume of tires to carry out its processing.
- The third logistic chain considers that the entire recovery process is carried out by a third-party company, which may be contracted by the manufacturer whose product is collected for the same purpose. This chain can also be extended for the creation of other products (diversification), either by means of this contracted company or by the manufacturing company itself. The third-party company can also perform the whole process or simply be a part of it.
- The fourth logistic chain is related to the previous chain and it represents the case in which the company that is collecting the product does it for a different manufacturing process than the production of tires. This leads to the diversification for new markets.

Regarding the specific stages of re-manufacturing and diversification, the alternatives reviewed and discussed in Sections 4.1 and 4.2 can be extended with the following novel approaches:

- Manufacturing processes focused on decreasing rolling resistance of tires. This can lead to improving fuel efficiency, reducing CO_2 emissions, and extending the useful life of the tire [54].
- Standard use of the "Reduce Index" (Re Index) to objectively measure and assess processes focused on achieving a longer wear life of tire for different vehicles [54].
- Use of recycled tire crumb (RTC) for the manufacturing of construction materials. The RTC has been identified as a promising approach to create lightweight cellular concrete (LCC) due to its insulation properties regarding sound, water, and temperature [25]. Countries with regions with extreme weather conditions such as Russia can get benefits from this proposal.
- Recycling processes based on state-of-the-art advances in chemical studies on the effects of amount, size and morphology of rubber granulate grains for the creation of modern polymer-rubber composites from used tires [26].

- Extend on thermo-mechanical devulcanization based on extrusion as a standard for recycling on an industrial scale. As discussed in [23], it is the most environmentally appropriate approach to tire recycling in comparison to chemical, ultrasound, microwave, and mechanical devulcanization.

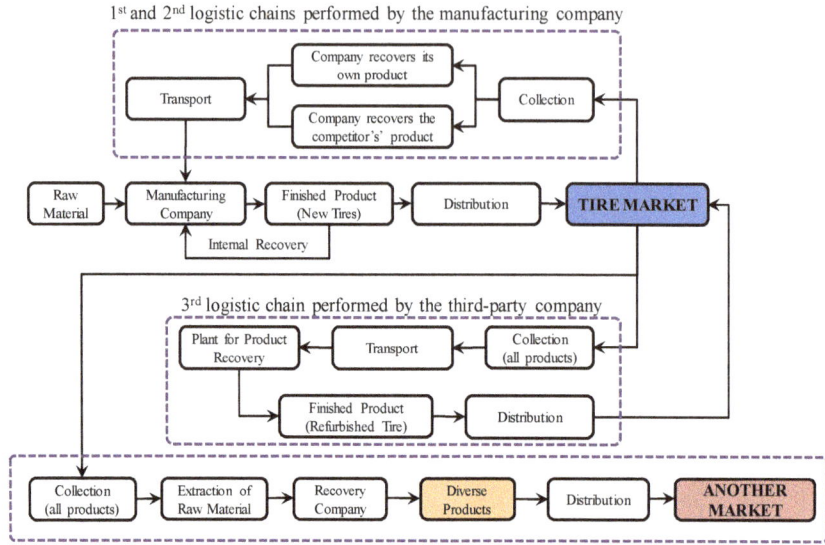

Figure 11. RL recovery process for tire waste (adapted from [61]).

5. Discussion on Findings

The current environment requires companies to increase the utilization of resources and optimize their practices through the incorporation of logistic processes. The consideration of a reverse flow in the logistic functions can support these objectives. Specifically, recycling of used or worn tires is the most urgent aspect to achieve sustainability within the tire industry.

According to Russian legislation, the organization of waste collection and recycling is under the responsibility of local authorities. However, classification and recovery are rarely performed by the Russian waste management system by the following situations: shortening of legislative regulation, the absence of strict requirements to separate waste, weak public awareness, and lack of collection stations and markets. Additionally, there is also a low efficiency in garbage trucks, and lack of transfer stations and incineration plants [62].

Not having an effective recycling process leads to the generation of waste. In Russia and Mexico, the most common form of waste treatment is landfill disposal. However, most landfills in operation are already overloaded and some constitute environmental and epidemiological hazards. Landfill problems in Russia and Mexico are due to non-compliance with environmental and sanitary standards, and closing of landfills without waste recovering. To reduce the dependence on landfills, the development of effective recycling processes can be established by reinforcing regulations, tax incentives, and improvement of disposal technologies (e.g., by-products or energy from waste). For this, business strategies can improve the economic outcomes of recycling. However, it is important to mention that, even with the latest recycling or re-manufacturing technologies, landfill is the main disposal strategy in the short and medium term.

In this regard, countries within the EU, such as Spain, Sweden and France, have achieved 100.0% recycling rates since 2011 [19], and the *Extended Producer Responsibility (EPR)* system has proved to be a suitable mean to accomplish this objective. This system can be adapted to Mexico and Russia as it was performed in Colombia [21]; however, it is important to consider the following difficulties:

- while incentives are considered within the EPR, and the company can rely on third-party companies for collection, recycling and re-manufacturing (including diversification), some processes may be too expensive and/or complex. This can limit the intended impact or commitment of all entities within the SC;
- implementation and compliance of new RL legislation may represent additional administrative and economic burden to local governments;
- among the critical factors for recycling are the management and transportation costs which must be performed by the responsible entity. In general, the collection of tire waste is not only carried out by the public (i.e., government) companies, but also by private companies that use this waste as raw material for their production processes (e.g., diversification). While transportation costs have reduced the collection tasks due to inefficient route planning, there is also an inefficient cooperation between public and private companies to accomplish these tasks.

The conceptual RL model presented in Figure 10 can support the decisions regarding the main processes to be considered for a waste management strategy. In addition, it can motivate the participation of companies in transportation and/or diversification through the appropriate visualization of the RL processes.

In this model, retreating or re-manufacturing have the following economic and environmental advantages for sustainability:

- savings in the energy used for production by 66.0%;
- 30.0–50.0% cheaper than the manufacturing cost of a new tire;
- a re-manufactured tire can have a useful life of 75.0–100.0% when compared to a new tire;
- very appropriate for heavy truck operators where the tire replacement rate is high.

On the other hand, market diversification represents an important investment opportunity that may be an option to close the SC cycle that includes both Direct and Reverse Logistics, as it would not only be paying attention to the process of production, but also to the problem of generated waste. This waste can be used as raw material within the SC's production process, or be used as raw material for new production lines designed to launch new products. This would help the companies to keep current customers or to obtain new customers by opening their doors to a secondary market. In this case, technologies such as those presented in [23,25,26] can make recycling and diversification more environmentally and economically viable.

However, as previously discussed, governmental and financial support is needed to invest in high-level recycling technology, and, at the same time, to obtain the highest rates. The attractiveness of secondary products and energy from waste also needs to improve economic and legal instruments, and public campaigns and procurement should be used to raise awareness of waste as a valuable resource. In addition, it is important to mention that landfills, even with future technologies and regulations, cannot (and will not) be eliminated completely in the short and medium term, independently of the RL strategy.

6. Conclusions

The present work defined a conceptual RL model for the tire waste management systems of Mexico and Russia due to an absence of work and consensus regarding RL strategies in these countries. For this purpose, a comprehensive review of tire waste management strategies in the EU and Japan, leading countries in recycling and sustainable practices, was performed. In addition, the current status in Mexico and Russia was reviewed considering the factors highlighted by the United Nations for the establishment of waste management systems.

This review provided the background to identify two specific RL processes: *re-manufacturing* and *diversification*, as a means to make a tire waste management system economically accessible and sustainable. These processes were integrated within an RL model considering the most standard processes performed in the EU where the Extended Producer Responsibility (EPR) is a recommended scheme that has provided benefits in countries such as Spain, France, Italy, and Colombia. Finally, a review on recycling technologies for diversification was discussed to support economic benefits.

Although a better strategy over the reviewed strategies cannot be assured due to the different regional conditions of financial mechanisms, regulations, involving entities and technologies established in each country, within the context of Mexico and Russia the proposed RL model can provide the guidelines to incorporate *re-manufacturing* and *diversification* in their waste management systems as no previous works concerning these countries have been reported. In addition, the present work can be improved regarding its limitations as no empirical research has been performed in Mexico and Russia regarding quantitative assessment of RL processes. This assessment represents also an approach to improve other strategies as those performed in the EU, Japan, Colombia, Italy and Romania.

Thus, as future work, the following aspects are considered:

- design of a Green-Reverse logistic model (e.g., by means of mixed integer linear programming) to improve the distribution network required for the collection, recycling, and diversification tasks;
- incorporation of economic variables within the RL structure to establish an efficient business model between all entities involved in the recycling process;
- incorporation of inventory supply strategies as a mean to optimize tire waste processing.

Author Contributions: All authors contributed to the research presented in this paper. M.-L.U.-M., S.-O.C.-M., J.-L.M.-F., P.C.-O. and A.-A.A. developed the main research under a double-degree program between UPAEP A.C. (Mexico) and UrFU (Russia) for the master's thesis of M.-L.U.-M.; S.-O.C.-M., J.-L.M.-F. and P.C.-O. revised the managerial and logistical aspects of the main research and of the master's thesis at UPAEP A.C.; M.-L.U.-M., S.-O.C.-M., PCO and J.-L.M.-F. wrote the paper version of the revised master's thesis.

Funding: This research received no external funding.

Acknowledgments: The authors acknowledge the scholarship provided by Consejo Nacional de Ciencia y Tecnologia—CONACyT (Mexico) and the support of the dual-degree program between Universidad Popular Autonoma del Estado de Puebla A.C.—UPAEP A.C. (Mexico) and Ural Federal University—UrFU (Russia). The authors also acknowledge the funding of UPAEP A.C. (Mexico) for publishing in open access.

Conflicts of Interest: The authors declare no conflict of interest.

References

1. Rutner, S.M.; Langley, C.J. Logistics Value: Definition, Process and Measurement. *Int. J. Logist. Manag.* **2000**, *11*, 73–82. [CrossRef]
2. Council of Logistics Management. *Council of Logistics Management 1998 Annual Conference—Logistics Excellence: Vision, Processes, and People*; Council of Logistics Management: Anaheim, CA, USA, 1998.
3. Zhao, C.; Liu, W.; Wang, B. Reverse Logistics. In Proceedings of the 2008 International Conference on Information Management, Innovation Management and Industrial Engineering, Taipei, Taiwan, 19–21 December 2008; pp. 349–353.
4. Barker, T.J.; Zabinsky, Z.B. A multicriteria decision making model for reverse logistics using analytical hierarchy process. *Omega* **2011**, *39*, 558–573. [CrossRef]
5. Gomez-Montoya, R.A.; Correa-Espinal, A.A.; Vasquez-Herrera, L.S. Reverse Logistics. An Approach with Business Social Responsibility (In Spanish). *Criterio Libre* **2012**, *10*, 143–158.
6. Rubio, L.S. El Sistema de Logística Inversa en la Empresa: Análisis y Aplicaciones. Ph.D. Thesis, Universidad de Extremadura, Badajoz, Spain, 2003. (In Spanish)
7. Guirong, Z.; Qing, G.; Bo, W.; Dehua, L. Green logistics and Sustainable development. In Proceedings of the 2012 International Conference on Information Management, Innovation Management and Industrial Engineering, Sanya, China, 20–21 October 2012; pp. 131–133.

8. Seroka-Stolka, O. The development of green logistics for implementation of sustainable development strategy in companies. *Procedia Soc. Behav. Sci.* **2014**, *151*, 302–309. [CrossRef]
9. Rodrigue, J.P.; Slack, B.; Comtois, C. Green logistics. In *Handbook of Logistics and Supply Chain Management*; Brewer, A.M., Button, K.J., Hensher, D.A., Eds.; Emerald Group Publishing Limited: Bingley, UK, 2008; pp. 339–350.
10. Sbihi, A.; Eglese, R.W. Combinatorial optimization and Green Logistics. *Ann. Oper. Res.* **2009**, *175*, 159–175. [CrossRef]
11. Scott, C.; Lundgren, H.; Thompson, P. *Guide to Supply Chain Management*; Springer: Berlin, Germany, 2011.
12. Nylund, S. Reverse Logistics and Green Logistics. Master's Thesis, Vaasan Ammattikorkeakoulu Vasa Yrkeshogskola University of Applied Sciences, Vaasa, Finland, 2012.
13. Kumar, A. Green Logistics for Sustainable Development: An Analytical Review. *IOSRD Int. J. Bus.* **2015**, *1*, 7–13.
14. Secretaría del Medio Ambiente (SEDEMA). *Inventario de Residuos Sólidos*; SEDEMA: Ciudad de México, Mexico, 2015. (In Spanish)
15. Adhikari, B.; De, D.; Maiti, S. Reclamation and recycling of waste rubber. *Prog. Polym. Sci.* **2000**, *25*, 909–948. [CrossRef]
16. World Business Council for Sustainable Development (WBCSD). *Managing End-of-Life Tires*; World Business Council for Sustainable Development: Geneva, Switzerland, 2008.
17. Tomenco, L.; Zagodyakin, G. Analysis of the return supply chain for the processing of used automobile tires with the use of agent modeling. *Successes Chem. Chem. Technol.* **2013**, *27*, 51–55. (In Russian)
18. UN-UNEP. *Revised Technical Guidelines for the Environmentally Sound Management of Used and Waste Pneumatic Tyres*; United Nations (UN): Basel, Switzerland, 2011.
19. Sebola, M.R.; Mativenga, P.T.; Pretorius, J. A Benchmark Study of Waste Tyre Recycling in South Africa to European Union Practice. *Procedia CIRP* **2018**, *69*, 950–955. [CrossRef]
20. Molino, A.; Donatelli, A.; Marino, T.; Aloise, A.; Rimauro, J.; Iovane, P. Waste tire recycling process for production of steam activated carbon in a pilot plant. *Resour. Conserv. Recycl.* **2018**, *129*, 102–111. [CrossRef]
21. Park, J.; Diaz-Posada, N.; Mejia-Dugand, S. Challenges in implementing the extended producer responsibility in an emerging economy: The end-of-life tire management in Colombia. *J. Clean. Prod.* **2018**, *189*, 754–762. [CrossRef]
22. Aoudia, K.; Azem, S.; Hocine, N.; Gratton, M.; Pettarin, V.; Seghar, S. Recycling of waste tire rubber: Microwave devulcanization and incorporation in a thermoset resin. *Waste Manag.* **2017**, *60*, 471–481. [CrossRef] [PubMed]
23. Asaro, L.; Gratton, M.; Seghar, S.; Hocine, N. Recycling of rubber wastes by devulcanization. *Resour. Conserv. Recycl.* **2018**, *133*, 250–262. [CrossRef]
24. Pavlovich, L.; Solovyova, N.; Strakhov, V. Utilizing waste tires with steel cord in coke production. *Coke Chem.* **2017**, *60*, 119–126. [CrossRef]
25. Kashani, A.; Ngo, T.D.; Mendis, P.; Black, J.R.; Hajimohammadi, A. A sustainable application of recycled tyre crumbs as insulator in lightweight cellular concrete. *J. Clean. Prod.* **2017**, *149*, 925–935. [CrossRef]
26. Sienkiewicz, M.; Janik, H.; Borzedowska-Labuda, K.; Kucinska-Lipka, J. Environmentally friendly polymer-rubber composites obtained from waste tyres: A review. *J. Clean. Prod.* **2017**, *147*, 560–571. [CrossRef]
27. Derakhshan, Z.; Ghaneian, M.; Mahvi, A.; Conti, G.; Faramarzian, M.; Dehghani, M.; Ferrante, M. A new recycling technique for the waste tires reuse. *Environ. Res.* **2017**, *158*, 462–469. [CrossRef] [PubMed]
28. Blanco-Machin, E.; Travieso-Pedroso, D.; Andrade-de-Carvalho, J. Energetic valorization of waste tires. *Renew. Sustain. Energy Rev.* **2017**, *68*, 306–315. [CrossRef]
29. Tsai, W.-T.; Chen, C.-C.; Lin, Y.-Q.; Hsiao, C.-F.; Tsai, C.-H.; Hsieh, M.-H. Status of waste tires' recycling for material and energy resources in Taiwan. *J. Mater. Cycles Waste Manag.* **2017**, *19*, 1288–1294. [CrossRef]
30. Torretta, V.; Rada, E.; Ragazzi, M.; Trulli, E.; Istrate, I.; Cioca, L. Treatment and disposal of tyres: Two EU approaches. A review. *Waste Manag.* **2015**, *45*, 152–160. [CrossRef] [PubMed]
31. Price, W.; Smith, E. Waste tire recycling: Environmentally benefits and commercial challenges. *Int. J. Environ. Technol. Manag.* **2006**, *6*, 362–374. [CrossRef]

32. Connor, K.; Cortesa, S.; Issagaliyeva, S.; Meunier, A.; Bijaisoradat, O.; Kongkatigumjorn, N.; Wattanavit, K. Developing a Sustainable Waste Tire Management Strategy for Thailand. Bachelor's Thesis, Worcester Polytechnic Institute (WPI), Worcester, MA, USA, 2013.
33. Stock, J.R. The 7 Deadly Sins of Reverse Logistics. *Material Handling Logistics* **2001**, *56*, MHS5.
34. Padilha, F.; Hilmola, O. Green Transport Practices and Shifting Cargo from Road to Rail in Finnish-Russian Context. Available online: https://dialnet.unirioja.es/servlet/articulo?codigo=3438108 (accessed on 6 September 2018).
35. Zueva, O.; Shakhnazaryan, S. Logistics of Return Flows of Secondary Resources. *Bull. Balt. Fed. Univ. Hum. Soc. Sci.* **2014**, *9*, 140–147. (In Russian)
36. Zueva, O.; Shakhnazaryan, S. Peculiarities of Introducing Reverse Logistics in Supply Chains. *J. Ural State Univ. Econ.* **2016**, *4*, 108–116.
37. Ovezov, B.; Fen, C. Reverse Logistics. *Young Sci. Econ. Manag.* **2016**, *1*, 441–446. (In Russian)
38. Gromov, V. Green Logistics in Russia. *Russ. J. Logist. Transp. Manag.* **2014**, *1*, 36–44. [CrossRef]
39. Shahnazaryan, S.; Potapova, S. The problem of defining the concept of "returnable logistics" and its role in supply chain management. *J. Ural State Univ. Econ.* **2013**, *46*, 123–128. (In Russian)
40. Weibold: Tyre Recycling Consulting. Tire Recycling Development in Crimea. Available online: https://weibold.com/tire-recycling-development-in-crimea/ (accessed on 21 November 2016).
41. Akishin, A. *Organizational and Economic Mechanism and Risk Management Tools in the Supply Chain of Enterprises in the Tire Industry*; Russian Chemical and Technical University: Moscow, Russia, 2011. (In Russian)
42. Recycling Points. Recycling of Automobile Tires. Available online: http://punkti-priema.ru/drugoe-vtorsiryo/utilizaciya-pokrishek (accessed on 3 May 2014).
43. Technological Resources. Features of Grinding Domestic Tires and Differences between Imported and Domestic Tires. Available online: http://www.stanki-ru.ru/poleznaya-informatsiya/osobennosti-pererabotki-shin-v-rossii-i-sng.html (accessed on 1 December 2017). (In Russian)
44. Nevyadomskaya, A.I.; Deriglazov, A.A. Utilization and Processing of Tires. In Proceedings of the ECONOMIC SCIENCES: XXV International Scientific and Practical Conference, Novosibirsk, Russia, 23 October 2014; pp. 230–237. (In Russian)
45. COLESAU.RU. Analysis of the Market of Processing of Rubber Products in Russia. Available online: http://colesa.ru/news/23249 (accessed on 7 July 2014). (In Russian)
46. Severa-Francés, D. Concept and evolution of the logistics function. *INNOVAR Rev. Cienc. Adm. Soc.* **2010**, *20*, 217–234.
47. Reynaldo, C.R.; Ertel, J. Reverse logistics network design for collection of End-of-Life Vehicles in Mexico. *Eur. J. Oper. Res.* **2009**, *196*, 930–939.
48. Ortiz, S. Logística Inversa: Al revés no es Igual. Available online: https://expansion.mx/manufactura/2009/05/06/logistica-inversa-al-reves-no-es-igual?internal_source=PLAYLIST (accessed on 6 May 2009). (In Spanish)
49. Jimenez, M.N. La gestión integral de residuos sólidos urbanos en México: entre la intención y la realidad. *Letras Verdes Revista Latinoamericana de Estudios Socioambientales* **2015**, *17*, 29–56. (In Spanish)
50. Instituto Nacional de Econología y Cambio Climático (INECC); Secretaría de Medio Ambiente y Recursos Naturales (SEMARNAT). *Diagnóstico Básico para la Gestión Integral de los Residuos 2012*; Instituto Nacional de Econología y Cambio Climático (INECC); Secretaría de Medio Ambiente y Recursos Naturales (SEMARNAT): Mexico City, Mexico, 2012. (In Spanish)
51. Rosagel, S. México se Rezaga en Reciclaje de Llantas. Available online: https://expansion.mx/manufactura/2011/07/25/mexico-se-rezaga-en-reciclaje-de-llantas (accessed on 25 July 2011). (In Spanish)
52. Cámara Nacional de la Industria Hulera (CNIH). *Plan de Manejo de Neumáticos Usados de Desecho*; Asociación Nacional de Distribuidores de Llantas y Plantas Renovadoras A.C. (ANDELLAC); Asociación Nacional de Importadores de Llantas A.C. (ANILLAC); Cámara Nacional de la Industria Hulera (CNIH): Mexico City, Mexico, 2013. (In Spanish)
53. Irevna. *Tire Recycling Industry: A Global View*; Global Research & Analytics—Irevna: Chennai, India, 2016.
54. JATMA. *Tire Industry of Japan 2018*; The Japan Automobile Tyre Manufacturers Association, Inc.: Tokyo, Japan, 2018.

55. ETRma. *End-of-Life Tyre: Report 2015*; EUROPEAN TYRE & RUBBER Manufacturers' Association: Brussels, Belgium, 2015.
56. Fleischmann, M.; Krikke, H.; Dekker, R.; Flapper, S.P. A characterisation of logistics networks for product recovery. *Omega* **2000**, *28*, 653–666. [CrossRef]
57. Soltani, S.; Naderi, G.; Ghoreishy MHR. Second Life: Never waste a tire fiber again. *Tire Technol. Int.* **2010**, *10*, 52–54.
58. Boustani, A.; Sahni, S.; Gutowski, T.; Graves, S. *Tire Remanufacturing and Energy Saving*; Environmentally Benign Manufacturing Laboratory, Massachusetts Institute of Technology: Cambridge, MA, USA, 2010.
59. Ferrer, G. The Economics of Tire Remanufacturing. *Resour. Conserv. Recycl.* **1997**, *19*, 221–255. [CrossRef]
60. SAyDS. *Resolución No 523 Sobre Manejo de Neumáticos*; Secretaría de Ambiente y Desarrollo Sustentable de la Nación (SAyDS): Mexico City, Mexico, 2013. (In Spanish)
61. Monroy, N.; Ahumada, M.C. Logística Reversa: Retos para la Ingeniería Industrial. *Rev. Ing.* **2006**, *23*, 23–33. (In Spanish).
62. Piippo, S. Municipal Solid Waste Management (MSWM) in Sparsely Populated Northern Areas: Developing an MSWM Strategy for the City of Kostomuksha, Russian Federation. Master's Thesis, University of Oulu, Oulu, Finland, 2012.

© 2018 by the authors. Licensee MDPI, Basel, Switzerland. This article is an open access article distributed under the terms and conditions of the Creative Commons Attribution (CC BY) license (http://creativecommons.org/licenses/by/4.0/).

Review

Sustainable Supply Chain Management Practices and Sustainable Performance in Hospitals: A Systematic Review and Integrative Framework

Verónica Duque-Uribe [1], William Sarache [1,*] and Elena Valentina Gutiérrez [2]

1. Departamento de Ingeniería Industrial, Universidad Nacional de Colombia, 170003 Manizales, Colombia; vduqueu@unal.edu
2. Departamento de Ingeniería Industrial, Universidad de Antioquia, 050010 Medellín, Colombia; elena.gutierrez@udea.edu.co
* Correspondence: wasarachec@unal.edu.co; Tel.: +57-68-879400

Received: 27 September 2019; Accepted: 22 October 2019; Published: 25 October 2019

Abstract: Hospital supply chains are responsible for several economic inefficiencies, negative environmental impacts, and social concerns. However, a lack of research on sustainable supply chain management specific to this sector is identified. Existing studies do not analyze supply chain management practices in an integrated and detailed manner, and do not consider all sustainable performance dimensions. To address these gaps, this paper presents a systematic literature review and develops a framework for identifying the supply chain management practices that may contribute to sustainable performance in hospitals. The proposed framework is composed of 12 categories of management practices, which include strategic management and leadership, supplier management, purchasing, warehousing and inventory, transportation and distribution, information and technology, energy, water, food, hospital design, waste, and customer relationship management. On the other side, performance categories include economic, environmental, and social factors. Moreover, illustrative effects of practices on performance are discussed. The novelty of this document lies in its focus on hospital settings, as well as on its comprehensiveness regarding the operationalization of practices and performance dimensions. In addition, a future research agenda is provided, which emphasizes the need for improved research generalizability, empirical validation, integrative addressing, and deeper analysis of relationships between practices and performance.

Keywords: supply chain management; hospital supply chain; sustainable supply chain; hospital logistics; hospital sustainability; healthcare logistics; sustainable hospital management

1. Introduction

Supply Chain Management (SCM) is a field of growing academic interest, as reflected in the increase in related literature [1]. The Council of Supply Chain Management Professionals [2] (p. 187) not only defines SCM as the "planning and management of all activities involved in sourcing and procurement, conversion, and all logistics management activities," but also emphasizes its role in the integration between players involved in the entire supply chain. SCM interest lies in its contribution to a competitive advantage, in terms of differentiation and the reduction of operating costs, especially in the current context of intense competition, globalization, and active consumer participation [3]. It is argued that better SCM results in superior performance, through the adoption of exemplary practices [4]. A wide range of publications support the existence of significant relationships between SCM practices and organizational performance, especially from the economic perspective [5–8].

Beyond the previously mentioned economic focus, a recent trend in SCM study points to the consideration of its link to sustainability, which incorporates the environmental and social dimensions, for two reasons. First, global poverty, health, working conditions, and climate change indicators [9], among others, have aroused worldwide interest in the promotion of sustainable development, defined as "the development that meets the needs of the present without compromising the ability of future generations to meet their own needs" [10] (p. 41). Second, given that organizations are often responsible for both environmental and social problems including pollution and unacceptable working conditions, they also have a duty to help mitigate such effects, as well as contribute to economic development.

The concepts of the Triple Bottom Line (TBL) and Sustainable Supply Chain Management (SSCM) have become significant. The former was coined by Elkington [11], and aims to consider the economic, environmental, and social dimensions to be equally important, since the economy is fundamental to support society, but doing business can become unfeasible in a depleted global ecosystem. The latter refers to the inclusion of environmental and social dimensions in the conventional notion of SCM, as proposed by Seuring and Müller [12] (p. 1700), who define SSCM as, "the management of material, information, and capital flows as well as cooperation among companies along the supply chain while taking goals from all three dimensions of sustainable development, i.e., economic, environmental, and social, into account, which are derived from customer and stakeholder requirements."

Hospital supply chains are often confronted by several economic, environmental, and social problems. From the economic point of view, increasing healthcare expenditures demand greater efficiency in the delivery of services [13,14]. The Organization for Economic Co-operation and Development has estimated that hospitals account for approximately 40% of total health expenditures [15]. Between 30% and 40% of a hospital's budget is dedicated to supply chain costs [16], which can be reduced by up to 8% through the use of best practices [17]. In addition, said best practices allow clinical personnel to focus on their core mission of caring [16].

Regarding the environmental dimension, hospital processes and services are intensive in terms of material, energy, and water consumption, generate significant amounts of waste (especially toxic waste, as compared to other sectors), and account for a large carbon footprint [14,18,19]. In England, for instance, the Sustainable Development Unit of the National Health Service has calculated that healthcare's footprint represents 39% of public sector emissions, from which procurement contributes 57%, energy contributes 18%, travel contributes 13%, and others account for 11% [20]. Moreover, acute services are responsible for the largest portion, which is approximately 50% of the total.

Social problems related to hospital supply chains are also tangible. From an internal perspective, although hospitals are large-scale employers, non-standard forms of employment are frequent, pay levels have decreased in comparison to other economic sectors, women are compensated worse and recognized less often than men, daily working hours exceed legal limits, and safety considerations are often neglected [21,22]. Work characteristics such as shift work and long working hours not only increase the likelihood of occupational accidents, and developing burnout and additional psychological stress than in other jobs [14,23–25], but also impact the quality of patient care [26–28]. From an external standpoint, hospitals have a deep impact on the population because health services influence, in one way or another, peoples' quality of life. Nevertheless, reported global problems include unsatisfactory health service coverage for the needs of the population, in terms of access and delivery [21].

Therefore, the goal of accomplishing the triple challenge of being more efficient, more environmentally-friendly, and offering better conditions to both workers and communities served, leads to the subjects of SSCM practices and sustainable performance. No matter the way that practices are defined, whether as organizational routines, rules, or standard procedures [29], best practices are linked to the objective of that which is recognized as superior by a majority [16]. In other words, poor performance can be considered a consequence of a lack of best practices [30].

Numerous publications demonstrate that SSCM is a field of increasing interest. As Carter and Washispack assert in a review, "we have reached a point of saturation" [31] (p. 242), in terms of appraising the structure and main themes of SSCM literature. However, specific relationships between constructs remain unexplored. Some empirical studies stress that SSCM practices and sustainable performance constructs have not been clearly or consistently defined [32–34]. Besides the primacy in the study of operational and economic topics, the environmental dimension has been more often addressed than the social dimension [35–39]. Moreover, the integration of the three sustainability dimensions has not been sufficiently robust [39–41], and industry-specific issues have not been elucidated to the extent to which they could be [34,42–44].

Despite dramatic growth in the SSCM literature [31], this is not the case when delimited to hospital settings. Academic database searches yield results on hospital SCM or hospital sustainability, but almost none appear to address hospital SSCM as such. Therefore, both SCM and sustainability may be relevant for hospitals, but they have likely been addressed in a fragmented manner in the literature. To the authors' knowledge, there are no existing reviews which address the intersection between hospital SCM and sustainability. Reviews focused on the healthcare supply chain [45–49] have not explicitly considered environmental and social issues, whereas reviews on hospital sustainability [50] have highlighted the environmental dimension.

In response, the aim of this article is to present a systematic review and an integrative framework for SSCM practices that can contribute to sustainable performance in hospital settings. Three research questions are specifically addressed: (1) What are the main SSCM practices applied by hospitals? (2) What are the main sustainability performance metrics used by hospitals? (3) How can the relationships between SSCM practices and performance be framed in the hospital setting?

This paper is organized as follows. In Section 2, the methodology is presented and explained. Section 3 discusses the main findings, considering two main components. The SSCM practices applied by hospitals and sustainable performance metrics used by hospitals. In Section 4, an integrative framework, derived from the systematic literature review, is developed. Section 5 examines future avenues for research. Lastly, a relevant set of conclusions are presented in Section 6.

2. Materials and Methods

In order to address the proposed research questions, this study is based on a systematic literature review. Contrary to narrative reviews, systematic reviews are characterized by their explicitness and transparency regarding the methods used to find reasonable evidence on a given topic [51]. In management, it has been increasingly asserted that systematic reviews are a useful way to identify relevant scientific contributions, inform research and practice, and enhance a field's body of knowledge, by applying rigorous principles that have been traditionally used in medical research [52].

As shown in Figure 1, the methodology implemented to undertake this review involves three stages: planning, conducting, and reporting, which is in line with several suggestions [51–53]. The planning stage was accomplished through the identification of need, based on the research questions proposed, as well as through the definition of the search strategy, the selection criteria, the quality assessment criteria, the data extraction strategy, and the data synthesis approach. The conducting and reporting stages were accomplished from the contents of findings and discussion sections. In parallel, the Preferred Reporting Items for Systematic Reviews and Meta-Analyses (PRISMA) checklist [54] was also followed to ensure rigor of the review process.

Search strategy: Scopus and Web of Science (WoS) were selected for the search, due to their strengths in terms of extension, coverage, and the possibility of classifying sources in accordance with impact criteria [55]. For Scopus, publications throughout history, up to February 2019, were considered. For WoS, the time horizon was set between 2001 and February 2019, as the core collection of this database was available beginning in 2001. Based on the intersection between the topics addressed and the research questions, the executed string was as follows: (TITLE-ABS-KEY ("supply chain management" OR "healthcare logistics") AND practice AND hospital) OR (TITLE (sustainab* AND hospital)).

The term "performance" was excluded from the search string, as some publications only address practices, irrespective of their link to performance. Along with supply chain management, the term "healthcare logistics" was employed, considering that both have been used interchangeably [56]. Regarding the connection between sustainability and hospitals, as keywords cover broad and diverse sustainability subtopics, the search was performed by the title, in order to ensure enhanced delimitation. The publications selected for this study were primarily in English, since the intention was to explore the topic globally [57]. Database search result duplicates were eliminated.

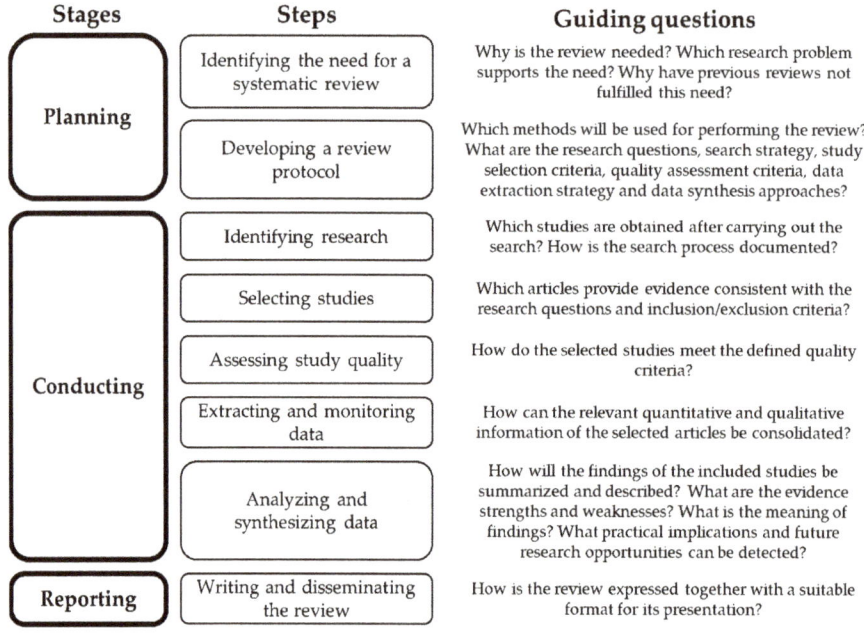

Figure 1. Methodology used for the review. Adapted from References [51–53].

Inclusion, exclusion, and quality assessment criteria: Publications with direct applicability to hospitals, from a comprehensive perspective, were included. Those that moved away from these entities as focal organizations, or on the contrary, focused on very specific chains, such as blood or laboratory, were excluded. Studies were also filtered based on their relationship to the TBL approach. Thus, contributions that referred to sustainability as the continuity of the specific health programs implemented, in order to analyze the effectiveness of such programs, were excluded. Articles and reviews from peer-reviewed journals were primarily considered. However, by review of publications' references, additional studies and international guidelines were considered suitable, such as References [58,59], since they specify SSCM practices applied in hospitals worldwide. Co-authors acted as coders to decide whether each publication retrieved from the search should be included or not. In cases of disagreement, these were discussed until consensus was achieved.

Data extraction strategy and synthesis approach: In accordance with the structure employed by most articles, as well as the information provided, the variables selected for data extraction, analysis, and synthesis were as follows: sustainability dimensions addressed, practices identified, performance metrics identified, and research suggestions. Concerning the data synthesis method, a mixture of interpretative and explanatory approaches was adopted, in an attempt to exceed description [51], as the pursued goal, being conceptual in nature, was the development of an integrative framework to facilitate the understanding of what, how, and why SSCM practices influence economic, environmental, and social performance in hospitals.

3. Results

As a result of the search strategy and selection criteria application, 58 documents were retained for analysis and synthesis (Figure 2). Out of 278 publications encountered in the databases, 80 were duplicates, 10 were added manually, and 150 were excluded, in accordance with the inclusion and exclusion criteria described in the methodology. This section is divided into three parts: in the first two, identified hospital SSCM practices and performance metrics are presented, respectively, in accordance with their categories. In the third section, specific practices and their effects on performance are discussed, to illustrate the relationships based on quantitative and qualitative findings reported in the reviewed literature.

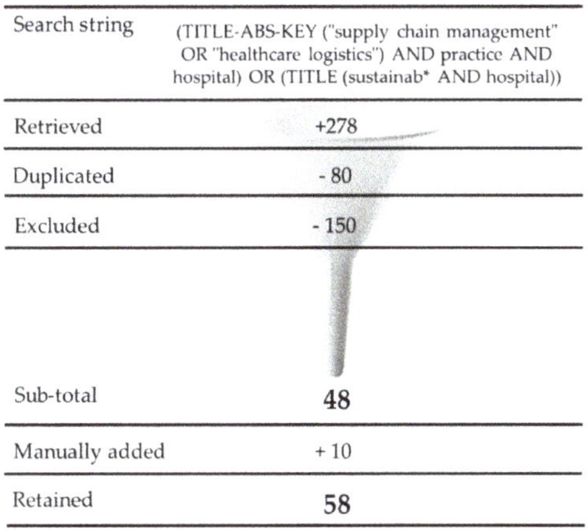

Search string	(TITLE-ABS-KEY ("supply chain management" OR "healthcare logistics") AND practice AND hospital) OR (TITLE (sustainab* AND hospital))
Retrieved	+278
Duplicated	- 80
Excluded	- 150
Sub-total	48
Manually added	+ 10
Retained	58

Figure 2. Search strategy results.

3.1. SSCM Practices

Different approaches may be found in the literature regarding the concept of Hospital SCM. Reference [60] divides this concept into internal management (material and information flows), and external management (material, information, financial, knowledge flows, and relationships). This differentiation is also addressed by other authors, who refer to the concept of healthcare logistics. For example, Reference [61], based on Reference [62], identifies an external chain composed of manufacturers, distributors, purchasing groups, providers, and users, as well as an internal chain that includes supply management, inventory management, replenishment, and utilization. In addition to medical products, hospital logistics include the management of support services required for care. These may encompass food, laundry, patient transportation, information technology management, waste management, and home care services [63,64]. In fact, the concept of healthcare logistics is also meant to include operations such as care units and operating rooms [30].

In addressing the first research question, 13 categories emerged from this review for the classification of SSCM practices, which include: (1) strategic management and leadership, (2) supplier management, (3) purchasing management, (4) warehousing and inventory management, (5) transportation and distribution management, (6) information and technology management, (7) energy management, (8) water management, (9) food management, (10) hospital design, (11) waste management, and (12) staff and community behavior. A final category called "others" was designated to include contributions that were problematic to fit into the above-mentioned topics, despite their potential to provide important insights for practices. The rationale for establishing these categories emerged on examination of the ways in which practices have been classified previously in the literature, which Figure 3 illustrates. Some examples of practices, in accordance with the above-defined categories, are shown in Table 1.

Table 1. Summary of categories and examples of identified sustainable supply chain management practices in hospitals.

Categories	Examples of Practices	Author(s)
Strategic management and leadership	1. Establishment of a strategic plan for supply chain management.	[16,17,65,66]
	2. Development of green and healthy policies and plans.	[58,59]
	3. Executive support for supply chain management processes.	[17,58,59,61,67]
	4. Use of indicators and measurement systems to assess total supply chain costs and performance.	[17,19,60–62]
	5. Involvement of clinical and non-clinical staff in supply chain decision-making.	[17,65,68]
Supplier management	1. Supplier base rationalization.	[62,65,69–71]
	2. Sharing information with suppliers related to material flow management (forecasts, planned consumption, inventory, costs, promotions, and performance).	[46,60,72]
	3. Inclusion of environmental, economic, and social dimensions in supplier arrangements.	[58,59,69]
	4. Selection of ISO 14000-certified suppliers.	[58,59,69]
	5. Work with suppliers to innovate and improve availability of sustainable products.	[58,59,69,72]
	6. Assessment of suppliers' sustainability and ethical practices.	[58,59,73]
	7. Knowledge sharing and transfer (improvements, special handling requirements, good practices, technical issues, management solutions, and new product planning and development).	[60]
	8. Payment control (enhanced control of payments to suppliers focused on preventing delays).	[60,65]
Purchasing management	1. Supply standardization.	[17,46,62,65,70,74]
	2. Use of purchasing groups.	[17,46,61,62,65,70,75]
	3. Alliances between hospitals for the purchase of common items (aggregating purchasing volumes) to attain lower prices and avoid monopolies.	[46,64]
	4. Use of the life cycle analysis to assess the environmental impacts of procured items.	[19,50,76]
	5. Considering the environmental and human rights impact of procured products.	[58,59,73,77]
	6. Purchasing of reusable, rather than disposable, products.	[50,58,59,76]
	7. Eliminating, minimizing, and substituting chemicals for safer alternatives.	[58,59,78]
	8. Coordination between hospitals to increase buying power for economic, environmental, and ethical purposes.	[58,59]
Warehousing and inventory management	1. Determination of quantity to order and reorder points based on information systems.	[61]
	2. Development of collaborative arrangements with trading partners to manage inventory of functional products (non-critical medical supplies) with high and stable demand (vendor-managed inventory, CPFR - collaborative planning, forecasting and replenishment, outsourcing).	[46,60,64,66,70,79,80]
	3. Use of hybrid stockless systems (high-volume products are delivered directly to points of care and low-volume products are delivered to the central store).	[46,64,79,81]
	4. Store consolidation and deployment of a centralized replenishment system for nursing units.	[16,62–65,74]
	5. Deployment of a two-bin system.	[16,65,68]
Transportation and distribution management	1. Consolidation of inter-site transport system.	[16]
	2. Consolidation of external transport.	[16,70]
	3. Promotion of active travel.	[50,58,59]
	4. Promotion of public transport use.	[50,58,59]
	5. Promotion of shared occupancy vehicle use.	[50,58,59]
	6. Use of alternative fuels and technologies.	[58,59]
	7. Development of services to minimize travel (e.g., telehealth, home healthcare, and videoconferencing).	[58,59]
Information and technology management	1. Use of information systems and technologies in interactions between hospital departments.	[17,60,65,67,82,83]
	2. Internal joint initiatives regarding product availability improvement and logistics cost reduction.	[60]
	3. Deployment of an e-commerce system.	[16,60,62,63,70]
	4. Use of track-and-trace systems (e.g., barcodes, Radio Frequency Identification).	[16,18,46,60,63,66,67,70,84,85]
	5. Collaboration among supply chain partners using pharmacy information systems.	[84]
	6. Automated central stores.	[16,66]
	7. Use of automated guided vehicle systems for the transportation of pharmaceuticals, meals, linen, waste, patient files, tests results, lab tests, blood samples, and non-stock purchases.	[64,65,68]
	8. Use of supplier relationship management system for the interaction between hospitals and their suppliers.	[60]

Table 1. *Cont.*

Categories	Examples of Practices	Author(s)
Energy management	1. Implementing initiatives for saving (e.g., conducting periodic audits, installing variable-speed drive fans for operating theatres, automatic lighting timers, and sensors, updating lighting to LED).	[19,50,58,59,76,78]
	2. Use of alternative technologies (e.g., cogeneration – combined heat and power).	[58,59,78]
	3. Shifting to cleaner fuels.	[58,59,78]
	4. Applying Lean Six Sigma approach to optimize a hospital linen distribution system.	[18]
	5. Implementing social marketing interventions (turning off machines, lights out when unnecessary, closing doors when possible).	[86]
Water management	1. Implementing initiatives for saving (auditing usage, controlling leaks, installing flow restrictors and dual-flush toilets, use of drought-resistant plants, reclaiming water from services such as dialysis and sterilization, harvesting rainwater).	[50,58,59,78,87]
	2. Switching from film-based radiology to digital imaging.	[58,59]
Food management	1. Serving locally grown and organic food.	[58,59,88,89]
	2. Integrating the nutritional care pathway, nutritional standards, and regional menu framework.	[90]
	3. Purchasing sustainable products (rBGH-free, cage-free eggs, meat produced without hormones or antibiotics, certified organic and fair-trade coffee).	[58,59,89]
	4. Identifying and working with small, local vendors to achieve healthy food goals.	[58,59,90,91]
	5. Limiting meat consumption.	[58,59,92]
	6. Applying tariffs to reduce prices for more sustainable choices (e.g., vegetarian meals) and maintaining higher prices for less-sustainable options (e.g., high-fat dishes).	[91,93]
	7. Recycling (fat, oil, grease, cardboard, paper, batteries, plastic, aluminum, newspaper, and tin cans).	[58,59,88,93]
	8. Composting.	[58,59,88,89]
Hospital design	1. Flow-through design (design for product, information, and people flow).	[62,65,68]
	2. Integrated nursing workstations.	[62,65]
	3. Building and adapting facilities considering sustainability criteria (using safer materials, local and regional materials, locating hospitals near public transportation routes, planting trees on site, incorporating design components such as day lighting, natural ventilation, and green roofs).	[19,58,59,78,94,95]
	4. Application of sustainability healthcare-building assessment tools (e.g., BREEAM, LEED, and CASBEE).	[50,94–96]
Waste management	1. Addressing over treatment and implementing methods like social prescribing.	[58,59,97]
	2. Development of processes that use less material and improved technology.	[67,78,83]
	3. Proper segregation.	[58,59,78,98,99]
	4. Recycling.	[58,59,78,98,99]
	5. Use of alternatives to incineration.	[58,59,78]
	6. Setting of criteria and procedures regarding reverse logistics.	[71]
	7. Take back programs of pharmaceuticals for patients and communities.	[58,59,71]
	8. Applying Lean Six Sigma.	[18,30,100]
Staff and community behavior	1. Hire/train well-qualified supply chain professionals.	[17,61]
	2. Encouraging critical thinking within the community to understand, adopt, and promote sustainability initiatives.	[50,58,59]
	3. Education of staff and community on sustainability.	[58,59,71,72,91,93]
	4. Joint initiatives with the community for disease prevention and environmental health.	[58,59]
	5. Collaboration with stakeholders to address environmental problems and develop plans to improve sustainability.	[58,59]
Other practices	Quality management practices (quality policy, employee training, product/service design, supplier quality management, process management/operating procedures, quality data and reporting, employee relations).	[83,101]
	Patient flow logistics (cross-functional or cross-organizational teams, information technology support, format standardization for information sharing, meetings focused on both medical and inter-organizational integration issues, and application of lean and agile approaches).	[14,102–104]

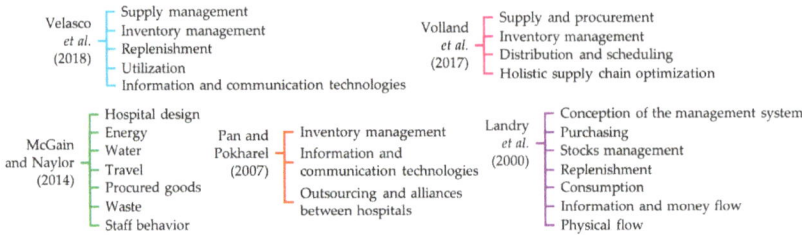

Figure 3. Categories identified for hospital sustainable supply chain management practices classification.

Strategic management and leadership practices are identified as a starting point to map out and control the resources, responsibilities, and implementation of other practices. Other categories, such as supplier management, purchasing management, warehousing and inventory management, transportation and distribution management, and information and technology management have traditionally been discussed from operational and economic perspectives. However, sustainability has contributed to the integration of environmental and social aspects into these topics, such as by considering the environmental and human rights impacts of procured products. Likewise, energy, water, food, hospital design, waste, and staff and community behavior have had a primarily environmental focus, as addressed by Reference [50].

"Other practices" comprises contributions that did not completely fit into the preceding categories. This holds true for studies that analyze the effect of quality management practices on non-financial and financial performance [101], and the application of an SCM perspective in patient flow logistics [14,102–104].

3.2. SSCM Performance

In addressing the second research question, SSCM performance metrics used by hospitals were identified and split into economic, environmental, and social factors (Table 2), as is customary in the SSCM literature [32,105,106], which is in line with the TBL approach. When applicable, metrics were also grouped into categories. For example, Reference [32] includes operation, market, and finance metrics as part of the economic dimension, pollution control, and resource utilization metrics as part of the environmental performance, and enterprise and employee perspective metrics as part of the social performance. Reference [105] divides performance into competitiveness, environmental, operations, and employee-centered and community social performance, while [106] groups metrics included economic, environmental, and social factors.

Table 2. Sustainable supply chain management performance metrics in hospitals identified in the literature.

Dimensions	Categories	Metrics	Author(s)
ECONOMIC	Purchasing and supplier management	Categories of items handled, percentage of purchases using contracts, contract renewal, percentage of purchases using purchasing groups, number of employees dedicated to supply management, percentage of purchases made directly from manufacturers, percentage of purchases using consignment, level of sophistication of the purchasing planning process, total number of products per order, total number of purchasing orders, percentage of complete orders, percentage of urgent orders, number of indicators used in supply management, demand and forecast accuracy, delivery reliability, percentage of perfect orders delivered by suppliers, quick response, lead time from suppliers, and number of active suppliers.	[18,60,61,63,70,72,107]
	Warehousing and inventory management	Space utilization, order sorting, receiving completeness, cross-docking, service levels in the central warehouse (internal customers), inventory policies (manual/information system), number of Stock Keeping Units (SKU), order delivery (planned/not planned), number of indicators used in inventory management, inventory visibility, inventory availability, number of items in inventory, inventory levels, rupture rate, medical devices and pharmaceuticals stockouts, inventory accuracy, inventory turnover, reduction in stock variety, and reduction of time spent by clinical staff to perform logistics tasks.	[16,18,30,60,61,63,68,70]
	Transportation and distribution management	Perfect delivery condition, order delivery in full, delivery performance to customer commit date, on-time delivery, service speed, overall average delivery lead times for formal orders, urgent delivery, number of transactions (inputs-outputs), utilization of transport services, and medication delivery trips.	[18,61,63,67,70,72]
	Information and technology management	e-procurement (extent to which it is implemented), ease of use and usefulness, product identification, accurate and reliable tracking, information availability, information accuracy, information kept up to date, adherence to standards and rules, communication among parties, and amount of information sharing.	[18,63,72]
	Market	Market share, capacity to develop a unique competitive profile, market growth, market development, and market orientation.	[30,101]
	Processes and capacity	Perceived operation processes standardization, procedure preparation time and waste, service capacity, and increase in efficiency due to visual work standards.	[18,67,70,72]
	Financial	Purchasing costs for medical devices and pharmaceuticals, value of orders coming from tender processes, value of orders chosen without tender processes, administration costs for medical devices and pharmaceuticals flows, value of discounts and rebates, supply expense as a percentage of total hospital expense, supply expense per patient admission, supply expense per case-mix index adjusted admission, inventory value, value of inventory lost, inventory carrying costs and stocking requirements, transportation costs, revenue growth, profitability, net profits, return on investment, profit to revenue ratio, cash flow from operations, cash flow rate, share of net patient revenue, patient profitability, cost of services, operating costs, cost reduction due to the quality of service delivery, maintenance costs, savings due to efficiency and conservation improvements in energy, water, waste, travel, and food, and social investment volume.	[19,30,50,58–61,67,70,72, 86–88,91,92,98,101,107]
ENVIRONMENTAL	Procurement	Reduction of material consumption, drugs and packaging, decrease in consumption of hazardous/harmful/toxic materials, reduction in air emission/pollution from procurement, and reduction in air emission/pollution from anesthetic gases.	[50,58,59,78,100]
	Energy	Reduction in energy consumption, increase in energy efficiency, reduction in air emission/pollution from energy consumption, energy usage per unit area, and increase in the use of clean and renewable energy.	[18,19,50,58,59,78,86]
	Water	Water consumption, water footprint.	[50,58,59,78,87]
	Travel	Reduction in air emissions/pollution from business travel, patient transportation services, staff and community travel, increase in fully electric fleet and pool vehicles, reduction in fuel consumption, decrease in staff car use, and proportion of journeys made by a car.	[50,58,59]

261

Table 2. Cont.

Dimensions	Categories	Metrics	Author(s)
SOCIAL	Food	Percentage of locally and sustainably sourced foods procured, reduction in air emission/pollution from food supply, reduction in nutritional waste, and patient and staff satisfaction with healthy food choices provided.	[50,58,59,88–91,93]
	Hospital design and buildings	Compliance with environmental and social value certification standards.	[58,59,94]
	Waste	Decrease in waste generation from pharmaceuticals, chemicals, materials (e.g., products and equipment, packaging), and food, perceived waste reduction in processes, avoidance of improper waste mixing and incineration, proper waste disposal, percentage of toxic waste, decrease in incineration waste as a percentage of the total, improvement in ability to reuse/recycle/compost, and a reduction in waste disposal sent to a landfill.	[50,58,59,67,72,78,98,100]
	Quality of patient care	Death rate, timely provision of healthcare, length of stay, improvement in patient experience (quality of sleep, level of privacy, thermal comfort, service quality as perceived by customers, overall satisfaction with hospital experience), perceived care quality compared to other hospitals, service level, and perceived service level compared to other hospitals.	[30,60,72,83,86,101]
	Employee	Improvement in worker safety and health at work, improvement in employee awareness and education, improvement in worker efficiency, employee satisfaction, employee work life quality, proportion of working hours to that planned, staff absenteeism, employee privacy, and staff utilization.	[19,70,72,86]
	Community	Job creation, image/reputation among major customer segments, reduction in corruption and bribes, increase in population well-being, and stakeholder satisfaction.	[30,72,87,101]

Findings show that most of the identified metrics are economic, which is coherent with the prominent attention that this dimension has received in the literature over time. Both recent and older publications that address the effects of SCM on performance, without explicit consideration of a holistic sustainability approach, have defined performance through a competitive advantage [7], operational [5] market, and financial constructs [108].

Conversely, the social dimension is that which contains the smallest number of identified indicators, which is consistent with the lesser recognition of this dimension in the literature [109]. On the one hand, social issues are considered difficult to measure, since they involve subjective, complex, and dynamic factors of human nature [110]. On the other hand, the literature shows significant advances in the identification of social issues of interest for supply chain management, but slow progress in their operationalization [109].

Environmental indicators are in a halfway position between economic and social indicators, pertaining to quantity. In the reviewed literature, efforts to measure natural resource consumption and waste generation, as well as the economic projection attributable to practice implementation, are evident. Predominance of this dimension over the social one may be explained by the considerable availability of publications and empirical results, mainly on green supply chain management [37,38].

3.3. Analysis of SSCM Practices and Illustrative Effects on Sustainable Performance

3.3.1. Strategic Management and Leadership

The SCM strategy has become a prerequisite for practice deployment [16]. However, it appears that strategy and organizational changes are hardly successful if there are no responsible and trained leaders who establish and control SCM priorities, plans, work teams, and performance measurement [17]. Clinical staff participation on logistics work teams is commendable, as it may help to solve natural conflicts between stakeholders [17,65], such as the product variety desired by physicians, in contrast to the economies of scale pursued by pharmacy managers [80]. To counteract patient demand uncertainty, high inventory levels are often seen as favorable by clinicians [17] and about 60% of supply spending is influenced [74,111]. Thus, the incorporation of clinical perspectives facilitates a consensus about purchasing decisions, in order to reduce costs without detriment to quality.

From a sustainability perspective, which not only includes economic aspects, certain matters become relevant in advocacy for green and healthy hospitals. According to References [58,59], the organizational culture needs to be changed through practices like the development of green and healthy policies and plans, upper management and staff support for environmental and health issues, and the dedication of human and financial resources to green initiatives.

3.3.2. Supplier Management

It seems clear that organizational performance depends upon the way in which suppliers are managed [112,113]. Practices such as supplier rationalization are often suggested in the reviewed literature, as it lends not only the possibility of ordering higher volumes that generate financial savings, but also of building long-term relationships that enhance trust and enable the implementation of collaborative initiatives [65,69,70]. For example, Reference [70] suggests that the implementation of vendor-managed inventory arrangements is easier after having reduced the supplier base.

Reference [60] found that, in addition to having high levels of cross-departmental interaction, leading hospitals embark on joint initiatives with their suppliers to improve product availability and reduce costs, as well as extensively share information and knowledge with them regarding forecasts, consumption plans, inventory levels, costs, joint efforts, technical information, good material flow management practices, and new products and services. Concerning financial flow management, leading hospitals tend to keep payments under control, in order to prevent delays. As a result, these hospitals value the effects of this external integration positively.

Given that quality of care, health, and hospital reputation can be compromised by problems related to procured products. Hospital sustainability implies supplier sustainability as well. In manufacturing contexts, a typical example is when the procurement of harmful materials takes place, which can cause adverse events and lead to recall products from the market, as well as other consequences including criticism, damage to hospital reputation, and economic losses [114]. Therefore, the selection of certified suppliers [58,59,69], supplier sustainability reporting [73], supplier audit programs [62], and assessment of suppliers' environmental and ethical practices [58,59,73] emerge as important practices in the arena of supplier management, as a way to ensure compliance with economic, environmental, and social standards. Furthermore, suppliers are uniquely poised to contribute to sustainability, as the development of more innovative and sustainable products largely depends on their capacity, readiness, and time invested therein [69].

3.3.3. Purchasing Management

Unsurprisingly, purchasing management is among the categories with the highest number of practices, as it represents a large portion of hospital budgets. On average, the share of supply expense, in reference to tangible supplies, is 15%, and can reach 40% in hospitals with high clinical complexity [107]. Among the most commonly mentioned practices are product standardization, purchasing group use, and creation of alliances with other hospitals. These practices have an essential economic orientation. For example, product standardization decreases item variety, and, therefore, the obtention of better prices and inventory reductions [62,65,70]. The use of purchasing groups has also turned out to be a beneficial practice for the achievement of more competitive prices and economies of scale, through the purchasing power acquired by these groups, as a consequence of volume consolidation [17,62,84]. Said effects also pertain to the creation of alliances with other hospitals [46,64].

Instead, practices like the consideration of the environmental and human rights impacts of procured products have more visible environmental and social backgrounds. Only in the United Kingdom does 57% of the healthcare footprint comes from procurement [20]. Consequently, criteria that refer to greater product durability, reduced waste generation, and less packaging and hazardous material use are recommended [58,59,75,77]. As Reference [75] demonstrates, purchasing of bundled new and refurbished products may result in significant economic and environmental gains, if properly combined. Similarly, Reference [77] sheds light on how the packaging design needs to be considered in purchasing processes in order to improve logistic efficiency.

In addition to economic and environmental motives, the decision to source from suppliers that offer the most competitive prices must not be at the expense of unethical conditions and human rights violations, as in the case reported by Reference [58], in which ten-year-old children worked in the street to produce surgical scissors.

3.3.4. Warehousing and Inventory Management

Because about one-fifth part of healthcare revenue is attributed to inventory management [80], implementation of practices to improve the reception, warehousing, and control of supplies can be more than justifiable. The reviewed literature reports, for example, how the determination of quantities to order and reorder points based on information systems, in contrast to manual processes, has aided in the prevention of stock-out and overstock [61]. It further reports the ways in which the development of collaborative arrangements can be effective, depending upon contingent factors. As found by Reference [79], vendor-managed inventory is likely to work well for products with high and stable demand, which are not subject to highly-regulated environments, and when spatial complexity is low, the distance between organizations is not excessive and does not put supply at high risk of breakdown.

Similarly, the use of hybrid stockless systems has been recommended. This involves the delivery of high-volume products directly to points of care and low-volume products to central stores [64,79,81]. A completely stockless system is ideal for removing central stores and releasing space [68], but may fail in a hospital environment that deals daily with unpredictable emergencies, or in remotely-located hospitals in which response times might be significant [79]. Related to points of care, the findings of Reference [16] suggest that centralization of replenishment systems for nursing units is a practice that results in reducing surplus inventory, as well as administrative time for nursing, which works in favor of their dedication to delivery of care.

3.3.5. Transportation and Distribution Management

Case studies discussed by Reference [16] include practices related to transportation. In the Canadian hospitals explored by these authors, it was shown that the consolidation of inter-site transport produced economic savings of up to 35%, substantially reduced delivery times, and positively impacted customer service, which are a basis for the consolidation of external transport. Reference [70] emphasizes the need in the health sector for such external consolidation, as the large number of transport providers and their independent operations create valuable opportunities for capacity and routing optimization, which reduces both time and costs.

It is important to mention that a stream of practices pushes toward transport minimization for environmental reasons, given its high impact on CO_2 emissions [50]. From this perspective, the avoidance, or at least reduction of travel, is a primary goal, through the encouragement of active travel and the promotion of the use of public transport, shared occupancy vehicles, and electric vehicles [50,58,59]. Virtual solutions have proven valuable for the replacement of face-to-face meetings and appointments, since they avoid unnecessary patient and staff travel, both in administrative and clinical environments, through solutions such as tele-conferencing and telehealth, respectively [50,58,59]. In addition, the exploration of regulatory mechanisms, based on incentives and fees, is proposed, to stimulate the adoption of travel options with lower environmental impact and discourage those with the highest impact.

3.3.6. Information and Technology Management

Sharing information with suppliers was discussed in the subsection referring to supplier management. Some of the internal practices applied by leading hospitals mention the use of electronic communication tools and information systems, such as Electronic Patient Record (EPR), bar codes, and Enterprise Resource Planning (ERP) systems [17,60,64]. The relevance of sharing information regarding forecasts, planning, inventory visibility, and delivery dates, as well as the establishment of cross-functional teams that encourage joint initiatives for product selection and standardization, inventory classification, and the discussion of performance metrics has been acknowledged [60].

A bundle of practices is concentrated on supply, inventory, and transport. These consider the implementation of electronic commerce or e-procurement, Radio Frequency Identification (RFID), the integration of medical and administrative information systems, and automation of warehouses and transportation systems. Some outcomes of e-procurement implementation include the reduction of clerical tasks, errors, use of paper, and associated costs [16,70]. RFID, along with barcodes, are part of track and trace systems, which identify medicines, individuals, supplies, or equipment. The identification of products, in particular, generates numerous advantages, in terms of a visibility increase and inventory cost reduction, manual task reduction, patient safety improvement, and support for reverse logistics [85]. By way of a case study about the location of infusion pumps with RFID, Reference [84] reached similar conclusions about the benefits of implementing this technology and even suggesting the integration of medical and administrative applications used by pharmacies to improve SCM agility.

Some experiences regarding the automation of central stores have come into being through the acquisition of carousels [16], while the use of automated guided vehicles has been suggested as a technology practice for transport [64,65,68]. Such vehicles are scheduled for the transportation of multiple items, such as pharmaceuticals, meals, linen, waste, patient files, tests results, lab tests, blood samples, and non-stock purchases. Although the investment payback has totaled approximately five years in hospitals that implemented automated guided vehicles, it has been considered a meaningful practice, given the minimal added value of conventional transportation jobs [68].

3.3.7. Energy Management

Several estimates provide notions of the high amount of energy consumed by hospitals. For example, it is calculated that these comprise 10% of total national consumption in the United States of America [18] and 20% of consumption in the Spanish tertiary sector [115]. Identified practices regarding energy mainly point to conservation measures, the use of alternative energy technologies and fuels, the application of lean six sigma, and behavior change interventions.

Motivated by facts such as the annual premature death of three million people due to air pollution, the University Health Network of Canada put a systemic approach into action that includes initiatives for energy use efficiency improvements. Some of these refer to the optimization of ventilation systems and replacement of existing lighting with LED, which resulted in quantifiable financial and consumption savings, and improved patient and staff comfort [19]. Similarly, Reference [76] suggests that variable-speed drive fans, lighting timers, and sensors have been effective in the reduction of energy consumption by up to 50% in operating rooms, while Reference [78] reports that cogeneration plants have allowed some hospitals to generate over half of their own energy.

By applying analytical tools derived from Six Sigma, Reference [18] proposed a future state to optimize a hospital linen distribution system, which led to improvements in communication, demand forecast accuracy, effectiveness, responsiveness, and reliability, which increased energy consumption efficiency. Reference [86] showed that turning off machines and lights when unnecessary, and closing doors when possible, as part of a social marketing intervention, proved successful not only in the reduction of energy consumption and carbon, but also in the improvement of the work environment and patient experience indicators such as quality of sleep and overall satisfaction.

3.3.8. Water Management

Hospitals use substantial amounts of water, which accounts for approximately 7% of the total water consumed in the tertiary sector in some countries [116,117]. According to the reviewed literature, auditing, controlling for leaks, and installing more efficient fixtures in both toilets and showers can lead to savings of up to 25% [50,118], while more complex solutions might imply transformations in clinical services operation. The latter choice refers to options including switches from conventional radiology to digital imaging, which not only reduces water use, but also reduces harmful radiological chemicals [58].

Another focus of practices involves recycling water from sterilization, dialysis, and other processes [50,118] for use in non-potable needs [87]. To examine the impact of different policies related to water management in hospitals, Reference [87] proposes a causal model and studies two scenarios by using system dynamics. Simulation results indicate that a 15% water reduction policy leads to a reduction of 12% in the water footprint, savings in cost of services up to 14%, and a population well-being increases from 1.116% to 1.117%. In contrast, a 20% water reuse policy leads to a reduction of 16% in the water footprint, savings in the cost of services at 19%, and a population well-being increases from 1.116% to 1.117%. The water footprint denotes water consumption, cost of services refers to daily average cost of resources per patient, and population well-being is measured in terms of patient admittance.

3.3.9. Food Management

High-fat processed food, the use of non-nutritive additives, meat produced using antibiotics and hormones, obesity, antibiotic resistance, diabetes, cancer, food waste, and pollution caused by food transport are among the problems that current food systems face [88,89,91,92]. Hospitals have the potential to impact sustainability by addressing food issues, given their role as intermediaries in the market, their buying power, their responsibility for the promotion of proper nutritional habits, and the large number of people who frequent these organizations, between patients, visitors, employees, and the community [93].

In the reviewed literature, publications that focus on food sustainability show that recycling and avoiding the sale of bottled water are common practices, in contrast to composting and serving organic and locally grown food [88]. Reference [93] identifies 12 opportunities through which food practices may be addressed: procurement, catering contracts, menu development, pricing, waste management, infrastructure, staff training, information, education, communication and feedback, partnerships, and special events. Similarly, Reference [91] suggests practices that range from the participative design of new options with staff and customers to behavioral initiatives that encourage the consumption of healthier food, while Reference [92] shows that reducing meat consumption by up to 20% and substituting it for vegetarian or alternative proteins from local sources is feasible for hospitals, without a detriment to budgetary increases.

A case presented by Reference [90] provides insights regarding the improved fulfilment of patients´ nutritional needs, their increased satisfaction, waste reduction, and local economy enhancement, by sourcing from a single and local supplier and articulating nutritional standards with regional menu frameworks. However, unlike Reference [90], positive outcomes in all sustainability dimensions are sometimes mixed. For Reference [88], the implementation of food sustainability practices overrides their costs, whereas Reference [93] finds cost to be an obstacle. Reference [91] concludes that not only is price a restriction on healthier food, but so too is the difficulty of preparation, staff involvement, and creativity to promote said meal in a market that is accustomed to and satisfied with fries and sugar-sweetened beverages.

3.3.10. Hospital Design

For any hospital, fluid architecture is desirable to facilitate logistics, which, in turn, assists with people, material, and information flows [65,68]. Coherently, one of the practices implemented in some hospitals has been the integration of nursing stations, through a design that groups the elements of information, medicines, and materials required for care, and which not only contributes to ergonomic improvements, but also contributes to reducing the distances travelled by nursing staff [65].

In addition to making flows more effective, sustainability raises challenges that generate the need for more complex planning for future facilities, as well as adaptation of existing facilities [96]. These challenges are aimed to ensure more efficient use of resources such as energy, water, and waste management, better social conditions, in terms of accessibility, safety, comfort, and patient experience, and improved economic outcomes with a reference to life cycle costs and contribution to local economies [94]. Specific recommendations for building and adapting facilities, considering sustainability criteria, include using safer and local materials, sitting hospitals near public transportation routes, planting on-site trees, and incorporating design components like day lighting, natural ventilation, and green roofs [58,59].

One limitation of current sustainability demands, however, many of today's hospitals operate in old buildings that consume large amounts of resources, and whose design is not carefully planned to favor aspects such as those mentioned above [96]. Furthermore, trade-offs can arise between dimensions. For example, larger patient rooms generate greater comfort, but consume additional environmental resources [94,119]. An initial step for hospital building modification lies in the application of a sustainability assessment tool. For this purpose, different options, such as the Building Research Establishment Environmental Assessment Method (BREEAM), Leadership in Energy and Environmental Design (LEED) for Healthcare, Green Star Healthcare, and Comprehensive Assessment System for Built Environment Efficiency (CASBEE) have been developed [95].

3.3.11. Waste Management

Significant volumes of waste are generated by hospitals. In Victoria, Australia, for example, public hospitals generate as much waste as 200,000 households [76]. Besides environmental motives, waste management is important for public health reasons. In countries like India, Reference [98] found that regulation is still weak while non-hazardous and hazardous waste are often mixed together, and large amounts of waste are unnecessarily incinerated, which causes avoidable toxic air pollution. Consequently, Reference [98] proposes a system that encompasses reduction strategies, segregation, and recycling of non-hazardous waste. After conducting a pilot study at one hospital, they projected quantities that can be prevented from being improperly disposed and incinerated, in addition to the economic benefits that this would bring, together by increasing recycling.

Even more effective practices refer to avoiding waste generation altogether, which has been made achievable by addressing overtreatment, instigating methods such as social prescribing, development of processes where less material is necessary, waste stream analysis, review of waste-generation processes, selecting safer chemicals, purchasing environmentally-friendly products, purchasing reusable rather than disposable products, and acquisition of improved technologies [58,76,78,97]. Consequently, some hospitals have reported decreased use of hazardous chemicals like mercury, and reduced waste generation, which translate into financial savings [78]. As stated by Reference [76], single-use plastic trays double the cost of reusable trays, which means that only the reuse of these elements would represent annual savings of $5,000 for a 300-bed hospital.

Case studies regarding the application of Lean Six Sigma in medication processes [18,100] and sterile processing [18] illustrate the way in which it can lead to improved medication availability, fewer missing medications, reduced medication delivery trips, less kit variety for sterile processing, less waste, and financial savings. Similarly, despite the fact that waste cannot be completely avoided, Reference [71] demonstrates that reverse logistics processes offer significant opportunities for hospitals and healthcare systems as a whole. An intervention undertaken in one hospital allowed the value to return to stock, and it was found that recycled and disposed drugs represented around 3% of total drug expenditures. This was a starting point for establishing responsibilities, criteria, procedures, and schedules for the collection, review, and classification of returned items [71].

3.3.12. Staff and Community Behavior

Several examples show that capabilities, culture, and psychological factors are key determinants for the successful implementation of SCM practices. In other words, lack of training and education of supply chain professionals and executives is a common barrier [17,61], as is proneness to issues like sharing information, which pivots on organizational culture [17]. Variables like shared values have proven to play a mediating role between other variables, such as the use of electronic medical records and physicians´ performance [120], while willingness to support the implementation of practices depends on individual interests and the degree to which they make true sense for employees, patients, and the community [50].

From an economic perspective, Reference [66] includes the institutionalization of training and development as a best practice for the improvement of hospital supply chains. Reference [17] draws attention to the existing need for training in analytical skills, SCM best practices, leadership, communication, and financial themes, among others. From the environmental and social perspectives, Reference [50] asserts that critical thinking needs to be fostered, and employees must be supported in the process of making ethical decisions that they consider to be coherent with their personal beliefs, if indifference toward and myths about sustainability that constrain action are to be dismantled.

3.3.13. Other Practices

Several studies have found positive and significant links between organizational performance and quality management practices. Reference [101] highlights the importance of encouraging staff involvement, managers development, and strengthening information and statistics tools, since employee relations, training, role of top management, and quality data and reporting, were found to prevail over practices related to factors such as service design, supplier quality management, and process management. Another interesting result is the concluded influence of quality practices on financial performance through non-financial performance, which might indicate the pertinence of investment in quality practices even if it does not result in better financial performance in the short run, but indirectly through market share gains, increased service quality, and other outcomes. [83,101]. Such findings provide helpful evidence for prioritizing these kinds of practices and support of their implementation, since they contribute to market share gains, increased service quality, reduced waste, higher speed, improved quality of care, a superior competitive position, and financial performance.

Lastly, attention should be called to certain publications that are connected with the realm of operations management, which employ the SCM concept as a guide to study planning processes, which assumes that patient flow can be more efficient if a rationale similar to that of product flow is applied. References [103,104] suggest enhancing cross-functional or cross-organizational teams as well as information technology support, and format standardization for information sharing and meetings focused on both medical and inter-organizational integration issues, to address communication, patient safety, waiting times, and integration problems that arise when manifold healthcare providers are involved in patient care. Similarly, Reference [102] discusses the applicability of lean to SCM in combination with agility, while Reference [14] argues that a socio-ecological approach can be applied in hospitals by moving sustainability into the core business, which requires that decisions be made about care planning and service design.

4. A Proposed Framework for Hospital SSCM

In response to the third research question, this section presents an integrative framework for SSCM practices that may impact sustainable performance in hospital settings (see Figure 4). This can be considered innovative in at least three ways. First, as found in the reviewed literature, several publications have outlined relevant sustainability issues, but little attention has been given to the amalgamation of scattered practices and performance measures in a single and articulated framework. Most of the previous research on hospital supply chain management focuses on logistics from a cost reduction perspective, which is indisputably crucial for sustainability, but is insufficient from the TBL approach. Moreover, the publications identified with the sustainability label pivot primarily on the environmental dimension and leave aside the social dimension.

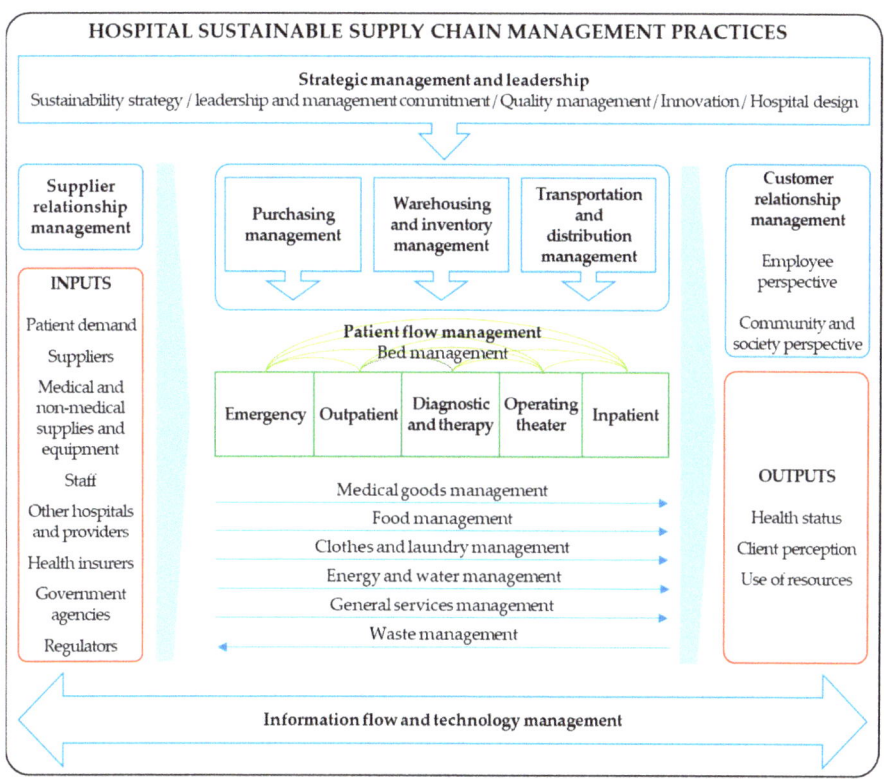

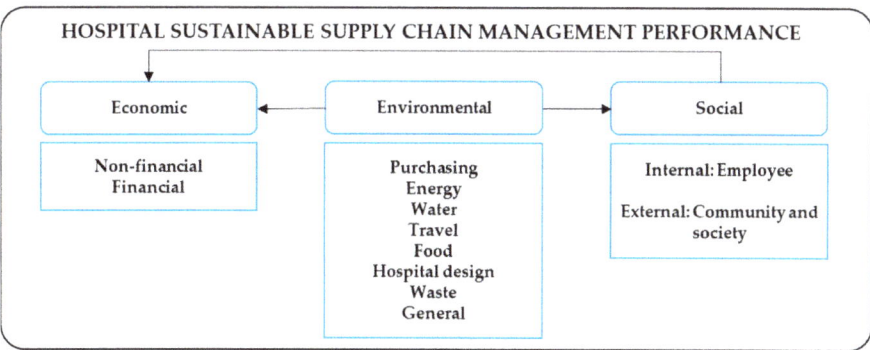

Figure 4. Conceptual framework for hospital sustainable supply chain management.

Second, the wide-ranging identification of practices and performance metrics achieved in the literature review, which gave rise to the proposed framework, likely allows to delineate a clear path toward empirical validation and the managerial implications of practice implementation and performance measurement. While a considerable number of frameworks provide valuable insights on interactions in sustainable supply chain management [40], healthcare supply chain management in the emerging economy with the sustainable lenses [121], and supply chain sustainability in the service industry [122], the degree of operationalization of the categories and exemplified relations presented in this case has not been detected in previous reviews.

A third contribution to highlight is the worldwide applicability of the proposed framework and its possible extension to other service sectors. On the one hand, health services are not new to humanity. Hospitals are necessary in any country and have always existed. Similarly, sustainability issues are of global concern. On the other hand, although the framework was developed from the hospital perspective, this does not prevent it from being used as a reference for other service sectors, if properly adapted. Just as it is extremely important for hospitals to adopt supply chain management concepts and practices that have proved successful in other sectors such as food, research focused on hospitals can be a source of learning [13].

The proposed framework is composed of two main blocks: practices and performance, whose corresponding exploded views are depicted in Figures 5 and 6. For practices, contributions that conceptualize the logistics management process and supply chain integration [3], health care operations management [123], and hospital logistics [124] were considered. Accordingly, components traditionally related to internal supply chains, namely purchasing, warehousing and inventory, and transportation and distribution management, are placed at the center, and serve the care units through which patients flow, which include emergency, outpatient, diagnostic and therapy, operating theater, and inpatient [123]. Undirected arcs connect these units, which means the multiple directions in which patient flow occurs, since varied medical needs create customized sequences [125]. Clothes and laundry management as well as general services management are included, along with medical goods, food, energy, water, and waste management, since they are considered hospital logistic fields [124], account for resource consumption, and influence healthcare delivery, quality of patient care, and patient satisfaction [123]. In addition, strategic management and leadership, as well as information flow and technology management are key constituents of the framework, as they can influence and support supply chain relationships.

Upstream and downstream linkages are represented by Supplier Relationship Management (SRM) and Customer Relationship Management (CRM), respectively. Patient demand heads the list of inputs, as internal operations depend thereupon [123], and healthcare demand has unique characteristics. Rather than desire, healthcare services are grounded on necessity [126], which implies that typical marketing approaches to stimulate demand are minimal, if at all applicable, in healthcare. Other framed inputs include suppliers, medical and non-medical supplies and equipment, staff, other hospitals and providers, health insurers, government agencies, and regulators, which is in line with previous healthcare operations definitions and the numerous players that provide goods, services, and information to make operations possible [123,127]. Regarding outputs, these comprise the health status that reflects in clinical indicators, the client perception that indicates how well staff and patient expectations are met, and the use of resources that denote operation efficiency [123].

In accordance with the scope of the CRM concept, identified practices in the reviewed literature may fall short. It was struggling to identify it as a clear construct because these practices are scarce in healthcare [128], despite the fact that their influence on performance has been widely studied in manufacturing [5,7,129–131]. In such publications, CRM practices have been operationalized into management of customer complaints, evaluation of customer satisfaction, determination of customer expectations, frequent interaction with customers to set standards, and consideration of information from customers for business design and planning. To fill the existing void in healthcare, Reference [128] emphasizes the need to fortify the adoption of CRM practices, considering that they can lead to better understanding of patient profitability, and that there is some evidence of their contribution to patient health and loyalty.

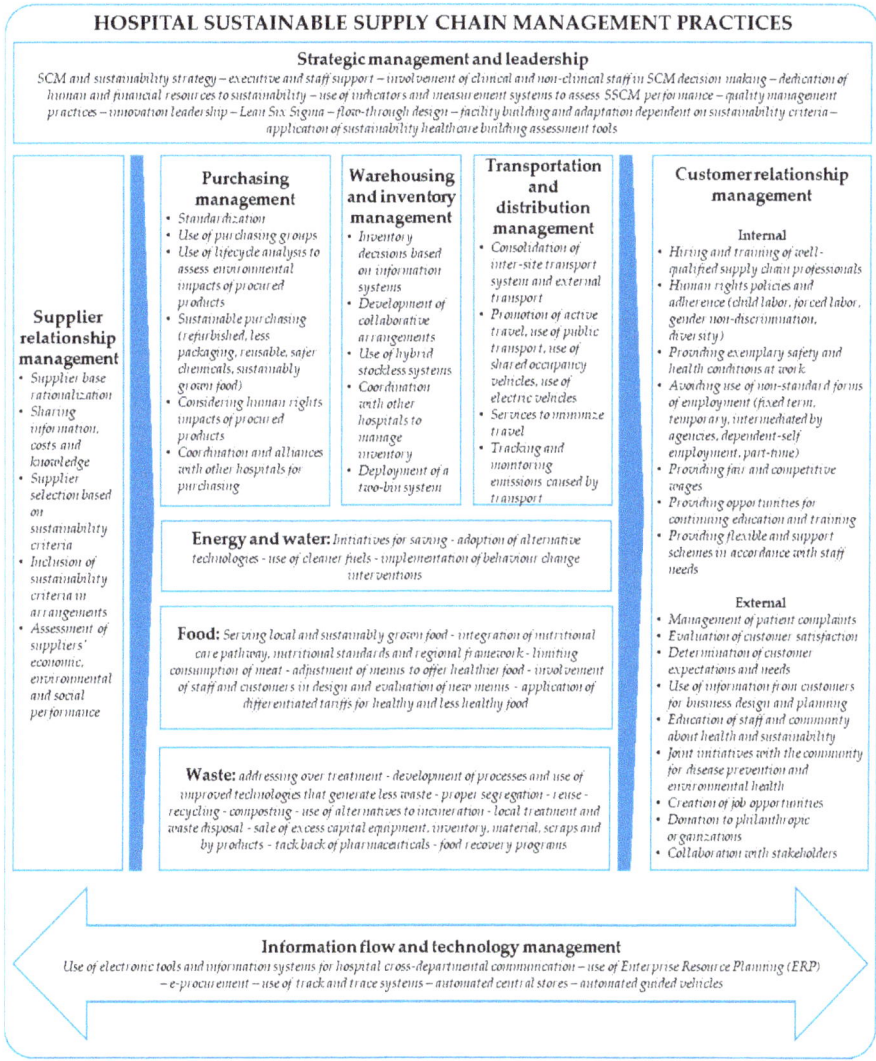

Figure 5. Conceptual framework for hospital sustainable supply chain management practices.

Recent studies on SSCM and specific social healthcare problems also lead to the inclusion of additional practices in the proposed framework. On the one hand, among the issues addressed by employee-centered social practices are wages, worker safety, and occupational health working conditions, employee participation, career planning for staff development, and the provision of opportunities for employees continuing their education [33,39,42,105,106,132–134]. On the other hand, community-centered social practices encompass labor laws, no child labor and human rights compliance, environmental awareness training, promotion of corporate social responsibility in the industry, sustainability reporting, donation to philanthropic organizations, provision of employment or business opportunities to the surrounding community, support of local health, educational, and cultural development, and volunteering at local charities [33,105,106,133,135].

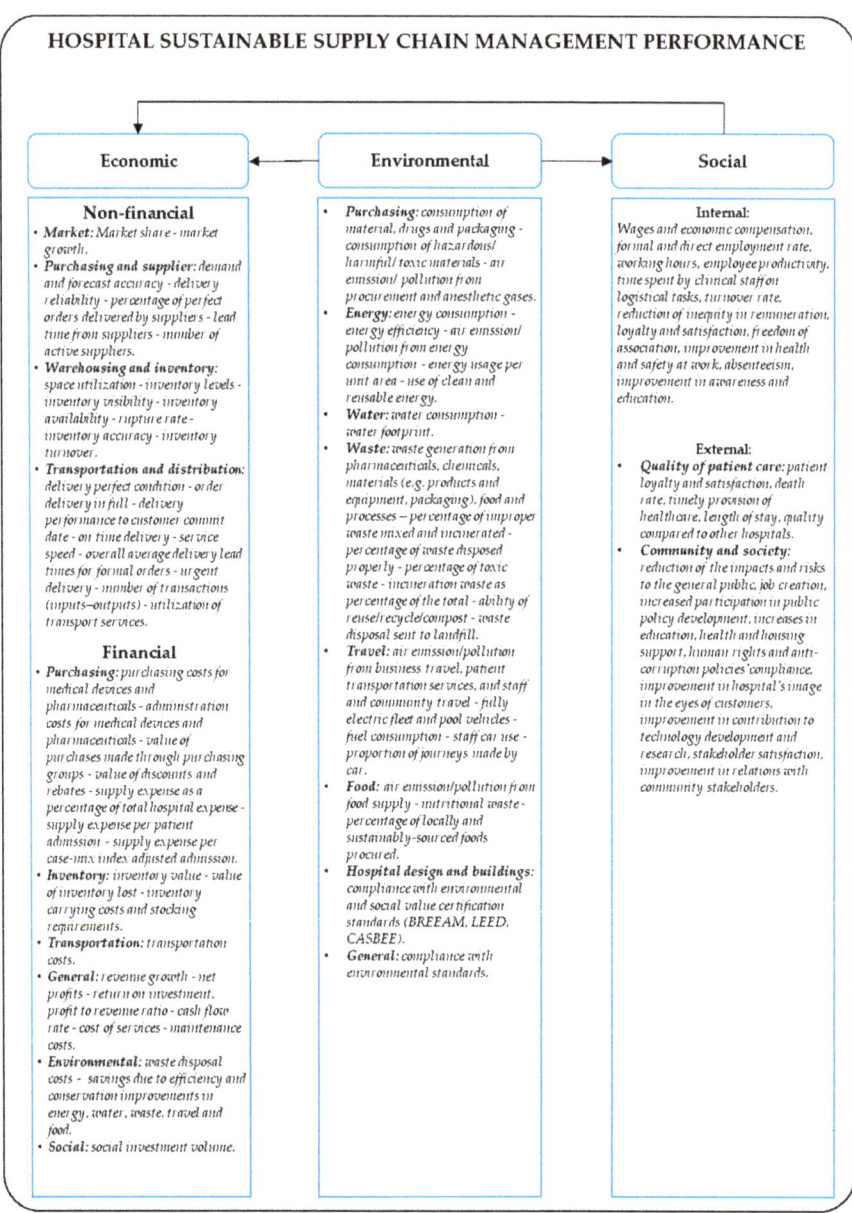

Figure 6. Conceptual framework for hospital sustainable supply chain management performance.

In alignment with the TBL approach, performance components are framed in terms of economic, environmental, and social dimensions. From both non-financial and financial perspectives, economic metrics point to measure operational issues and the costs of classical logistics processes such as purchasing, warehousing, inventory, transportation, and distribution management. However, market-specific metrics are also included in non-financial metrics, since they were reiterative in the reviewed literature. Analogously, general metrics intended to reflect outcomes of the entire hospital, such as profitability, cost of services, and return on investment, are included in financial metrics. Moreover, environmental performance is sometimes converted into financial terms by quantifying actual and potential savings due to efficiency and conservation improvements in energy, water, travel, food, and waste [19,50,58,59,86,88,91,92,98], while social investment volume is used to economically measure social performance [72].

Environmental performance metrics are classified in accordance with the topics of purchasing, energy, water, travel, food, hospital design, and waste. To a certain extent, it could be said that carbon emissions reduction is the last target of environmental interventions, since it depends upon other metrics such as the reduction of resources, materials, drugs, and packaging consumption, increases in the use of clean and renewable energy, decreases in car use, percentage of locally and sustainably sourced procured foods, and avoidance of improper waste mixing. For example, it has been found that 5% of the carbon footprint of acute organizations comes from anesthetic gases [136]. Similarly, waste incinerators emit toxic air pollutants, such as dioxin and mercury [19,98,99]. Therefore, emissions can be reduced as consumption and waste generation decrease.

Except a few data regarding percental decrease in injuries caused by improper disposal [19] and improvements in awareness, education, and efficiency [86], hospital social performance metrics are scarce or vaguely addressed in the reviewed literature. For example, Reference [72] mentions employee satisfaction, work life quality, proportion of working hours to those planned, staff absenteeism, and employee privacy as concepts of hospital SCM, without distinguishing between practices and performance variables. Thus, contributions that focus on SSCM and manufacturing settings are taken as a basis for the framework [32,43,105,106,137–140]. In particular, metrics are divided into internal and external ones, and the latter, in turn, into community and society [137].

Metrics derived from the quality of patient care are included in the social external perspective, since hospital supply chains are social by nature, and failures in service provision may have fatal consequences on health and life [18]. Reference [60] found significant differences between leading, developing, and under-developed hospitals, in terms of performance on quality indicators, which means that a lower death rate and higher timely provision of healthcare are perceived by those hospitals with the greatest extent of applying healthcare SCM practices. In the same way, quality of patient care has been operationalized in terms of patient experience criteria [72,86,101] and perceptions in comparison with other hospitals [30,83].

In the proposed framework, relationships between sustainable performance dimensions are drawn. First, it is noted that the environmental dimension influences the economic and the social dimensions. Second, the social dimension influences the economic dimension. Those cases reported by References [19,78,87] are only a few of the studies that highlight specific economic outcomes of implementing initiatives through which the consumption of resources, such as energy and water, is reduced. According to Reference [58], improved environmental performance prevents health systems from incurring costs, and positively impacts the social dimension by reducing diseases caused by climate change.

Lastly, the economic dimension is thought to be influenced by the social one. Despite the lack of financial indicators that reflect the management of social issues in the supply chain, a significant number of studies (albeit not focused on hospitals) have concluded that social performance positively impacts economic performance [110]. In particular, Reference [110] argues that the implementation of SCM practices that seek to enhance social issues results in greater loyalty, legitimacy, socially responsible investment, and trust, as well as in lower stakeholder criticisms and risk, which, in turn, lead to cost reduction and increased economic benefits.

5. Further Research Agenda

In accordance with the reviewed literature, avenues worthy of future research comprise both methodological and conceptual issues. Most suggestions are concerned with limitations of generalizability, research methods, and scope.

- **Generalizability.** Some contributions recommend using wider samples [50,60,67,92] and replicating studies in other cities and countries [16,60,61,64,69,83,92,96,101], since more information from different populations and geographical areas might help validate existing research and explain heterogeneities. The broadening of moderating variables is also emphasized. Reference [141], for instance, found that different priorities are held by public and private hospitals in terms of sustainability dimensions, since pressures undergone appear to be dissimilar for both organization types. Apart from hospital type and size [67,83], suggested moderators include operations outsourcing [83], information applications by type [67], forms of technology [63], nature of purchases [69], and contingent factors that affect the inventory [79].
- **Research methods.** Directions for the research methods employed depend largely on the types of studies covered in the reviewed literature. For instance, papers with an analytical and mathematical foci advocate addressing parameters that allow the simplification and improvement of proposed models [75,84,87]. Similarly, other studies posit that qualitative data is desirable to complement quantitative results [60], whereas those based on qualitative data require empirical validation through quantitative tools, as mentioned by Reference [69]. Moreover, some researchers point out the limitations of cross-sectional studies, and, therefore, recommend the use of longitudinal designs, in order to learn about supply chain relationships over time [69], and to unveil the effects of these practices on performance in the long run [39,83]. Ultimately, the concept of being sustainable implies a long-term vision and a strategic approach [39].
- **Scope.** The need to dig deeper into what is meant by hospital SSCM practices and their influence on sustainable performance is brought to light in several ways. Technological, clinical, and organizational innovations that help hospitals be more sustainable are bound to being more explored [50]. In addition, the documentation of less successful practices, in contrast with the most successful ones, is stressed as an issue that needs additional attention [61], albeit more dissemination of exemplar cases is also required to encourage the adoption of practices [89]. Furthermore, much can be said about the impacts of hospital supply chains, but the measurement of the effects themselves represents a challenge for hospitals. As Reference [61] found, few hospitals use a wide range of indicators for purchase and inventory management. Reference [63] recommends including patient safety as a performance dimension. From an environmental standpoint, Reference [50] highlights the measurement of footprints across internal hospital supply chains as imperative.

In addition, further analysis of the influences of practices on performance is outlined. It is important to disclose the ways in which specific practices affect specific performance indicators [60], at the time that the incorporation of sectorial, social, and cultural issues into hospital SSCM research becomes prominent. While it can be a good practice to hire and train well-qualified supply chain professionals [17], it can be equally vital to know which concrete skills are required by supply chain managers in hospital settings [16]. While the relevance of promoting active travel is almost indisputable, the determinants of travel behavior remain unclear [50]. While adjusting menus to offer healthier dishes in hospital cafeterias is urgent, preference for less healthy food is rooted in the mindsets of the majority [91]. Consequently, since social and cultural factors can hinder or facilitate the implementation of practices [89], it is of paramount importance for sustainability improvement to gain understanding about the ways in which behaviors and culture need to change [50], and what kind of incentives and motivations lead staff and communities to demand, adopt, and promote better practices [91,93].

Lastly, the extant need of additional integrative research on hospital SSCM and sustainable performance merits mention. Most studies address the economic dimension, whereas few address the environmental one, and fewer yet address the social one. Unfortunately, although the under-representativeness of the social dimension is not unusual in the field of SCM [109], and it is difficult to ignore the economic rationality on which SCM research is based, it is clear that hospitals have social concerns that, if ignored, will make a growing healthcare deterioration more evident. This is more than serious, which takes into account the interdependence between health and sustainable development, since one of the goals of sustainable development is oriented toward health improvement, but health is a condition for sustainable development [14].

6. Conclusions

Framing both SSCM practices and sustainable performance metrics at once is not an easy task. The concept of practice, per se, is difficult to define. Practices take various forms and can represent technologies, processes, ways of doing things, or ways of organizing work [65,68]. In addition, they can have different meanings or rationales from a sustainability approach and, for this reason, can overlap whichever categories have been established for their classification. In this way, a practice such as serving locally grown food can be conceived to improve food freshness and nutritional quality, favor the environment by avoiding transport activities, or strengthen local economies. Multiple purposes and interconnections among practices are more than visible and demonstrate the massive opportunities for action and impact that an integrated approach for sustainability provides, as well as its complexity.

Regarding performance, the main difficulty is that many effects of practices are not completely clear because there have not been enough empirical studies completed, and even less so regarding the interactions and trade-offs that may arise between dimensions. Moreover, indicator operationalization and validation are still incipient. For instance, not all the items encompassed by the review and the proposed framework have been measured in the literature. On one hand, it might be indicative of the exploratory status of current research, and the nascent interest in disclosing the elements that make up hospital SSCM. On the other, this could be interpreted as a symptom of the low level of adopting metrics and measurement systems, to such an extent that it would be more important to learn whether hospitals use indicators to measure performance than to calculate the values of such indicators.

The proposed framework can serve as a starting point for studying SSCM practices implementation in hospitals, in order to improve performance in this type of organizations, from a holistic sustainability approach. However, it needs to be validated and refined by using both quantitative and qualitative research methods. The practices and performance metrics covered are examples extracted from the literature to allow for a complete overview, rather than an instruction manual to be followed uncritically, since hospitals vary in accordance with their range of services, capacities, types, complexities, technologies, problems, impacts, needs, and more. Furthermore, it would be useful to prioritize elements of the framework, such as through multiple-criteria decision analysis techniques.

Apart from further validation required of the proposed framework, this review has several limitations. Additional databases and languages could be used. Since a search strategy that separately includes each of the topics of SCM or sustainability in hospitals was not formulated, the resulting analysis is comprehensive, but leaves room for improvements in exhaustiveness. The identification of categories of practices and performance could be an input with which to carry out a more thorough, detailed search for evidence, and enhance forthcoming debates on existing relationships. In addition, an interesting way to refine the definition of SSCM practices could be by covering literature that addresses drivers, barriers, and enablers. These were not fully or directly considered in the paper at hand, due to the early development of the proposed framework, but these could delineate a way for, or even help to explain which practices are or should be adopted, and why. Another limitation refers to subjectivity regarding the selection of keywords and paper classification, as well as in terms of established categories for practices and performance metrics, despite three researchers that have been involved throughout the review process. Recognized methodology guidelines have been referred to and followed.

Author Contributions: Conceptualization, V.D.-U., W.S., and E.V.G. Methodology, V.D.-U., W.S., and E.V.G. Writing—original draft preparation, V.D.-U. Writing—review and editing, V.D.-U., W.S., and E.V.G. Supervision, W.S. and E.V.G.

Funding: This research received no external funding.

Acknowledgments: This work is part of the doctoral study of the first author, which is financially supported by the Departamento Administrativo de Ciencia, Tecnología e Innovación – COLCIENCIAS, Colombia.

Conflicts of Interest: The authors declare no conflict of interest.

References

1. Shiau, W.L.; Dwivedi, Y.K.; Tsai, C.H. Supply chain management: Exploring the intellectual structure. *Scientometrics* **2015**, *105*, 215–230. [CrossRef]
2. CSCMP—Council of Supply Chain Management Professionals Supply. Chain Management Definitions and Glossary. Available online: http://cscmp.org/imis0/CSCMP/Educate/SCM_Definitions_and_Glossary_of_Terms/CSCMP/Educate/SCM_Definitions_and_Glossary_of_Terms.aspx?hkey=60879588-f65f-4ab5-8c4b-6878815ef921 (accessed on 8 February 2019).
3. Christopher, M. *Logistics & Supply Chain Management*; Pearson Education Limited: Dorchester, UK, 2011.
4. Paulraj, A.; Chen, I.J.; Lado, A.A. An empirical taxonomy of supply chain management practices. *J. Bus. Logist.* **2012**, *33*, 227–244. [CrossRef]
5. Truong, H.Q.; Sameiro, M.; Fernandes, A.C.; Sampaio, P.; Duong, B.A.T.; Duong, H.H.; Vilhenac, E. Supply chain management practices and firms' operational performance. *Int. J. Qual. Reliab. Manag.* **2017**, *34*, 176–193. [CrossRef]
6. Shi, M.; Yu, W. Supply chain management and financial performance: Literature review and future directions. *Int. J. Oper. Prod. Manag.* **2013**, *33*, 1283–1317. [CrossRef]
7. Li, S.; Ragu-Nathan, B.; Ragu-Nathan, T.S.; Subba Rao, S. The impact of supply chain management practices on competitive advantage and organizational performance. *Omega* **2006**, *34*, 107–124. [CrossRef]
8. Tan, K.C. Supply Chain Management: Practices, Concerns, and Performance Issues. *J. Supply Chain Manag.* **2002**, *38*, 42–53. [CrossRef]
9. United Nations General Assembly. Transforming Our World: The 2030 Agenda for Sustainable Development. Available online: http://www.un.org/ga/search/view_doc.asp?symbol=A/RES/70/1&Lang=E (accessed on 28 August 2018).
10. United Nations General Assembly. Report of the World Commission on Environment and Development: Our Common Future. Available online: http://www.un-documents.net/our-common-future.pdf (accessed on 22 March 2019).
11. Elkington, J. *Cannibals with Forks: The Triple Bottom Line of 21st Century Business*; Capstone Publishing Limited: Oxford, UK, 1997.
12. Seuring, S.; Müller, M. From a literature review to a conceptual framework for sustainable supply chain management. *J. Clean. Prod.* **2008**, *16*, 1699–1710. [CrossRef]
13. Kumar, S.; Blair, J.T. U.S. healthcare fix: Leveraging the lessons from the food supply chain. *Technol. Heal. Care* **2013**, *21*, 125–141. [CrossRef]
14. Weisz, U.; Haas, W.; Pelikan, J.M.; Schmied, H. Sustainable Hospitals: A Socio-Ecological Approach. *GAIA* **2011**, *20*, 191–198. [CrossRef]
15. OECD—Organization for Economic Co-operation and Development. *Health at a Glance 2017: OECD Indicators*; OECD Publishing: Paris, France, 2017.
16. Landry, S.; Beaulieu, M.; Roy, J. Strategy deployment in healthcare services: A case study approach. *Technol. Soc. Chang.* **2016**, *113*, 429–437. [CrossRef]
17. McKone-Sweet, K.E.K.E.; Hamilton, P.; Willis, S.B.S.B. The Ailing Healthcare Supply Chain: A Prescription for Change. *J. Supply Chain Manag.* **2005**, *41*, 4–17. [CrossRef]
18. Zhu, Q.; Johnson, S.; Sarkis, J. Lean six sigma and environmental sustainability: A hospital perspective. *Supply Chain Forum* **2018**, *19*, 25–41. [CrossRef]
19. Pisters, P.; Bien, B.; Dankner, S.; Rubinstein, E.; Sheriff, F. Supporting hospital renewal through strategic environmental sustainability programs. *Healthc. Manag. Forum* **2017**, *30*, 79–83. [CrossRef] [PubMed]
20. SDU NHS—Sustainable Development Unit National Health Service. Carbon Update for the Health and Care Sector in England. 2015. Available online: https://www.sduhealth.org.uk/policy-strategy/reporting/hcs-carbon-footprint.aspx (accessed on 28 August 2018).
21. ILO—International Labour Organization. Improving Employment and working conditions in health services: Report for discussion at the Tripartite Meeting on Improving Employment and Working Conditions in Health Services. Available online: https://www.ilo.org/wcmsp5/groups/public/---ed_dialogue/---sector/documents/publication/wcms_548288.pdf (accessed on 21 October 2017).

22. Manyisa, Z.M.; van Aswegen, E.J. Factors affecting working conditions in public hospitals: A literature review. *Int. J. Afr. Nurs. Sci.* **2017**, *6*, 28–38. [CrossRef]
23. da Silva, A.A.; Sanchez, G.M.; Mambrini, N.S.B.; De Oliveira, M.Z. Predictor variables for burnout among nursing professionals. *Rev. Psicol.* **2019**, *37*, 319–348. [CrossRef]
24. Dilig-Ruiz, A.; MacDonald, I.; Demery Varin, M.; Vandyk, A.; Graham, I.D.; Squires, J.E. Job satisfaction among critical care nurses: A systematic review. *Int. J. Nurs. Stud.* **2018**, *88*, 123–134. [CrossRef] [PubMed]
25. Wagstaff, A.S.; Lie, J.A.S. Shift and night work and long working hours—A systematic review of safety implications. *Scand. J. Work. Env. Heal.* **2011**, *37*, 173–185. [CrossRef]
26. Weigl, M.; Schneider, A. Associations of work characteristics, employee strain and self-perceived quality of care in Emergency Departments: A cross-sectional study. *Int. Emerg. Nurs.* **2017**, *30*, 20–24. [CrossRef]
27. Aiken, L.H.; Sloane, D.M.; Bruyneel, L.; Van den Heede, K.; Sermeus, W. Nurses' reports of working conditions and hospital quality of care in 12 countries in Europe. *Int. J. Nurs. Stud.* **2013**, *50*, 143–153. [CrossRef]
28. Stone, P.W.; Mooney-Kane, C.; Larson, E.L.; Horan, T.; Glance, L.G.; Zwanziger, J.; Dick, A.W. Nurse working conditions and patient safety outcomes. *Med. Care* **2007**, *45*, 571–578. [CrossRef]
29. Parmigiani, A.; Howard-Grenville, J. Routines revisited: Exploring the capabilities and practice perspectives. *Acad. Manag. Ann.* **2011**, *5*, 413–453. [CrossRef]
30. Adebanjo, D.; Laosirihongthong, T.; Samaranayake, P. Prioritizing lean supply chain management initiatives in healthcare service operations: A fuzzy AHP approach. *Prod. Plan. Control Control* **2016**, *27*, 953–966. [CrossRef]
31. Carter, C.R.; Washispack, S. Mapping the Path Forward for Sustainable Supply Chain Management: A Review of Reviews. *J. Bus. Logist.* **2018**, *39*, 242–247. [CrossRef]
32. Hong, J.; Zhang, Y.; Ding, M. Sustainable supply chain management practices, supply chain dynamic capabilities, and enterprise performance. *J. Clean. Prod.* **2018**, *172*, 3508–3519. [CrossRef]
33. Mathivathanan, D.; Kannan, D.; Haq, A.N. Sustainable supply chain management practices in Indian automotive industry: A multi-stakeholder view. *Resour. Conserv. Recycl.* **2018**, *128*, 284–305. [CrossRef]
34. Marshall, D.; Mccarthy, L.; Heavey, C.; Mcgrath, P. Environmental and social supply chain management sustainability practices: Construct development and measurement. *Prod. Plan. Control* **2015**, *26*, 673–690. [CrossRef]
35. Khan, M.; Ajmal, M.; Hussain, M.; Helo, P. Barriers to social sustainability in the health-care industry in the UAE. *Int. J. Organ. Anal.* **2018**, *26*, 450–469. [CrossRef]
36. Asgari, N.; Nikbakhsh, E.; Hill, A.; Farahani, R.Z. Supply chain management 1982-2015: A review. *Ima J. Manag. Math.* **2016**, *27*, 353–379. [CrossRef]
37. Morioka, S.N.; de Carvalho, M.M. A systematic literature review towards a conceptual framework for integrating sustainability performance into business. *J. Clean. Prod.* **2016**, *136*, 134–146. [CrossRef]
38. Ashby, A.; Leat, M.; Hudson-Smith, M. Making connections: A review of supply chain management and sustainability literature. *Supply Chain Manag.* **2012**, *17*, 497–516. [CrossRef]
39. Carter, C.R.; Rogers, D.S. A framework of sustainable supply chain management: Moving toward new theory. *Int. J. Phys. Distrib. Logist. Manag.* **2008**, *38*, 360–387. [CrossRef]
40. Yun, G.; Yalcin, M.G.; Hales, D.N.; Kwon, H.Y. Interactions in sustainable supply chain management: A framework review. *Int. J. Logist. Manag.* **2019**, *30*, 140–173. [CrossRef]
41. Morali, O.; Searcy, C. A Review of Sustainable Supply Chain Management Practices in Canada. *J. Bus. Ethics* **2013**, *117*, 635–658. [CrossRef]
42. Mathivathanan, D.; Haq, A.N. Comparisons of sustainable supply chain management practices in the automotive sector. *Int. J. Bus. Perform. Supply Chain Model.* **2017**, *9*, 18–27. [CrossRef]
43. Mitra, S.; Datta, P.P. Adoption of green supply chain management practices and their impact on performance: An exploratory study of Indian manufacturing firms. *Int. J. Prod. Res.* **2014**, *52*, 2085–2107. [CrossRef]
44. Hassini, E.; Surti, C.; Searcy, C. A literature review and a case study of sustainable supply chains with a focus on metrics. *Int. J. Prod. Econ.* **2012**, *140*, 69–82. [CrossRef]
45. Yanamandra, R. Development of an integrated healthcare supply chain model. *Supply Chain Forum* **2018**, *19*, 111–121. [CrossRef]
46. Volland, J.; Fügener, A.; Schoenfelder, J.; Brunner, J.O. Material logistics in hospitals: A literature review. *Omega* **2017**, *69*, 82–101. [CrossRef]

47. Leaven, L.; Ahmmad, K.; Peebles, D. Inventory management applications for healthcare supply chains. *Int. J. Supply Chain Manag.* **2017**, *6*, 1–7.
48. Kwon, I.W.G.; Kim, S.H.; Martin, D.G. Healthcare supply chain management; strategic areas for quality and financial improvement. *Technol. Soc. Chang.* **2016**, *113*, 422–428. [CrossRef]
49. Dobrzykowski, D.; Saboori Deilami, V.; Hong, P.; Kim, S.-C. A structured analysis of operations and supply chain management research in healthcare (1982–2011). *Int. J. Prod. Econ.* **2014**, *147*, 514–530. [CrossRef]
50. McGain, F.; Naylor, C. Environmental sustainability in hospitals—A systematic review and research agenda. *J. Heal. Serv. Res. Policy* **2014**, *19*, 245–252. [CrossRef]
51. Briner, R.B.; Denyer, D. Systematic Review and Evidence Synthesis as a Practice and Scholarship Tool. In *The Oxford Handbook of Evidence-Based Management*; Rousseau, D.M., Ed.; Oxford University Press: New York, NY, USA, 2012; pp. 112–129.
52. Tranfield, D.; Denyer, D.; Smart, P. Towards a Methodology for Developing Evidence-Informed Management Knowledge by Means of Systematic Review. *Br. J. Manag.* **2003**, *14*, 207–222. [CrossRef]
53. Kitchenham, B. *Procedures for Performing Systematic Reviews*; Keele University: Keele, UK; Empirical Software Engineering National ICT Australia Ltd.: Eversleigh, Australia, 2004.
54. Moher, D.; Liberati, A.; Tetzlaff, J.; Altman, D.G. Preferred reporting items for systematic reviews and meta-analyses: The PRISMA statement. *J. Clin. Epidemiol.* **2009**, *62*, 1006–1012. [CrossRef] [PubMed]
55. Aghaei Chadegani, A.; Salehi, H.; Md Yunus, M.M.; Farhadi, H.; Fooladi, M.; Farhadi, M.; Ale Ebrahim, N. A comparison between two main academic literature collections: Web of science and scopus databases. *Asian Soc. Sci.* **2013**, *9*, 18–26. [CrossRef]
56. Landry, S.; Beaulieu, M. The Challenges of Hospital Supply Chain Management, from Central Stores to Nursing Units. In *Handbook of Healthcare Operations Management: Methods and Applications*; Denton, B.T., Ed.; Springer: New York, NY, USA, 2013; pp. 465–482.
57. La Rotta, D.; Pérez Rave, J. A relevant literary space on the use of the European Foundation for Quality Management model: Current state and challenges. *Total Qual. Manag. Bus. Excel.* **2017**, *28*, 1447–1468. [CrossRef]
58. HCWH—Health Care Without Harm. A Comprehensive Environmental Health Agenda for Hospitals and Health Systems Around the World. Available online: http://greenhospitals.net/wp-content/uploads/2011/10/Global-Green-and-Healthy-Hospitals-Agenda.pdf (accessed on 4 March 2019).
59. WHO—World Health Organization; HCWH. Health Care Without Harm Healthy Hospitals, Healthy Planet, Healthy People: Addressing Climate Change in Healthcare Settings. Available online: http://www.who.int/globalchange/publications/climatefootprint_report.pdf?ua=1 (accessed on 4 March 2019).
60. Rakovska, M.A.; Stratieva, S.V. A taxonomy of healthcare supply chain management practices. *Supply Chain Forum* **2018**, *19*, 4–24. [CrossRef]
61. Velasco, N.; Moreno, J.P.; Rebolledo, C. Logistics practices in healthcare organizations in Bogota. *Acad. Rev. Lat. Adm.* **2018**, *31*, 519–533. [CrossRef]
62. Beaulieu, M.; Landry, S.; Roy, J. La productivité des activités de logistique hospitalière. Available online: http://cpp.hec.ca/cms/assets/documents/recherches_publiees/CE_2011_06.pdf (accessed on 5 February 2019).
63. Kritchanchai, D.; Hoeur, S.; Engelseth, P. Develop a strategy for improving healthcare logistics performance. *Supply Chain Forum* **2018**, *19*, 55–69. [CrossRef]
64. Pan, Z.X.; Pokharel, S. Logistics in hospitals: A case study of some Singapore hospitals. *Lead. Heal. Serv.* **2007**, *20*, 195–207. [CrossRef]
65. Landry, S.; Beaulieu, M.; Friel, T.; Duguay, C.R. *Étude Internationale des Meilleures Pratiques de Logistique Hospitalière, Cahier de Recherche 00-05*; Groupe de recherche Chaîne sur l'intégration et l'environnement de la chaîne d'approvisionnement, École des HEC: Montreal, QC, Canada, 2000.
66. Böhme, T.; Williams, S.; Childerhouse, P.; Deakins, E.; Towill, D. Squaring the circle of healthcare supplies. *J. Heal. Organ. Manag.* **2014**, *28*, 247–265. [CrossRef] [PubMed]
67. Yoon, S.N.; Lee, D.H.; Schniederjans, M. Effects of innovation leadership and supply chain innovation on supply chain efficiency: Focusing on hospital size. *Technol. Soc. Chang.* **2016**, *113*, 412–421. [CrossRef]
68. Landry, S.; Philippe, R. How Logistics Can Service Healthcare. *Supply Chain Forum Int. J.* **2004**, *5*, 24–30. [CrossRef]

69. Oruezabala, G.; Rico, J.C. The impact of sustainable public procurement on supplier management - The case of French public hospitals. *Ind. Mark. Manag.* **2012**, *41*, 573–580. [CrossRef]
70. Brennan, C.D. Integrating the healthcare supply chain. *Healthc. Financ. Manag.* **1998**, *52*, 31–34.
71. Ritchie, L.; Burnes, B.; Whittle, P.; Hey, R. The benefits of reverse logistics: The case of the Manchester Royal Infirmary Pharmacy. *Supply Chain Manag.* **2000**, *5*, 226–233. [CrossRef]
72. Mirghafoori, S.H.; Sharifabadi, A.M.; Takalo, S.K. Development of causal model of sustainable hospital supply chain management using the intuitionistic fuzzy cognitive map (IFCM) method. *J. Ind. Eng. Manag.* **2018**, *11*, 588–605. [CrossRef]
73. Schieble, T.M. Advertised sustainability practices among suppliers to a University Hospital operating room. *J. Hosp. Mark. Public Relat.* **2008**, *18*, 135–148. [CrossRef]
74. Budgett, A.; Gopalakrishnan, M.; Schneller, E. Procurement in public & private hospitals in Australia and Costa Rica—A comparative case study. *Heal. Syst.* **2017**, *6*, 56–67.
75. Ross, A.D.; Jayaraman, V. Strategic purchases of bundled products in a health care supply chain environment. *Decis. Sci.* **2009**, *40*, 269–293. [CrossRef]
76. McGain, F. Sustainable hospitals? An Australian perspective. *Perspect. Public Health* **2010**, *130*, 19–20. [CrossRef] [PubMed]
77. Kumar, S.; Degroot, R.A.; Choe, D. Rx for smart hospital purchasing decisions: The impact of package design within US hospital supply chain. *Int. J. Phys. Distrib. Logist. Manag.* **2008**, *38*, 601–615. [CrossRef]
78. Ashourian, K.T.; Young, S.T. Greening Healthcare: The Current State of Sustainability in Manhattan's Hospitals. *Sustainability* **2016**, *9*, 73–79. [CrossRef]
79. Bhakoo, V.; Singh, P.; Sohal, A. Collaborative management of inventory in Australian hospital supply chains: Practices and issues. *Supply Chain Manag.* **2012**, *17*, 217–230. [CrossRef]
80. Kelle, P.; Woosley, J.; Schneider, H. Pharmaceutical supply chain specifics and inventory solutions for a hospital case. *Oper. Res. Heal. Care* **2012**, *1*, 54–63. [CrossRef]
81. Rivard-Royer, H.; Landry, S.; Beaulieu, M. Hybrid stockless: A case study. Lessons for health-care supply chain integration. *Int. J. Oper. Prod. Manag.* **2002**, *22*, 412–424. [CrossRef]
82. Dobrzykowski, D.D.; Tarafdar, M. Understanding information exchange in healthcare operations: Evidence from hospitals and patients. *J. Oper. Manag.* **2015**, *36*, 201–214. [CrossRef]
83. Lee, S.M.; Lee, D.H.; Schniederjans, M.J. Supply chain innovation and organizational performance in the healthcare industry. *Int. J. Oper. Prod. Manag.* **2011**, *31*, 1193–1214. [CrossRef]
84. Nabelsi, V.; Gagnon, S. Information technology strategy for a patient-oriented, lean, and agile integration of hospital pharmacy and medical equipment supply chains. *Int. J. Prod. Res.* **2017**, *55*, 3929–3945. [CrossRef]
85. Romero, A.; Lefebvre, E. Combining barcodes and RFID in a hybrid solution to improve hospital pharmacy logistics processes. *Int. J. Inf. Technol. Manag.* **2015**, *14*, 97–123. [CrossRef]
86. Manika, D.; Gregory-Smith, D.; Wells, V.K.; Comerford, L.; Aldrich-Smith, L. Linking environmental sustainability and healthcare: The effects of an energy saving intervention in two hospitals. *Int. J. Bus. Sci. Appl. Manag.* **2016**, *11*, 32–54.
87. Faezipour, M.; Ferreira, S. A System Dynamics Approach for Sustainable Water Management in Hospitals. *IEEE Syst. J.* **2018**, *12*, 1278–1285. [CrossRef]
88. Huang, E.; Gregoire, M.B.; Tangney, C.; Stone, M.K. Sustainability in hospital foodservice. *J. Foodserv. Bus. Res.* **2011**, *14*, 241–255. [CrossRef]
89. Dauner, K.N.; Lacaille, L.J.; Schultz, J.F.; Harvie, J.; Klingner, J.; Lacaille, R.; Branovan, M. Implementing healthy and sustainable food practices in a hospital cafeteria: A qualitative look at processes, barriers, and facilitators of implementation. *J. Hunger Env. Nutr.* **2011**, *6*, 264–278. [CrossRef]
90. Bloomfield, C. Putting sustainable development into practice: Hospital food procurement in Wales. *Reg. Stud. Reg. Sci.* **2015**, *2*, 552–558. [CrossRef]
91. Pitts, S.J.; Schwartz, B.; Graham, J.; Warnock, A.L.; Mojica, A.; Marziale, E.; Harris, D. Best practices for financial sustainability of healthy food service guidelines in hospital cafeterias. *Prev. Chronic Dis.* **2018**, *15*, 1–8.
92. Ranke, T.D.; Mitchell, C.L.; St. George, D.M.; D'Adamo, C.R. Evaluation of the Balanced Menus Challenge: A healthy food and sustainability programme in hospitals in Maryland. *Public Health Nutr.* **2015**, *18*, 2341–2349.

93. Goggins, G. Developing a sustainable food strategy for large organizations: The importance of context in shaping procurement and consumption practices. *Bus. Strat. Env.* **2018**, *27*, 838–848. [CrossRef]
94. de Fátima Castro, M.; Mateus, R.; Bragança, L. Healthcare Building Sustainability Assessment tool - Sustainable Effective Design criteria in the Portuguese context. *Environ. Impact Assess. Rev.* **2017**, *67*, 49–60. [CrossRef]
95. de Fátima Castro, M.; Mateus, R.; Bragança, L. A critical analysis of building sustainability assessment methods for healthcare buildings. *Environ. Dev. Sustain.* **2015**, *17*, 1381–1412. [CrossRef]
96. Buffoli, M.; Gola, M.; Rostagno, M.; Capolongo, S.; Nachiero, D. Making hospitals healthier: How to improve sustainability in healthcare facilities. *Ann. Ig.* **2014**, *26*, 418–425.
97. SDU NHS—Sustainable Development Unit National Health Service. Reducing the Use of Natural Resources in Health and Social Care: 2018 Report. Available online: https://www.sduhealth.org.uk/policy-strategy/reporting/natural-resource-footprint-2018.aspx (accessed on 11 February 2019).
98. Doiphode, S.M.; Hinduja, I.N.; Ahuja, H.S. Developing a novel, sustainable and beneficial system for the systematic management of hospital wastes. *J. Clin. Diagn. Res.* **2016**, *10*, LC06–LC11. [CrossRef] [PubMed]
99. Theofanidis, D.; Fountouki, A.; Vosniakos, F.; Papadakis, N.; Nikoalou, K. Sustainable management of hospital waste: The view of Greek nurses. *J. Environ. Prot. Ecol.* **2008**, *9*, 391–403.
100. De Oliveira Furukawa, P.; Cunha, I.C.K.O.; Da Luz Gonçalves Pedreira, M.; Marck, P.B. Environmental sustainability in medication processes performed in hospital nursing care. *Acta Paul. Enferm.* **2016**, *29*, 316–324.
101. Turkyilmaz, A.; Bulak, M.E.; Zaim, S. Assessment of TQM Practices as a part of supply chain management in healthcare institutions. *Int. J. Supply Chain Manag.* **2015**, *4*, 1–9.
102. Aronsson, H.; Abrahamsson, M.; Spens, K. Developing lean and agile health care supply chains. *Supply Chain Manag.* **2011**, *16*, 176–183. [CrossRef]
103. Meijboom, B.R.; Bakx, S.J.W.G.C.; Westert, G.P. Continuity in health care: Lessons from supply chain management. *Int. J. Health Plann. Manag.* **2010**, *25*, 304–317. [CrossRef]
104. Meijboom, B.; Schmidt-Bakx, S.; Westert, G. Supply chain management practices for improving patient-oriented care. *Supply Chain Manag.* **2011**, *16*, 166–175. [CrossRef]
105. Das, D. Development and validation of a scale for measuring Sustainable Supply Chain Management practices and performance. *J. Clean. Prod.* **2017**, *164*, 1344–1362. [CrossRef]
106. Wang, J.; Dai, J. Sustainable supply chain management practices and performance. *Ind. Manag. Data Syst.* **2018**, *118*, 2–21. [CrossRef]
107. Abdulsalam, Y.; Schneller, E. Hospital Supply Expenses: An Important Ingredient in Health Services Research. *Med. Care Res. Rev.* **2019**, *76*, 240–252. [CrossRef]
108. AL-Shboul, M.A.; Garza-Reyes, J.A.; Kumar, V. Best supply chain management practices and high-performance firms: The case of Gulf manufacturing firms. *Int. J. Prod. Perform. Manag.* **2018**, *67*, 1482–1509. [CrossRef]
109. Nakamba, C.C.; Chan, P.W.; Sharmina, M. How does social sustainability feature in studies of supply chain management? A review and research agenda. *Supply Chain Manag.* **2017**, *22*, 522–541. [CrossRef]
110. Yawar, S.A.; Seuring, S. Management of Social Issues in Supply Chains: A Literature Review Exploring Social Issues, Actions and Performance Outcomes. *J. Bus. Ethics* **2017**, *141*, 621–643. [CrossRef]
111. Schneller, E.; Smeltzer, L. *Strategic Management of the Health Care Supply Chain*; Jossey-Bass: San Francisco, CA, USA, 2006.
112. Salam, M.A.; Khan, S.A. Achieving supply chain excellence through supplier management: A case study of fast moving consumer goods. *Benchmarking* **2018**, *25*, 4084–4102. [CrossRef]
113. Forslund, H. Exploring logistics performance management in supplier/retailer dyads. *Int. J. Retail Distrib. Manag.* **2014**, *42*, 205–218. [CrossRef]
114. Gimenez, C.; Tachizawa, E.M. Extending sustainability to suppliers: A systematic literature review. *Supply Chain Manag.* **2012**, *17*, 531–543. [CrossRef]
115. González González, A.; García-Sanz-Calcedo, J.; Salgado, D.R. A quantitative analysis of final energy consumption in hospitals in Spain. *Sustain. Cities Soc.* **2018**, *36*, 169–175. [CrossRef]
116. Gómez-Chaparro, M.; García Sanz-Calcedo, J.; Armenta-Márquez, L. Study on the use and consumption of water in Spanish private hospitals as related to healthcare activity. *Urban Water J.* **2018**, *15*, 601–608. [CrossRef]

117. EPA—Environmental Protection Agency. Saving Water in Hospitals. Available online: https://www.epa.gov/watersense/types-facilities (accessed on 26 March 2019).
118. Victorian Government Department of Health Guidelines for Water Reuse and Recycling in Victorian Health Care Facilities. Available online: https://www2.health.vic.gov.au/about/publications/policiesandguidelines/Guidelines-for-water-reuse-and-recycling-in-Victorian-health-care-facilities-Nondrinking-applications (accessed on 11 February 2019).
119. Baum, M.; Shepley, M.; Rostenberg, B.; Ginberg, R. *Eco-Effective Design and Evidence-Based Design: Removing Barriers to Integration*; AIA Board Knowledge Committee: San Francisco, CA, USA, 2009.
120. Dobrzykowski, D.D. Linking Electronic Medical Records Use to Physicians' Performance: A Contextual Analysis. *Decis. Sci.* **2017**, *48*, 7–38. [CrossRef]
121. Scavarda, A.; Daú, G.L.; Scavarda, L.F.; Korzenowski, A.L. A proposed healthcare supply chain management framework in the emerging economies with the sustainable lenses: The theory, the practice, and the policy. *Resour. Conserv. Recycl.* **2019**, *141*, 418–430. [CrossRef]
122. Hussain, M.; Khan, M.; Al-Aomar, R. A framework for supply chain sustainability in service industry with Confirmatory Factor Analysis. *Renew. Sustain. Energy Rev.* **2016**, *55*, 1301–1312. [CrossRef]
123. Vissers, J.; Beech, R. *Health Operations Management: Patient Flow Logistics in Health Care*; Taylor & Francis Group: Abingdon, UK, 2005.
124. Kriegel, J.; Jehle, F.; Dieck, M.; Mallory, P. Advanced services in hospital logistics in the German health service sector. *Logist. Res.* **2013**, *6*, 47–56. [CrossRef]
125. Barnes, S.; Golden, B.; Price, S. Applications of Agent-Based Modeling and Simulation to Healthcare Operations Management. In *Handbook of Healthcare Operations Management. International Series in Operations Research & Management Science*; Denton, B., Ed.; Springer: New York, NY, USA, 2013.
126. Thompson, S.M.; Day, R.; Garfinkel, R. Improving the Flow of Patients Through Healthcare Organizations. In *Handbook of Healthcare Operations Management. International Series in Operations Research & Management Science*; Denton, B., Ed.; Springer: New York, NY, USA, 2013.
127. Langabeer, J.R.; Helton, J. *Health Care Operations Management: A Systems Perspective*; Jones & Bartlett Learning: Burlington, VT, USA, 2016.
128. Baltacioglu, T.; Ada, E.; Kaplan, M.D.; Yurt, O.; Kaplan, Y.C. A new framework for service supply chains. *Serv. Ind. J.* **2007**, *27*, 105–124. [CrossRef]
129. Gandhi, A.V.; Shaikh, A.; Sheorey, P.A. Impact of supply chain management practices on firm performance: Empirical evidence from a developing country. *Int. J. Retail Distrib. Manag.* **2017**, *45*, 366–384. [CrossRef]
130. Gawankar, S.A.; Kamble, S.; Raut, R. An investigation of the relationship between supply chain management practices (SCMP) on supply chain performance measurement (SCPM) of Indian retail chain using SEM. *Benchmarking* **2017**, *24*, 257–295. [CrossRef]
131. Li, S.; Rao, S.S.; Ragu-Nathan, T.S.; Ragu-Nathan, B. Development and validation of a measurement instrument for studying supply chain management practices. *J. Oper. Manag.* **2005**, *23*, 618–641. [CrossRef]
132. Wang, J.; Zhang, Y.; Goh, M. Moderating the role of firm size in sustainable performance improvement through sustainable supply chain management. *Sustainability* **2018**, *10*, 1654. [CrossRef]
133. Das, D. The impact of Sustainable Supply Chain Management practices on firm performance: Lessons from Indian organizations. *J. Clean. Prod.* **2018**, *203*, 179–196. [CrossRef]
134. Jia, P.; Diabat, A.; Mathiyazhagan, K. Analyzing the SSCM practices in the mining and mineral industry by ISM approach. *Resour. Policy* **2015**, *46*, 76–85. [CrossRef]
135. Köksal, D.; Strähle, J.; Müller, M.; Freise, M. Social Sustainable Supply Chain Management in the Textile and Apparel Industry—A Literature Review. *Sustainability* **2017**, *9*, 100. [CrossRef]
136. SDU NHS—Sustainable Development Unit National Health Service. Carbon Footprint from Anaesthetic Gas Use. Available online: https://www.sduhealth.org.uk/areas-of-focus/carbon-hotspots/anaesthetic-gases.aspx (accessed on 15 January 2019).
137. Henao, R.; Sarache, W.; Gomez, I. A Social Performance Metrics Framework for Sustainable Manufacturing. *Int. J. Ind. Syst. Eng.* (In press).
138. Paulraj, A.; Chen, I.J.; Blome, C. Motives and Performance Outcomes of Sustainable Supply Chain Management Practices: A Multi-theoretical Perspective. *J. Bus. Ethics* **2017**, *145*, 239–258. [CrossRef]
139. Esfahbodi, A.; Zhang, Y.; Watson, G. Sustainable supply chain management in emerging economies: Trade-offs between environmental and cost performance. *Int. J. Prod. Econ.* **2016**, *181*, 1–17. [CrossRef]

140. Zailani, S.; Jeyaraman, K.; Vengadasan, G.; Premkumar, R. Sustainable supply chain management (SSCM) in Malaysia: A survey. *Int. J. Prod. Econ.* **2012**, *140*, 330–340. [CrossRef]
141. Rodriguez, R.; Svensson, G.; Eriksson, D. Organizational logic to prioritize between the elements of triple bottom line. *Benchmarking* **2018**, *25*, 1626–1640. [CrossRef]

© 2019 by the authors. Licensee MDPI, Basel, Switzerland. This article is an open access article distributed under the terms and conditions of the Creative Commons Attribution (CC BY) license (http://creativecommons.org/licenses/by/4.0/).

MDPI
St. Alban-Anlage 66
4052 Basel
Switzerland
Tel. +41 61 683 77 34
Fax +41 61 302 89 18
www.mdpi.com

Sustainability Editorial Office
E-mail: sustainability@mdpi.com
www.mdpi.com/journal/sustainability

www.ingramcontent.com/pod-product-compliance
Lightning Source LLC
LaVergne TN
LVHW071938080526
838202LV00064B/6636